Spatial Uncertainty in Ecology

Springer-Science+Business Media, LLC

Carolyn T. Hunsaker   Michael F. Goodchild
Mark A. Friedl   Ted J. Case
Editors

Spatial Uncertainty in Ecology

Implications for Remote Sensing and GIS Applications

With 100 Illustrations, 17 in Full Color

 Springer

Carolyn T. Hunsaker
USDA Forest Service
Forestry Sciences Laboratory
Pacific Southwest Research Station
Fresno, CA 93710-4639
USA
chunsaker@fs.fed.us

Michael F. Goodchild
National Center for Geographic
   Information and Analysis
Department of Geography
University of California, Santa Barbara
Santa Barbara, CA 93106-4060
USA
good@ncgia.ucsb.edu

Mark A. Friedl
Department of Geography
Boston University
Boston, MA 02215-1401
USA
friedl@bu.edu

Ted J. Case
Department of Biology
University of California, San Diego
La Jolla, CA 92093-0116
USA
case@biomail.ucsd.edu

Cover illustration: A simulation: 200-m contours, 800-m baseline. See Figure 11.6.

Library of Congress Cataloging-in-Publication Data
Spatial uncertainty in ecology: implications for remote sensing and GIS applications/
edited by Carolyn T. Hunsaker ... [et al.].
       p. cm.
   Includes bibliographical references (p.).
   ISBN 978-0-387-98889-4   ISBN 978-1-4613-0209-4 (eBook)
   DOI 10.1007/978-1-4613-0209-4
       1. Spatial ecology—Mathematical models.   2. Uncertainty (Information theory)
I. Hunsaker, Carolyn T.
   QH541.15.S62 S63 2001
   577′.01′5118—dc21   00-061268

Printed on acid-free paper.

Production coordinated by Chernow Editorial Services, Inc., and managed by Francine
McNeill; manufacturing supervised by Erica Bresler.
Typeset by Scientific Publishing Services (P), Ltd., Madras, India.

9 8 7 6 5 4 3 2 1

ISBN 978-0-387-98889-4

Acknowledgment

Working Group Participant. This work was conducted as part of the *Spatial Uncertainty* Working Group supported by the National Center for Ecological Analysis and Synthesis, a Center funded by NSF (Grant #DEB-94-21535), the University of California at Santa Barbara, and the State of California.

Contents

Part I Introduction

Part II Dimensions of Spatial Ecological Data

Part III Methods

Part IV Epilog

Contributors

KATE BEARD
National Center for Geographic Information and Analysis, University of
Maine, Orono, ME 04469, USA

BARBARA P. BUTTENFIELD
Department of Geography, University of Colorado, Boulder, CO 80309,
USA

TED J. CASE
Department of Biology, University of California, San Diego, La Jolla, CA
92093-0116, USA

NOEL CRESSIE
Department of Statistics, The Ohio State University, Columbus, OH 43210-
1247, USA

RONALD EASTMAN
Department of Geography, Clark University, Worcester, MA 01610, USA

GEOFFREY EDWARDS
Department of Geomatic Sciences, Laval University, Ste. Foy, Quebec
G1K 7P4, Canada

PETER FISHER
Department of Geography, University of Leicester, Leicester LE1 7RH, UK

ROBERT N. FISHER
USGS Biological Research Division and Biological Field Stations, Depart-
ment of Biology, San Diego State University, San Diego, CA 92182-4614,
USA

MARIE-JOSÉE FORTIN
School of Resource and Environmental Management, Simon Fraser
University, Burnaby, British Columbia V5A 1S6, Canada

STEVEN E. FRANKLIN
Department of Geography, University of Calgary, Calgary, Alberta T2N
1N4, Canada

MARK A. FRIEDL
Center for Remote Sensing, Department of Geography, Boston University,
Boston, MA 02215-1401, USA

MICHAEL F. GOODCHILD
National Center for Geographic Information and Analysis and Department
of Geography, University of California, Santa Barbara, CA 93106-4060,
USA

CAROLYN T. HUNSAKER
USDA Forest Service, Forestry Sciences Laboratory, Pacific Southwest
Research Station, Fresno, CA 93710-4639, USA

PHAEDON C. KYRIAKIDIS
Department of Geography, University of California, Santa Barbara, CA
93610-4060, USA

KENNETH C. MCGWIRE
Desert Research Institute, Earth and Ecosystem Sciences, Reno, NV 89512,
USA

DOUGLAS K. MCIVER
Center for Remote Sensing, Department of Geography, Boston University,
Boston, MA 02215, USA

KEVIN S. MCKELVEY
USDA Forest Service, Forest Sciences Laboratory, Rocky Mountain
Research Station, Missoula, MT 59807, USA

BARRY R. NOON
Department of Fishery and Wildlife Biology, Colorado State University,
Fort Collins, CO 80523, USA

ASHTON SHORTRIDGE
Department of Geography, Michigan State University, East Lansing, MI
48824, USA

FRED H. SKLAR
South Florida Water Management District, Everglades Systems Research Division, West Palm Beach, FL 33406, USA

PETER A. STINE
USDA Forest Service, Pacific Southwest Research Station, Sierra Nevada Framework for Conservation and Colloboration, Sacramento, CA 95814, USA

JAY M. VER HOEF
Wildlife Conservation Division, Alaska Department of Fish and Game, Fairbanks, AK 99701, USA

A-XING ZHU
Department of Geography, University of Wisconsin, Madison, WI 53706, USA

Part I
Introduction

1
Introduction

Michael F. Goodchild and Ted J. Case

In 1650 Elsevier Press published *Geographia Generalis* by the young Bernhard Varenius. It represented one half of Varenius's view of the world, which was divided into general geography (i.e., the principles of the earth's geometry and the processes that are responsible for creating and modifying its human and physical landscape), and special geography (i.e., the description and inventory of the unique characteristics of places). The book came to the attention of Sir Isaac Newton, who had it translated, "improved" and illustrated, and republished in London (Warntz 1989). The prevailing view of the surface of the earth was then and remains today essentially Newtonian, with space and time as rigid frames, and quantum and relativistic effects well outside the range of relevant scales.

This book is about ecology, which focuses on one of the more important of the sets of dynamic processes that are responsible for the patterns and phenomena we find on the earth's surface. *Ecology* studies the dynamics of populations of living organisms, including the interactions that occur both within populations, and between populations and their environment. Ecological knowledge is thus part of Varenius's general geography because all other things being equal the behavior of a population of living organisms is invariant under relocation anywhere on the earth's surface.

1.1 Ecological Models

Dynamic processes are driven by interaction, and the opportunity for interaction typically declines with physical separation. Animals are able to overcome the effects of limited physical separation through motion, and plants through transport mechanisms such as wind, so early attempts to model ecological dynamics often ignored physical separation. Although each organism takes up a finite amount of space and interacts with other individuals only in some neighborhood defined by its movements, space itself and movements across it were largely left out of theoretical thinking. It was implicitly assumed that qualitative results would be insensitive to the

3

locations of individuals. The state variables in these theories were abundances of different species or levels of composite factors, without assigning individuals or populations to specific points in space. Models addressed the existence and stability of various species associations.

It was gradually recognized that several important conclusions were critically linked to the neglect of spatial movements. For example, poorly competitive species that could not coexist locally could nevertheless coexist globally in a wider region, if dispersal rates to unoccupied patches were high enough to compensate for their lack of competitive ability. An understanding of *fugitive species* thus requires a spatial backdrop and limits to dispersal, as well as an understanding of population extinction. For such a fugitive species to persist and colonize, the appearance of new vacant patches clearly must be a relatively common occurrence. Vacant patches can arise by extinctions of competitors in existing patches, or because physical processes (i.e., floods, earthquakes, landslides, fires, or hurricanes) create new openings in the landscape.

For purposes of discussion we can distinguish two types of models. *Patch occupancy models* deal with an archipelago of habitat islands surrounded by a sea of less hospitable terrain. Patch occupancy models use as the state variable the fraction of these patches that are occupied by a species, regardless of its density within. In fact, an assumption is made that the densities in each patch are equivalent, which might be the case if the species quickly reached a carrying capacity that was approximately the same for all patches.

A *meta-population* is simply a population of populations (Levins 1970). In the simplest meta-population model patches already occupied by a species are the source of colonists to unoccupied patches. Furthermore, another very restrictive assumption is typically invoked: The spatial arrangement of occupied and empty patches makes no difference to the colonization rate for empty patches. All empty patches are equally accessible to individuals from all occupied patches regardless of their distance away. Most animals have more limited dispersal abilities than this, and are more likely to find vacant patches nearby than far away.

Patch occupancy models can be modified to incorporate distance-mediated dispersal in an implicit way. Distance is irrelevant for each individual within a particular subpopulation of the meta-population, but movements between populations are affected by some simple function of the physical distance between them. Outcomes of these spatially implicit meta-population models are invariant under relocation of individuals within meta-populations, but not invariant under relocation of meta-populations on the earth's surface.

The second type of model we will call *spatially explicit*. Two subclasses can again be distinguished. In individual-based models, a finite number of individuals are tracked as they move, give birth, and die. Movement algorithms may range from a simple random walk to correlated walks or to more sophisticated learning using neural nets. Parameters of movements, births and deaths can be explicitly tagged to landscape and habitat features. On the other hand, the state variables may be the numbers of individuals at spatial points (continuous space) or in spatial grid cells (discrete space).

A *reaction-dispersal* equation results when space and time are continuous and movements are simple random walks. For two-dimensional space (the plane) the model for species I in space x, y takes the form:

$$\frac{dN_i(x, y, t)}{dt} = F_i(N_1(x, y, t), N_2(x, y, t), \ldots, N_n(x, y, t))$$
$$+ D_i \left[\frac{\partial^2 N_i(x, y, t)}{\partial x^2} + \frac{\partial^2 N_i(x, y, t)}{\partial y^2} \right]$$

In other words, the growth rate of species i at position x, y and at time t is equal to the sum of a growth function for species i that depends on the densities of all of the species at position x, y at time t (the *reaction*) part of the equation, and a term representing *diffusion* across two-dimensional space x, y with diffusion coefficient D_i. With several interacting species the reaction-diffusion equations are coupled through the reaction F_i functions, and spatial locations are coupled through the diffusion term. The analysis and solution of such equations can be quite formidable. The important feature to note, however, is that the F_i functions will typically contain parameters that affect birth and death rates, whose magnitude depends on spatial position x, y. Finally, the dispersal component may be more complicated than diffusion; for example, movements downhill or downwind may be more likely than uphill or upwind. Movement rules may change in different habitat patches, and obstacles on the landscape may prevent movements for some species but not for others.

By giving each organism a location and expressing movements directly through a spatial metric, both individual-based, patch-occupancy models with distance-mediated dispersal and reaction-dispersal models create situations where ecological outcomes are no longer invariant under relocation of the patch or organisms on the earth's surface.

In the history of dynamical models in ecology, spatially explicit models were born out of purely theoretical concerns, yet, as the models developed and as they stimulated more interest in the movement rules of real organisms and how the texture of the landscape influenced these movements, ecologists began applying these models to real species in real places, not just to abstract processes on hypothetical landscapes. Ecologists addressed environmentally relevant questions such as: How will human alteration of the landscape affect the persistence of legally protected sensitive species? Where are pollutants likely to end up on the terrain? How much soil and nutrients will be lost from a watershed and where will they end up? With sea-level rises due to global warming, which parts of the landscape may become submerged and when? The chapters in this book attest to the several real-world applications.

Two technical innovations also greatly accelerated the application of spatial models to landscape-level problems. Advances in remote sensing allowed a much higher resolution of variation across real landscapes to be provided in readily accessible data, and geographic information systems (GIS) provided a spatially tagged record-keeping system for managing, visualizing, and analyzing these data (for general introductions to GIS see, for example, Burrough and McDonnell 1998; Longley et al. 1999; and for their applications in ecology see, for example, Goodchild et al. 1993, 1996; Haines-Young et al. 1993; Johnston 1998). For the first time, ecologists could test the hypotheses that they had developed at the landscape scale. Ecology embraced geography: remote sensing and GIS allowed maps at large spatial scales to become dynamic in ecologically relevant time scales.

With the application of ecological research to legislatively driven environmental problems came a greater need to quantify the likelihood of alternative outcomes predicted by dynamic models. Error and uncertainty enter at various stages and mix into the synthesis of ecological thinking and into the formats of models. If ecology was to be relevant it would have to quantify its uncertainty about spatial dynamics, and do so urgently, within the time scales of environmental decision making. This book is a direct outgrowth of this urgent need.

1.2 Uncertainty in Spatial Processes

The outcomes of any dynamic ecological process depend directly on the initial or boundary conditions that provide its inputs. Even if the earth were a flat, rotationless object in a fixed position relative to a constant sun, ecological and other processes would have created nonuniformity, which would then influence the outcome of subsequent processes. Termites have to build their mounds somewhere, even on a flat, featureless plain; and the basic instability of many processes ensures, for example, that rivers will form and erode valleys in response to the slightest imperfection in the surface. In reality, however, the boundary conditions provided by the earth's surface are extremely diverse. Some processes serve to smooth, with outcomes that reduce the high-resolution variation in their inputs (i.e., glaciation removes much of the high-resolution variation in terrain). Other processes simply preserve the spatial patterns of their inputs; and many ecological processes create or enhance high-resolution variation, as in the case of the termite mounds or the landscape architecture of the beaver. Outcomes of one set of processes provide inputs to others, ensuring that any process on the earth's surface will have complex spatial variation in its boundary conditions.

The boundary or initial conditions of ecological processes are Varenius's special geography, captured today in the spatial databases that are increasingly available in digital form through such technologies as remote sensing, GIS, and the World Wide Web. To someone concerned with planning, managing, or conserving, knowledge of both forms of geography is essential

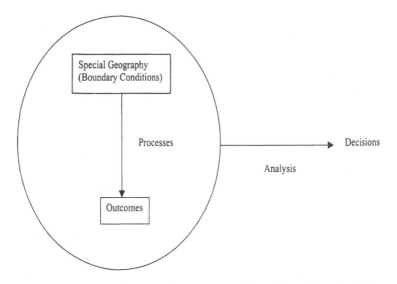

FIGURE 1.1. The role of spatial data, process models, and predictions in decision making.

in order to predict outcomes, and make rational decisions. In Figure 1.1, ecological processes operate under boundary conditions to create outcomes, which are then analyzed in order to support decisions. It would clearly be absurd to try to make decisions based on knowledge of processes alone without understanding what outcomes those processes caused under the boundary conditions present at some location on the earth's surface.

   This book is about the problems that uncertainties create in this simple framework. It is virtually impossible to have certainty about the analysis process that leads to decisions because in today's world decisions almost always involve multiple stakeholders with multiple viewpoints. Knowledge of ecological processes is similarly never perfect, so uncertainty is present whenever outcomes are predicted. Finally, we can never have perfect knowledge of the boundary conditions because the real world is far too complex to be fully measured, observed, or represented. This book covers all aspects of uncertainty, although its heaviest emphasis is on the third kind: uncertainty in the spatial data that provide the boundary conditions for ecological processes. We have chosen to place the emphasis there for several reasons: spatial data uncertainty is a rich area of research in several fields, as the chapters of this book illustrate, but this literature is relatively unknown in ecology, and the methods are not often applied; application in ecology will help to enrich the set of examples and motivating applications for the general study of spatial data uncertainty; and we believe the problems posed by spatial data uncertainty are among the most challenging, the most important for decision makers, and may in some cases be the largest in magnitude.

Spatial data uncertainty arises for many reasons. Complete measurement of the real world may be impossible because of cost, or because the capacity to store and process the resulting volume of data is simply not available. Generalization is almost inevitable, with high-resolution data either not being collected or being discarded at some point through a process of generalization. One of the largest sources of uncertainty lies in the lack of formal methods of generalization; instead, generalization in fields like cartography often has elements of subjectivity and art that have proven very resistant to formalization, and it may be difficult to determine exactly what objective relationship exists between a generalized map (i.e., a map of soils) and the reality that it is intended to represent. Uncertainty arises from measurement errors, notably measurements of position, and from the spurious changes introduced by format conversion and digitizing. It is also present in the essential fuzziness and vagueness associated with much geographic data. We define *uncertainty* for the purposes of this book as:

The state of knowledge about a relationship between the world, and a statement about the world.

Uncertainty defined in this way exists, for example, when an item of data indicates that the elevation at a given point is 200.5 m, but we know that the expected difference between that item of data and the true elevation is 2.4 m; or we may know that the difference between the true vegetation class in an area, and the database record for that area, is characterized by a probability of correct classification of 0.85.

1.3 The Book's Organization

The book has two major parts. The first section includes many demonstrations of the use of spatial data in ecological analysis, and instances of the framework shown in Figure 1.1. The chapters in this section demonstrate the importance of all three kinds of uncertainty and its essentially multidimensional nature.

The second section presents a collection of models of uncertainty, framed within a variety of paradigms that range from spatial statistics to cognitive science. It is clear to us that no single paradigm can fully frame the complex multidimensional issue of uncertainty, and that some paradigms may be more useful in some circumstances than others. Each of them, though, possesses elements of the four-stage structure shown in Table 1.1, which provides an expanded view of the processes implicit in Figure 1.1, and is commonly used to structure discussions of spatial data uncertainty and its impacts (see, e.g., Goodchild and Gopal 1989; Guptill and Morrison 1995; Burrough and McDonnell 1998; Heuvelink 1998; Goodchild and Jeansoulin 1998).

TABLE 1.1. Issues at various stages of application of spatial data.

Stage	Uncertainty issue
Data	Description of uncertainty
Display	Visualization of uncertainty
Analysis, modeling	Models of uncertainty
Decision making	Propagation of uncertainty, confidence limits

Attention initially focuses on the description of the uncertainty present in data. Although various generic methods of uncertainty description exist, and are reviewed in many of the chapters of this book, the problem is always to describe and measure uncertainty in generic ways that are not specific to applications. We hope, for example, that in describing the positional uncertainty of features in a data set we will later be able to analyze the impacts of this uncertainty on the results of analysis and modeling in numerous applications, but that may not always be true. In that context, the uncertainties described and measured by a collector or disseminator of data represent a compromise designed to satisfy as many as possible of the diverse needs of the data's users, just as the standard deviation represents a compromise among the numerous possible ways of describing variation in a sample of measurements of a scalar.

The second stage addresses the vital issue of communication of uncertainty to the users of data. Data increasingly pass between numerous custodians in their life cycle, and the person who eventually uses the data may be unable to contact or communicate with the person who created them. Moreover, that person may be a member of the general public or of a nontechnical discipline who is unfamiliar with the complexities of spatial statistics, and likely to reject any description that is couched in such terms. Visualization is one of the most important tools for communicating effective information on uncertainty and for giving the user some useful basis for assessing its impacts.

Models of uncertainty (e.g., the statistical models exemplified by the classical Gaussian distribution) are essential if we are to understand the impacts of uncertainty on the results of modeling and analysis. They provide the basis for an analytic derivation of confidence limits in a specific instance, or the basis of simulation models which will be used to explore the impacts of uncertainties, especially for users who are not technical experts.

Finally, in the best of all possible worlds every user of spatial data would be provided with a set of confidence limits on any results of analysis, predictions of models, or inputs to decisions. Those confidence limits might then be incorporated into decision-making processes to assess the risks associated with each option. Increasingly, the results of decisions made using spatial data, via such technologies as GIS, are being tested in court under legalistic standards of rigor. Today, maps of wetlands or vegetation cover are being used to implement regulations regarding land use, and those regulations are clearly open to challenge if it can be shown that the regulator knew about uncertainties in the maps, and failed to deal with them appropriately or responsibly. Even though uncertainties may be no more than an inconvenience to the scientific community, they may be the Achilles' heel of attempts to turn the results of science into public policy.

References

Burrough, P.A., and R.A. McDonnell. 1998. Principles of geographical information systems. Oxford University Press, New York.

Goodchild, M.F., and S. Gopal, eds. 1989. Accuracy of spatial databases. Taylor & Francis, London.

Goodchild, M.F., and R. Jeansoulin, eds. 1998. Data quality in geographic information. Editions HERMES, Paris.

Goodchild, M.F., B.O. Parks, and L.T. Steyaert, eds. 1993. Environmental modeling with GIS. Oxford University Press, New York.

Goodchild, M.F., L.T. Steyaert, B.O. Parks, C.A. Johnston, D. Maidment, M. Crane, et al., eds. 1996. GIS and environmental modeling: progress and research issues. GIS World, Fort Collins, CO.

Guptill, S.C., and J.L. Morrison, eds. 1995. Elements of spatial data quality. Elsevier Science, New York.

Haines-Young, R., D.R. Green, and S.H. Cousins, eds. 1993. Landscape ecology and GIS. Taylor & Francis, London.

Heuvelink, G.B.M. 1998. Error propagation in environmental modelling with GIS. Taylor & Francis, London.

Johnston, C.A. 1998. Geographic information systems in ecology. Blackwell Science, Oxford.

Levins, R. 1970. Extinction. Pages 77–107 in M. Gerstenhaber, ed. Some mathematical problems in biology. American Mathematical Society, Providence, RI.

Longley, P.A., M.F. Goodchild, D.J. Maguire, and D.W. Rhind, eds. 1999. Geographical information systems: principles, techniques, applications and management. John Wiley & Sons, New York.

Warntz, W. 1989. Newton, the Newtonians, and the Geographia Generalis Varenii. Annals of the Association of American Geographers 79(2):165–91.

Part II
Dimensions of
Spatial Ecological Data

One basic null hypothesis in ecology is that the future state of a particular piece of land does not depend upon the state of its neighbors. As naïve as this sounds, it has dominated the models in ecology. This is often referred to as a *mean-field assumption* in dynamical systems. It essentially assumes that each individual interacts with all others independent of their physical location. Each individual experiences the average, or mean-field environment across space—the particular geometric arrangement of these individuals over space is ignored. The inclusion of explicit spatial context into ecological dynamics adds realism, but at the cost of increasing model complexity with many more parameters to be fit from observational data. The reward is not just a matter of gaining a greater degree of spatial resolution and prediction for particular units of landscape; rather, it gives entirely new insights about potential ecological outcomes. The inclusion of spatial processes and neighborhood effects in modeled dynamics can qualitatively change the ecological outcome altogether.

The next four chapters in this book deal with some common uses of spatial data in ecology and resource management, particularly as they are affected by spatial uncertainty. There are two related issues here, one phenomenological and one epistemological: (1) How does spatial and temporal variation affect community function and composition? (2) How should we propagate positional uncertainty in our data gathering and through our calculations to avoid biased statistical inferences about such things as species habitat affinities, a population's likelihood of persisting in a particular patch, or the outcome of, for example, changing water regime management in the Everglades on wood stork recruitment?

Chapter 2, by Sklar and Hunsaker, reviews the many ways to include spatial information into dynamical models of ecological processes operating on real landscapes. They provide a framework for the systematics of these diverse models and go on to discuss five common sources of uncertainty in spatial data and models: data collection, data processing, model structure, human intervention, and natural variability. They suggest that despite a diversity of approaches, uncertainty will propagate through four components of model structure: inputs, initial conditions, calibration, and validation. They also show how some of these models can be used to evaluate the outcome of management decisions (e.g., water flow regulation in the Everglades), where consequences might only materialize 40 or 50 years later but have dramatic effects on sensitive species protected by legislative mandates.

Fisher and Case (Chap. 3) describe research in progress on the reptiles and amphibians in Southern California, one of the test grounds for new policy under the Federal Endangered Species Act. They find that the occurrence and probable persistence of several species depends on the state of a particular patch of land as well as on the patch location with respect to adjacent land use. Moreover, they wrestle with the problem of developing a predicative habitat affinity model when the spatial data from field studies contains autocorrelations that violate the principle of independence underpinning typical parametric statistics. This theme and some of the data are also analyzed later in this book by Ver Hoef et al. (Chap. 12).

Resource managers and decision makers must know the reliability of conclusions reached from an analysis using geographic information systems (GIS). For many GIS applications, it is not possible to compare these products with an independently derived "truth." The alternative is to conduct a sensitivity analysis based on randomization of the data set, assuming reasonable forms for the error structure at the various stages of data collection and analysis. This is the approach taken by McKelvey and Noon (Chap. 4) as they develop a habitat affinity model for the northern spotted owl and assess the effect of location and vegetation classification errors. Location errors arise from difficulties in pinpointing the exact location of radiocollared individuals by workers on the ground because owl positions can only be estimated approximately by triangulation. Their simulations show that more correct inferences (i.e., closer to the truth) are gained by keeping probability distributions for both types of errors in the analysis, rather than assigning simply the mean or modal value to location and vegetation state, as is often done. This is especially true for organisms with strong associations to fairly uncommon vegetation types.

The last chapter in this group, by Stine and Hunsaker forms a bridge to the next section of the book, which focuses on formal methods of analysis of uncertainty. Stine and Hunsaker review areas where spatial variation is necessary for an understanding of ecological function. They also briefly review developments in spatial statistics, remote sensing, and GIS to quantify and communicate spatial patterns. Finally, they foreshadow the next section of this book by summarizing ways that spatial uncertainty can be measured and analyzed to better inform scientists and decision makers.

An argument could be made that the order of this book's sections should be reversed (i.e., why have what amounts to the "results" section come before the "methods"?). This latter organization would make sense if we had agreed-upon methods for solving different spatial problems and fixed recipes to follow for handling and communicating uncertainty, but as pointed out in the introduction to the Methods section, the field has not yet reached this level of development. The editors felt that the nuts-and-bolts of quantitative analysis could not be fully appreciated without first convincing the reader that their effort would pay off; that a range of interesting and practical problems require that these methods be understood, accessible, and, we hope better tailored to their specific applications. These four chapters, although diverse, by no means exhaust the ways that ecologists think about spatial variation and uncertainty in nature and in their data. We hope, however, that they motivate the reader to consider the scope of activities in ecology, resource-management, and land-use planning that require a fuller arsenal of spatial metrics and analysis.

2
The Use and Uncertainties of Spatial Data for Landscape Models: An Overview with Examples from the Florida Everglades

FRED H. SKLAR and CAROLYN T. HUNSAKER

Models are usually developed for one of two purposes: to better understand ecological systems or to evaluate the influence of an altered condition to make a decision or set policy. A spatial model can loosely be defined as one that has either one or more state variables that are a function of space or can be related to other space-dependent variables. The information-richness of spatial data and models comes from their simultaneous depiction of temporal and spatial patterns. This richness is easily grasped by only the most powerful of all computers, the mind. For example, the mind can immediately compare a temperature map of today with the map from the day before and quickly develop a spatial trend analysis for areas of change. Put three maps together and the mind computer may even venture to predict areas cooling down or warming up. It does this easily, with no thought of the spatial distribution of weather stations, kriging techniques, data collection errors, the influence of scale, or knowledge of meteorological models, yet these things and more are all factors that contribute to the spatial uncertainty of this *mental model*. These uncertainties become extremely important when one tries to formalize and quantify this spatial modeling process for prediction.

Ecological modeling differs greatly among the subdisciplines of ecology, but it is developing in the same general direction: integration of temporal and spatial effects (Hunsaker et al. 1993). Most advances in the past concerned temporal effects, and considerable progress has been made. The next logical step was to consider spatial phenomena. The null hypothesis in spatial analysis is that the future state of a unit on a landscape is independent of adjacent units. This assumption was often implicit in many ecological models, but the degree to which it is true remains unclear.

Landscape models have become popular as ecological research and natural resource management tools (Sklar and Costanza 1991; Hunsaker et al. 1993; Sklar et al. 1999). These spatial models include transition probability models (Turner 1987; Wu et al. 1997), gradient models (Walker 1995), geostatistical models (Ripley 1981; Turner et al. 1991), distributional mosaic models (Costanza et al. 1990; Fitz and Sklar 1999), and individual-based models (DeAngelis and Gross 1992; Wolff 1994). They can be static (analytical) or dynamic (a function of time). They can be simple or complex. They all use spatial data such as, topography, rainfall, soil type, or land use type. These inputs, as part of the landscape modeling process, are either interpolated across space, analyzed for trends, or simulated to understand ecological interactions and predict spatial change. Fast number-crunching machines and huge datasets, made possible by advances in computer technologies, geographic information systems (GIS), and remote sensing have inspired a second quest by ecologists (the first followed Lotka 1925 and Volterra 1926) for ecological causality and mechanics. The paradigm of classical physics, particularly laws of conservation and thermodynamics, are now the *boundary conditions* within which all biological processes are constrained (Prigogine et al. 1972a, 1972b; Odum 1983). These new tools and laws, however, have also heightened scientific awareness of the irregularity, variability, and unpredictability of ecological systems (Breckling and Dong 2000) such that the quest now is for a better understanding and quantification of the *uncertainty* of ecological causality and mechanics.

A generalized understanding of different landscape models suggests that despite a diversity of approaches, which we will discuss, uncertainty resides within four components of model structure. These components are: (1) inputs (e.g., spatial interpolation of point data), (2) initial conditions (e.g., spatial interpretation of point data), (3) calibration (e.g., spatial interpretation), and (4) verification (e.g., spatial comparisons). Although all spatial models are subject to all the same four components of model error, some are more sensitive to one component than another. In this chapter we will also discuss the sources of uncertainty in spatial data and models that can be traced to problems associated with (1) data collection, (2) data processing, (3) model structure, (4) human intervention, and (5) natural variability. Reducing these uncertainties may be as simple as taking multiple samples or as complex as running a simulation 1000 times. Here, we present a descriptive overview of the types of uncertainties inherent to the field of landscape modeling. Other chapters in this book will demonstrate techniques for describing, quantifying, and/or reducing these uncertainties. For a more mathematical formulation of the errors attributed to modeling landscape attributes from fine-scale ecological data and knowledge we recommend Rastetter et al. (1992).

2.1 Data for Landscape Models and Some of Their Associated Uncertainties

All landscape models require data. These data are used as inputs, initial conditions, calibration parameters, and verification components. Stine and Hunsaker (see Chap. 5) discuss types and sources of error and uncertainty associated with spatial data. The modeling process is such that the scale of simulated events and states should match the scale of events and states of the data used by the model as closely as possible (Sklar et al. 1990). That is to say, in terms of both spatial and temporal organization, the scale of a habitat model is very different from that of a global model. For example, if one used a global temperature model as an input to a habitat model of California grape production, then the result would be economic collapse for the wine industry because the microclimate needed to estimate grape production is not captured at the global scale. This is not to say that a global temperature model could not be used to identify *potential* grape-producing regions. The mismatching of scales will most often produce uncertainties associated with model structure and conclusions. Errors associated with mismatching of scales are unfortunately not easy to identify. A landscape model operating at multiple scales creates the problem of possible internal scale inconsistencies. These types of uncertainties can only be avoided by careful peer-review and clearly defining and bounding the model objectives within a set of spatial, temporal, and structural constraints.

Spatial and temporal constraints on a landscape model are related to the model building process (Fig. 2.1). Every landscape model requires an initial condition: a point in time and a region in space where the model begins. It

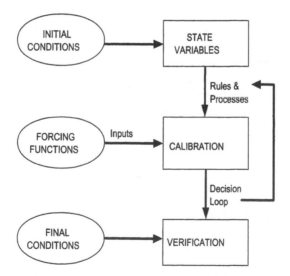

FIGURE 2.1. The model-building process (rectangles) and associated spatial data (ovals) has three distinct levels of interaction. Uncertainties exist at the initialization phase when spatial data are interpreted and configured to fit the structure of the model. Uncertainties exist at the calibration phase when spatial data are rescaled and interpolated to fit the time frame of the model. Uncertainties exist at the validation phase when spatial data are re-scaled, interpreted, and interpolated to fit the structure and time frame of the model output.

is rare to find a situation where spatial boundaries are clear and where a complete set of appropriately scaled landscape maps exist. In most cases, the number of state variables in the model requiring initialization is much greater than anything available. As a result, it is necessary to create maps by spatial interpolation and interpretation of point data. The most utilized of all such maps are topography/elevation quadrates, which are landscape models and are called digital elevation models (DEMs). The contour maps produced by DEMs are used by landscape models for either categorizing spatial features or simulating gravitational material flows. In addition to DEMs other coverages used to initialize landscape models include soil type (e.g., porosity, bulk density, nutrient content), vegetation (e.g., biomass, density, species), animals (e.g., species, distribution, abundance), water depth, and temperature. All of these coverages have a level of uncertainty that is affected by the number and distribution of point data, the complexity of the landscape, the interpolation method, and the resolution required for the landscape model. A very hetereogeneous landscape represented by only a few data points used as input to a high-resolution model are the ingredients for extreme initialization error. This situation fortunately may have a minimal effect on some landscape models, particularly if the modeling structure is found to be insensitive to initial conditions (relative to processes and rules) or the modeling objectives require a range of initial conditions. It is actually a test for initial conditions (e.g., percolation models, Gardner and O'Neill 1991).

Forcing functions, used by mechanistic models, are another important source of uncertainty in the landscape model building process (Fig. 2.1). Forcing functions are the inputs needed to move the simulation from time t_m to t_{m+1}. They are not modified or simulated by the modeling algorithms. Temperature is the best example. Temperature affects many biogeochemical processes, but it is very difficult to simulate. Most landscape modelers use weather station data as input rather than write complex heat budget equations (this is not true of atmospheric and climate modelers). The most significant uncertainty associated with temporally dynamic forcing functions, like temperature, is missing data. Accurate sensors exist, but 0% failure does not. To *correct* for missing temporal data, one can use averages, interpolation schemes, alternative data sources, or predictive models. None of these are 100% certain. Other common forcing functions include precipitation, wind, solar radiation, river discharge, water depth, and material loadings. These can be spatially continuous (e.g., precipitation) or spatially discrete (e.g., river discharge). Spatially continuous data require numerous data collectors throughout the landscape; otherwise, spatial uncertainties become significant. Spatially discrete input data only require that the locations of point sources be known and that the data collectors are working properly.

All landscape models have some kind of mathematical structure that defines rules, processes, statistical relationships, or state change. The net result is a time-series of simulated landscape variables. This calibration step in the model building process requires a comparison of the simulated state and vector variables with those actually measured and a re-adjustment of the mathematical constructs to maximize observed and simulated resemblance. This step, for the setting of parameter values, is usually devoted to quantifying flows, fluxes, and rates of change. There are numerous parameter estimation procedures (i.e., least-squares fitting, Snedecor and Cochran 1967; Kalman filtering, Cosby et al. 1984; Monte-Carlo fitting, Hornberger et al. 1985; Levenberg-Marguardt fitting, Press et al. 1986; maximum likelihood, Brown and Rothery 1993). It is not our purpose here to evaluate these procedures; however, we do want to point out that model calibration uncertainties are a function of errors associated with both observational and simulated data. The observational data have the same uncertainties as forcing functions and initial conditions plus the added uncertainties associated with the interpretation of experimental data. The simulated data have uncertainties associated with model structure that is a function of the type of landscape model (more on this point later in the chapter).

In general, the more complex the model structure in terms of temporal, spatial, and variable articulation, the greater the total variation of observational and simulated data, and the less accurate (more uncertain) the model (Costanza and Sklar 1985). Modelers tend to ask narrow questions in order to achieve higher accuracy. Models with low complexity can achieve higher accuracy because they say much about little; however, the real effectiveness or explanatory power of a model is a function of both how much it attempts to explain (complexity) and how well it explains what was attempted (accuracy). In a survey of wetland models with varying degrees of temporal, spatial, and structural complexity, Costanza and Sklar (1985) developed an index of effectiveness or explanatory power to measure the balance between accuracy and the broadness of the questions being asked (complexity). The result was Figure 2.2. The upper bound for effectiveness was low for low-complexity, high-accuracy models, increased to a maximum, and decreased again for high-complexity/low-accuracy models (those that say little about much). This implies that there is an optimum size or complexity beyond which the *benefits* of additional complexity are outweighted by the *costs* of lowered accuracy.

Landscape models tend to be very complex. An analysis of accuracy and effectiveness has not been done on these models, so it is difficult to say how much uncertainty has been added due to the higher complexity of the spatial component. One way to compensate for this structural uncertainty is to increase the number of state and vector variables in the calibration process. This requires more observational data but serves to decrease the structural uncertainty. For landscape models, the calibration process should be an iterative process of model reconstruction to simplify temporal and structural complexity and maximize coefficients of determination.

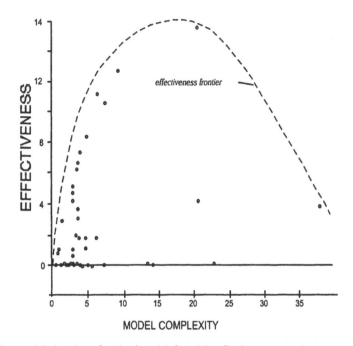

FIGURE 2.2. A review of wetland models found the effectiveness or explanatory power of a model was a function of both how much it attempts to explain (complexity) and how well it performs (accuracy). Models with low complexity can achieve higher accuracy because they say much about little, whereas models with high complexity achieve lower accuracy because they say little about a lot. The Effectiveness Index (the accuracy normalized for complexity) was plotted as a function of complexity. The result was a bell-shaped curve and a relationship between effectiveness (i.e., information) and model complexity that is suggestive of an *Effectiveness Frontier*, where both accuracy and information can be optimized (reprinted from Costanza and Sklar 1985, with permission from Elsevier Science).

Finally, there is the uncertainty associated with the verification process of model building that also involves the spatial and temporal comparisons of observational and simulated data. The difference here is that the observational data are usually spatial state variables that were not utilized in the calibration process. They may be population distributions, or vegetation types, or water quality data never seen by the modelers. Existing data can also be set aside specifically for the verification process. The degree to which these data are both temporally and spatially reproduced by the landscape model is a measure of model verification. The uncertainties of verification are the same as those for calibration except for one difference: verification errors increase with time (i.e., the prediction period). Predicting tomorrow's weather is much less uncertain than predicting the weather for next week. The validity of landscape models will diminish as prediction time increases because uncertainty is a cumulative component of models (Rastetter et al. 1992).

2.2 Landscape Models in Ecology

Ecological modeling has historically had a temporal emphasis (i.e., projections of state changes at one location over time). Different approaches were necessary for terrestrial, aquatic, and marine models (Hunsaker et al. 1993). Terrestrial plant models did not consider organisms moving over the ground, whereas animal models did. Aquatic models had to consider that organisms were embedded in a medium (water) that moved; only rarely was movement of the terrestrial medium (the soil) an important factor in terrestrial models. The notion that animals and plants in oceans and lakes are always immersed in some form of chemical solution was also considered in aquatic models. Aquatic models often incorporated chemical and toxicity submodels. It was necessary in aquatic models to integrate temporal and spatial processes (two dimensional and three dimensional) for many organisms and hundreds of model grid cells. Hunsaker et al. (1993) reviewed the current state of dynamic spatial models for terrestrial, freshwater aquatic, and marine ecosystems. It is interesting to note that even though three-dimensional models for freshwater aquatic systems existed prior to GIS being commonly available, models that incorporate spatial processes are more common for terrestrial ecosystems. Although marine ecosystems should lend themselves to spatial models, few spatially dependent marine ecosystem models existed as of 1993.

The expansion of ecological research during the last two decades of the twentieth century, from relatively small plots of land to large landscapes is reflected in the expansion of ecological models from a temporal emphasis (i.e., spatially implicit models) to ones that are spatially explicit (Sklar et al. 1994). The temporally dynamic, spatially implicit models extrapolate output to the entire landscape by assuming that the landscape is homogeneous. These models simulate spatial change and differences either by altering the forcing functions to represent different regional inputs or by altering the model structure to represent different regions or both. In the United States, in 1984, the inability of this modeling approach to describe landscape features or predict spatial change quantitatively across biogeochemical gradients gave rise to spatially explicit procedures and definitions for a new discipline (Risser et al. 1984). There have been hundreds of landscape models developed since 1984 for regions as different as deserts and oceans. Reviews of some of these can be found in Sklar and Costanza (1991) and Hunsaker et al. (1993).

We have, for this overview, divided dynamic landscape models in ecology into four general categories (Table 2.1). Although not all landscape models can be categorized into these designations this will be of value because most landscape models combine or integrate features from these approaches.

TABLE 2.1. The objectives, characteristics, and uncertainties associated with dynamic landscape models.

Model type	Objective	Scale*	Typical variables	Basic characteristics	Uncertainty characteristics
Transitional Probability (Whole mosiac)	Calculate a steady-state landscape	TF = 10–500 yrs dt = years SF = 100–5,000 km ds = 1–100 m	Land use type Vegetation type Neighborhood effects Density Distance	Probability of spatial change is based upon descrete time-series maps and neighborhood characteristics	Most susceptible to initialization and validation errors. Uncertainty arises from errors in spatial interpretation and interpolation of raster and point data.
Gradient (Mass balance)	Predict mass balance across an energy gradient	TF = 1–50 yrs dt = 1 min.–5 days SF = 1–10,000 km ds = 1–100 m	Pollutants Nutrients Seeds Water Sediments Diffusion Mass	Flow of matter is simulated as continuous function controlled by rates of diffusion, motion, momentum, and first-order kinetics	Most susceptible to forcing function and calibration errors. Uncertainty arises from natural variability and data collection errors along the boundaries, particularly at the "upstream" region of the gradient. Uncertainty associated with calibration is due to aggregation and simplification of model structure and parameters.
Process-based (Distributional mosaic)	Predict state change and simulate mechanisms of change across a spatial pattern	TF = 10–100 yrs dt = minutes–weeks SF = 100 m–global ds = 100–1,000 m	Water depth Vegetation type Biomass Density/diffusion Nutrients Birth/growth/death	A spatial pattern is the result of site-specific biogeochemical mechanisms that control energy and material flows	Most susceptible to initialization, forcing function, and calibration errors. Uncertainty arises from natural variability and data collection errors along the boundaries and across the landscape. Uncertainty associated with calibration is

			Transpiration Respiration Sediments	due to complexity of model structure and parameters.
Individual-based (Rule-based)	Predict population abundance and distribution in a spatial pattern	TF = 10–100 yrs dt = minutes–weeks SF = 100–10,000 km ds = 1–100 m	Species Migration Birth/growth/death Behavior Habitat Distance Density	Individuals within a landscape select behavioral responses to local conditions, neighborhood populations, and physiological requirements	Most susceptible to initialization and calibration errors. Uncertainty arises from natural variability, data collection errors, and spatial interpretation of raster and point data. Uncertainty is associated with the complexity of behavioral responses and the level of landscape aggregation.

*Scale (from Costanza and Sklar 1985): TF = temporal framework; dt = temporal resolution; SF = spatial framework; ds = spatial resolution.

2.2.1 Transitional Probability Models

Transitional probability models are a broad group of statistical approaches that simulate a spatial configuration as a function of discrete state space (i.e., GIS layers). These maps are often ordinal value mosaics of cells or polygons whose structure is determined by interactions with neighbors, local environmental conditions, and spatial dependencies (Hunsaker et al. 1993). The principle of their design is based upon the idea that it is possible to predict the *chances* that a system will be in a particular state at a later time if its present state is known. Mechanistic realities are ignored; however, causality can be inferred. A typical landscape transitional probability model requires a minimum of two maps; one at time t and another at time $t + m$. The landscape transition matrix is the probability of moving from the initial map to a map at time $t + m$. If each map has but two states, then there would be four transition probabilities:

1. The probability of a cell changing to State 2, given that the cell is currently in State 1.
2. The probability of a cell to remain as State 1, given that the cell is currently in State 1.
3. The probability of a cell changing to State 1, given that the cell is currently in State 2.
4. The probability of a cell to remain as State 2, given that the cell is currently in State 2.

For a transition matrix of all positive entries, it is possible to calculate the successive state matrixes (i.e., a Markov-chain process) until a steady-state matrix for the system is found. This final map is the long-term prediction of landscape structure. Most transitional probability models have broad temporal (10–500 years) and spatial extent (100–5000 km), long integration intervals ($dt = 1$ year or more), and fine spatial resolution ($ds =1$–100 m). State variables, which are typically community type, land-use type, or vegetation type, are simulated as either present or absent for each cell within the whole mosaic. A variety of descriptive statistics have been developed as matrix modifiers in an effort to increase the predictive accuracy of these models. These include indexes of spatial pattern such as fractal dimension (Milne 1991), adjacency (Turner et al. 1989), contagion (Riitters et al. 1995), and lacunarity (Plotnick et al. 1993), as well as measures of spatial concurrence such as multivariate regression, and spatial autocorrelation (Sokal and Oden 1978; Turner et al. 1991).

One can create a very complex set of landscape modifiers for adjusting a probability matrix. It is unlikely, however, that these numerous modifiers will alter the uncertainties associated with transitional probability models because they do not decrease initialization and verification errors. No matter how much one attempts to enhance probability values, they will only be as good as the input maps and GIS data layers. These input data are most susceptible to errors associated with spatial interpretation and interpolation of point data. The uncertainty of transitional probability models is decreased by having good field verification and by increasing the mapping accuracy of the initial and state matrixes. On-the-other-hand, uncertainties associated with the nature of statistical inferences for transitional probability models (e.g., linear dependencies, temporally constant probability functions, no ecological feedbacks, and the lack of causality) cannot be reduced by creating better maps or more significant regression coefficients.

An example of a well-designed Everglades transitional probability model can be found in a simulation of vegetation change by Wu et al. (1997). In this model, the proportion of the natural sawgrass landscape replaced by cattail vegetation, based upon classified SPOT imagery, increased from less than 5% in 1973 to more than 30% by 1991. From an analysis of patchiness and adjacency it was found that the number of cattail cells adjacent to a sawgrass cell can significantly influence spatial dynamics (P_{adj}). The average probability of a 20×20 m sawgrass cell changing into a cattail cell increased from 0.049 when adjacent to one cattail cell, to 0.094 when adjacent to eight cattail cells. It was also found that this conversion to cattail had a natural background probability (P_{nat}), an association with levees (P_{levee}), and a relationship with water depth and phosphorus ($P_{PW\&TP}$). The result was a transition probability for sawgrass conversion (P_{sawcat}) based upon:

$$\text{Prob}_{sawcat} = \varepsilon(P_{adj} + P_{nat} + P_{levee} + P_{W\&TP})$$

where ε is an effective coefficient used to adjust all probabilities to the amount of phosphorus added to the system each year. The result was a simulation of the number of hectares converted from sawgrass to cattail (Fig. 2.3a) and the degree of patchiness produced in the process (Fig. 2.3b). This steady-state model assumed no remote sensing errors, constant probability functions into the future, and no sawgrass recovery. It predicted a 50% conversion to cattail by the year 2000 and complete conversion after 2030. It was later discovered, however, that the remote sensing estimate of cattail abundance was higher than that estimated from aerial photos (Rutchey and Vilchek 1998). Transition probabilities, however, are easily recalculated and the new cattail expansion rate was found to be one third of what was previously calculated (Wu, personal communication). The model is now viewed as a relative index of change and as a tool for developing hypotheses about cause and effect. This is a good example of how improved vegetation mapping techniques can reduce the uncertainty of a landscape modeling approach. One must nevertheless be cautious when using these types of landscape models to predict too far into the future because techniques improve and assumptions of constant probabilities are likely to be violated.

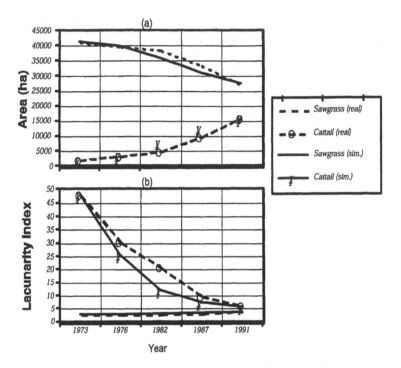

FIGURE 2.3. A transition probability landscape model of the Everglades was used to compare observed and simulated changes in (a) hectares of sawgrass and cattail and (b) degree of patchiness for sawgrass and cattail (from Wu et al. 1997).

2.2.2 Gradient Models

Gradient models are a class of landscape models that tend to focus on mass balance, input–output equations, and the aggregation of parameters. They consist of coupled mass–balance and water (or air)–balance equations that represent a steady-state system and first-order kinetics. These models are designed for landscapes with obvious upstream and downstream components (e.g., weather patterns, mountains, streams, watersheds, marshes, and aquifers). State variables can be pollen, water, nutrients, seeds, soil type, pollutants, biomass, or even species abundance. The extent of the gradient is usually defined by field studies and the equation:

$$A = W \cdot D$$

where A = cumulative downstream area; W = the width of the flow path; and D = the gradient distance. Lateral flows across W are assumed to be minimal. Differential equations are used to simulate the mass balance of some material (M) within an incremental area (dA) such that:

$$d(QM)/dA = \text{Inputs} - \text{Outputs}$$

and

$$Q = Q_{in} \text{ at } D = 0$$

$$M = M_{in} \text{ at } D = 0$$

where Q = flow rate of the medium; Q_{in} = inflow of the medium at the top of the gradient; and M_{in} = concentration of M at the top of the gradient. Inputs to a zone dA_j or a cell$_j$ are the sum of all upstream inflows (e.g., precipitation and runoff) and their respective flow-weighted concentrations (Walker 1995):

$$\text{Inputs} = \Sigma Q_{in} \cdot M_{in}$$

Outflows from a zone A_j or a cell$_j$ are similar but include a first-order removal coefficient (k):

$$\text{Outputs} = M_j \Sigma Q_j + k M_j$$

The removal coefficient represents the net consequence of many ecological processes within an incremental area with no attempt to represent specific mechanisms. For example, the accumulation of soil phosphorus (kM) in an incremental area of a soil phosphorus gradient in the Everglades (i.e., the removal of P from the water column) is a process that is affected by vegetation type, droughts, abiotic and biotic nutrient uptake rates, soil chemistry, decomposition, and peat accretion (Kadlec and Newman 1992; Reddy et al. 1999). Even though these processes are not explicitly simulated, gradient models for the Everglades, with a singular net removal coefficient for phosphorus, have been used effectively to design stormwater treatment areas (Walker 1995), evaluate the nutrient contribution of runoff (Moustafa 1998), and predict water quality impacts of various water management plans.

Spatial and temporal extent of gradient models can range from small, quickly changing systems (e.g., seed dispersal by wind) to very large, slowly changing systems (e.g., the Southern oscillation). Spatial and temporal resolution can similarly range two-to-three orders of magnitude. No matter what the resolution, however, the simplicity of these differential equations are such that they can be integrated analytically to solve for flow and mass. The analytical equations can then be used to predict M_j and kM_j at any distance downstream given estimates or measurements of inputs and outputs. Uncertainty in these models is due primarily to calibration and aggregation errors associated with the net removal parameter k and measurement errors associated with Q_{in} and M_{in}.

Calibration exercises for finding the best estimates of k are the most critical aspect of gradient model development. This involves knowing the structure of the gradient and being able to measure inputs at the top of the gradient and outputs at the bottom of the gradient. Measures of inputs and outputs are subject to the same uncertainties as any forcing function; namely, missing data and measurement error. Much of the uncertainty of gradient models is associated with the spatial and temporal variability of the gradient structure. As such, there can be much uncertainty associated with the observational data, including natural variability, measurement error, and scaling inconsistencies; however, most of the uncertainty is associated with model structure. Like biogeochemical structures, gradients have complex, nonlinear spatial interactions and ecological feedback mechanisms. There may be times when no gradient can be observed and other times when it looks more like a series of discrete step functions. It is difficult to select a removal coefficient that will accurately ($r^2 > 0.9$) reproduce gradient structure for a broad temporal and spatial extent. Stable gradients will have less uncertainty than ephemeral gradients and subdividing a landscape into multiple gradients and calculating site specific values for k can reduce spatial uncertainty.

An Everglades gradient model created by Walker (1995) is a good example of how these models function and deal with uncertainty. In this model, the removal coefficient k was calibrated to predict 26-year-average phosphorus soil accretion rates measured in a 10.5-km wide flow path (W) and a 39-km long gradient (D), downstream of a water control structure with a Q_{in} of 347 million m^3/year and a total phosphorus flow-weighted M_{in} of 122 ppb. To predict soil phosphorus (P), a water-column phosphorus equation was designed with an "effective settling velocity" (Ke) to account for net P accumulation in the soil:

$$d(QC)/dA = PCp - Ke \cdot Fw \cdot C$$

where Q = overland flow (m^3/year); C = water-column phosphorus concentration (ppb); A = area (m^2); P = rainfall (m/year); Cp = phosphorus concentration in rain (ppb); Ke = removal coefficient (m/year); and Fw = wet period fraction (fraction of time that water covers the marsh). Actual phosphorus accumulation rates were measured using the Cesium-137 dating technique and were found to vary from a high of 1200 mg/m^2-year at the top of the gradient to a low of 10 mg/m^2-year at the bottom of the gradient. A least-squares calibration was derived by comparing observed and predicted P accumulation rates for various values of Ke. The results indicate that with a Ke value of 10.2 m/year, the gradient model was able to account for almost 90% of the variation observed for P accumulation in area A (Fig. 2.4). Uncertainty associated with Cp (due to measurement problems) was evaluated by testing a broad range of values (i.e., a sensitivity analysis) and found to have a minor affect compared to Ke. Most of the uncertainty in this model appears to be within the 2–6-km interval at the top of the gradient, where differences between observed and predicted values of P accumulation suggest a more complex or different soil–water chemistry then at the end of the gradient.

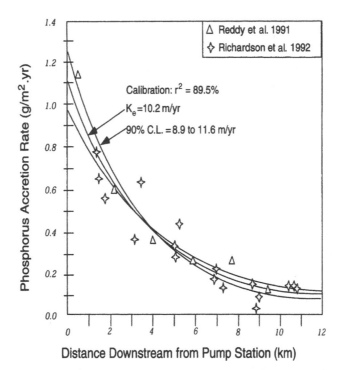

FIGURE 2.4. An example of a gradient model used to predict phosphorus removal by wetlands in the Everglades and design Stormwater Treatment Areas. Symbols depict measured, 26-year-average accretion rates. Center line depicts model predictions with a 10.2 m/year settling rate. Adjacent lines are the model's 90% confidence interval in settling rate (from Walker 1995).

The inherent structure of gradient models makes them difficult to apply across landscape mosaics with discrete structural or functional properties; however, one can effectively reduce this problem by grouping disparate temporal or spatial data into separate gradient models (Walker 1995).

2.2.3 Process-Based Mosaic Models

Process-based mosaic models are a very broad category of dynamic ecosystem simulations designed to distribute a geometric pattern (cells or polygons) across a landscape, obey the laws of thermodynamics, and use site-specific biogeochemical mechanisms to control energy and material flows. This category includes models of hydrodynamics and circulation (Kasahara and Washington 1967), climate and precipitation (Washington and Meehl 1984; Schlesinger and Zhao 1989), forest growth (Botkin et al. 1972; Shugart 1984), zooplankton movement (Show 1979), fisheries (Reyes et al. 1994), fire (Rothermel 1972; Wu et al. 1996), soil–water nutrient exchange (Kamp-Nielsen 1983; Cerco and Cole, 1993; Fitz and Sklar, 1999), and grassland succession (Sklar et al. 1985; Parton et al. 1987; Boumans and Sklar 1990; Costanza et al. 1990). These are some of the most complex landscape models in the world today. Most require large databases for initialization, numerical analysis of boundary conditions for calculating inputs, experimental observations for calibration, and numerous sampling sites for verification.

Hierarchy theory points out that at least two independent perspectives are needed for explaining the full range of ecological phenomena (MacMahon et al. 1978; O'Neill et al. 1986). Processed-based landscape models as well as individual-based landscape models (see Sec. 2.2.4) are examples of dual hierarchies (O'Neill et al. 1986) that solve problems from process-functional and population-community perspectives, respectively. Process-functional problems emphasize phenomena such as nutrient cycling, biomass accumulation, and productivity. Population-community problems emphasize age structures, abundance, and individuals organized into populations and guilds (O'Neill et al. 1986). State variables in process-functional models, are the end result of energy budgets (Odum 1983) and mass transformations (i.e., kinetic processes such as diffusion, friction, decomposition, and photosynthesis). In individual-based models state variables are the end result of population and behavioral rules. These dual perspectives are intertwined.

Population-community structure exists within process-based landscape models as independent variables that constrain function. For example, photosynthesis is a typical process-based function in a process-based landscape model. A change of plant species in this type of model is not typically as important as photosynthesis. Different species can perform the same function; however, species constrain or bound the function within realistic limits. Species-based tolerances to abiotic and biotic factors are used to set boundary conditions for processed-based simulations. Processes similarly set boundary conditions for populations in individual-based landscape models. For example, individual migration across a landscape is limited by the energy required for movement. This digression on hierarchy is meant to highlight how uncertainty in dynamic landscape models is often nested within layers of design and boundary conditions.

Uncertainty occurs in all four modeling components of process-based models, but it is greatest for calibration. Finding mathematical equivalencies for all relevant processes can be very difficult. Incorporating realistic mechanisms most often means adding feedback mechanisms, threshold functions, and nonlinear interactions. Reducing the uncertainty of this component is accomplished by following some simple guidelines. First, study previous models on the same subject and evaluate their effectiveness. Some mechanistic details have already been worked out and can be transferred to other models. Second, use experimental observations to constrain parameter values. For every transformation of energy and matter there is, at least, one rate coefficient (i.e., first-order kinetics) that describes how a process is controlled by inputs, state variables, and flow rates. Parameterization is the process of finding the appropriate combination of coefficient and controlling parameter. For purely physical processes, like evaporation, the parameters and their coefficients have been found (Thornthwaite and Mather 1955; Christiansen 1968) and should be used whenever possible. However, for biophysical processes, like transpiration, the parameters and their coefficients may need to be found for a specific situation. It is ideal to conduct site-specific experiments to calculate these process-based coefficients, but it is not necessary as only the range of values are needed, and is seldom practical. The uncertainty of calibration is significantly reduced if parameters and their coefficients have been found for similar ecosystems or biological entities. The most important data for parameterization is often the knowledge that a process is constrained. Another way to reduce calibration uncertainty for process-based landscape models is to compare simulation output with large quantities of observed data. A map time-series of model state variables from imagery or sensor arrays is the most effective control on uncertainty. These maps can show spatial and temporal discontinuities, the need for nonlinear functions, degrees of uncertainty in model design, and the need for specific information and experiments.

Process-based landscape models suffer the same uncertainties discussed for transitional and gradient landscape models; namely, problems with mapping accuracy for initialization and verification, and problems with accurately measuring inputs at the top of a gradient and outputs at the bottom of a gradient. Why create models that have the added uncertainty of a complex calibration component? When dealing with discrete spatial change or ephemeral gradients, the advantages of process-based models includes the flexibility of model structure to incorporate ecological feedback mechanisms and response thresholds, and the ability to evaluate critical ecological pathways (i.e., sensitivity analysis), test hypotheses, and recommend management options.

The Everglades landscape model (ELM), which is a spatial model of wetland nutrient cycling, plant growth, and hydrology (Fitz et al. 1996; Fitz and Sklar 1999), is one of the more complex process-based models for water management and as such, is a good example for discussion of structure and uncertainty. The ELM divides the Everglades into 10,000 square, 1×1–km grid cells. Each cell is a "unit model" (Fitz et al. 1996): This is a repeated description of the growth process for plant and algae communities (i.e., habitat types) in response to nutrients, water, sunlight, and temperature. Complexity is in the form of realistic feedback mechanisms.

Horizontal transport of water and nutrients is a fundamental component of any process-based landscape model, but it is especially important in wetlands. The flows of surface water among ELM grid cells were calculated using a mass balance, finite difference algorithm solving two-dimensional diffusion equations. The realistic feedback to this explicit solution for flow

(Q; cubic meters per day) was the extent of frictional drag (n) from plants such that:

$$Q = (h^{5/3}L^{1/2}\Delta H^{1/2})/n$$

where h is the water depth, L is the length of a grid cell, ΔH is the difference in water stage between source and destination cells, and n (the Manning's roughness coefficient; $\sec/m^{1/3}$):

$$n = n_{max} - |(n_{max} - n_{min})(2^{(1-h/mac)} - 1)| \tag{2.1}$$

was modified to be a dynamic function of vegetation type and height relative to water depth (Fitz et al. 1996) such that n_{min} and n_{max} are the respective minimum and maximum roughness coefficients associated with a cell's habitat type and mac is the plant height. Thus, the physical process of flow decreases as a biological process (plant growth) increases. Uncertainty in this case is not in the reality of the process, which is known to exist, but in the accuracy of the estimates for the coefficients and the sensitivity of the model to these errors. In general, the more sensitive the coefficient, the less certain it will be. If a 100% change in n or mac only alters Q by 1%, then the accuracy, hence its uncertainty, is not critical—at least, not in the short term.

Nutrient cycling and associated soil processes in the ELM also have feedback mechanisms. Process-based accumulation of soil in the ELM is very different from what was discussed earlier for gradient models. Here, accreted organic matter (AOM) is the balance between the amount of plant and animal detritus added to the soil from above- and below-ground production, and the amount of microbial decomposition (AOM_{dec}) below ground. This AOM decomposition is described by:

$$AOM_{dec} = AOM \cdot k_{ar}TP[U_{sat}(H_{ar}/H + k_{an}(H_{an}/H)]$$

where k_{ar} is the maximum rate of aerobic decomposition; T is an optimum temperature control function; P is a phosphorus control function (impacted by the nutrient content and growth of plants); U_{sat} is the percent moisture in the upper 30 cm of soil (impacted by n and mac; Eq. 2.1); and the H variables (H_{ar}, H_{an}, and H) are habitat specific heights (indirectly impacted by n and mac) of the aerobic decomposition zone, the anaerobic decomposition zone, and the total of the two, respectively. The active zone of decomposition (H) is clearly a nonlinear function constrained by minimums and maximums for a multitude of parameters. The equation for AOM_{dec} attempts to capture cause and effect relationships; however, it can never completely succeed because these types of multiplicative, process-functional equations are such that a high degree of uncertainty in one coefficient can offset a high degree of certainty in all the others. As a result, it is more vulnerable to criticism and uncertainty than transitional, gradient, or nonspatial models. To offset this calibration uncertainty, it is sometimes best to sacrifice a little realism by dropping unreliable coefficients (go with what you know) for the sake of increased certainty of output (Costanza and Sklar 1985). If the parameters of greatest uncertainty are of theoretical or managerial significance, however, then they should remain in the model, and their range of influence evaluated as part of the modeling exercise.

Despite the inherent high uncertainties, particularly in terms of structure and calibration, process-based spatial simulations are more likely to be used to manage and predict landscape structure than any other type of model. This is because process-based modeling has a long history of success. It began with Newton's differential calculus and has developed into very sophisticated simulations of weather patterns (Kaufmann and Smarr 1993). The output of the ELM builds upon this history (Sklar and Costanza 1991) and the quest for causality in landscape ecology. Typical output of the ELM (Fig. 2.5; see color insert) demonstrates how three variables (depth of water in the unsaturated zone, phosphorus remineralization rates, and macrophyte biomass) are the result of spatial differences in rates of plant-water-soil interactions. The black dots in Figure 2.5, across a diversity of habitats and nutrient gradients were areas of synoptic investigations used to calibrate the ELM. In this example, potential evapotranspiration (ET) was increased, thus altering the water depths without changing atmospheric inputs and phosphorus loading from water control structures. Compared with the nominal case of average ET, this test simulation maintained a deeper unsaturated zone for longer periods (Fig. 2.5a), which in turn produced a feedback of increased decomposition, which in turn again increased soil P remineralization (Fig. 2.5b) and increased macrophyte biomass (Fig. 2.5c). For the restoration of the oligotrophic Everglades, this increase in plant biomass as a result of lower water tables is not desirable.

2.2.4 Individual-Based Models

Individual-based models are a group of landscape models that focus on behavioral rules for an individual or an assemblage of individuals as a function of spatial constraints and opportunities (DeAngelis and Gross 1992; Wolff 1994). Individual-based models are modifications to the intrinsic growth (r) and carrying capacity (K) parameters of the logistic equation used in the discipline of population ecology. These models, with a hierarchical system of rules, allow various behaviors (e.g., foraging, mating, nesting, and social functions) to interact with individual physiological needs and environmental resources. They have been used to represent moose migration (Saarenmaa et al. 1988), reproduction of Wood Storks (Wolff 1994), and winter foraging by large ungulates (Turner et al. 1994). They are potentially the most complex of all landscape models because individual-based models can easily incorporate all three previously mentioned modeling approaches.

The time frame for individual-based models can be hundreds of years. Most simulate only a few years or some critical life stage [e.g., nesting (Wolff 1994) or winter survival (Turner et al. 1994)]. The time step (dt) can be minutes to weeks and depend upon the mobility of the individuals in relation to the spatial resolution of the model. For a cell size of one hectare, birds may have a dt of 15 minutes, while alligators would have a dt of 2–3 days. The spatial framework of these models, in theory, can be as small as a hectare (e.g., snail movements) and as large as a hemisphere (e.g., bird migration). The finer the spatial resolution, the smaller the geographic extent usually is.

Spatial uncertainty in these models arises from data collection and interpretation errors, as well as natural variability. These uncertainties are most manifest during initialization and calibration. Initialization requires a set of base maps of environmental resources and physical attributes at a scale consistent with individual behavioral rules. These maps may be descriptions of vegetation type, topography, snow depth, rainfall patterns, or indexes of habitat suitability. If these are static maps, ones that do not change with time or as a function of feedback mechanisms, then spatial data collection and interpolation errors will contribute most significantly to model uncertainty. In this case, the static structure of the landscape, in combination with behavioral rules, can dictate travel routes, resources, or refugia. As such, it can control population survival. As a result, it is common in these models to create idealized initial conditions as a way to explore impacts of habitat change or test hypotheses (Fleming et al. 1994). If the set of initial base maps change as a function of feedbacks or as a result of a separate modeling effort, however, then uncertainty arises from the spatial data needed for calibration as well as, spatial data collection and interpolation errors.

Individual-based models that incorporate two-way interactions between the landscape and the individuals can have all the same uncertainties associated with forcing functions and calibration as process-based models. Reyes et al. (1994) developed a combined habitat suitability index (CSI) and a general diffusion model (Okubo 1980) to describe planktivorous fish movement according to:

$$\text{CSI}_{ij} = (S \cdot T \cdot F^2)_{ij}^{1/4} \quad \text{and,}$$

$$F_{ij} = \text{Phyto}_{ij} - \alpha P_{ij} + \beta P_{ij}$$

where CSI in each ij cell was a function of salinity (S), temperature (T), and food availability (F). Food (F) was in turn, a function of the phytoplankton biomass (Phyto), fish feeding (αP_{ij}), and fish feedback to water nutrient quality (βP_{ij}). In other words, landscape change (fish abundance in cell ij) within this model required forcing functions (S and T) and model parameters for simulating abiotic and biotic causal mechanisms (β and α, respectively), just as for process-based models. These models can also have some of the same uncertainties as transitional probability models because behavioral rules can be stochastic. In the Reyes et al. (1994) model, rules were not discrete. The CSI was used as a coefficient that modified the probability of general diffusion to neighboring cells. Like transitional probability models, verification errors were the result of uncertain maps of fish distributions.

An excellent example of an Everglades individual-based model is that for wood storks (Fleming et al. 1994; Wolff 1994). Designed to capture the effects of water management on reproductive success for colonies of 50 or 250 mating pairs of birds, this model had separate submodels for hydrology, prey biomass, adult behavior, and nestling growth. The hydrology component, which is subject to data collection and interpolation uncertainty, predicted water depth change as a function of elevation, precipitation, and ET for each $1/4 \times 1/4$ km grid cell. The prey component, which is subject to initialization uncertainty, assigned average densities (10 prey in shallow habitats and 50 prey in deep habitats) of all edible fish for each cell. Droughts (no water) and birds (many predators) were able to set fish density to zero because fish recovery was not possible within the modeling time frame of 150 days (an extended nesting season). Adult behavior, which is most susceptible to calibration errors, was based upon energetic requirements of movement and rules for foraging efficiency according to the following:

1. Average adults must ingest 500 g of fish daily to survive.
2. Average adults can carry 300 g of fish back to the nest.
3. Adults fly 25 km/hour.
4. Foraging is only in cells with water depth of 10×40 cm.
5. Foraging in a particular cell is either random or a function of flock size.
6. Adults have thresholds for giving up foraging, remaining at a site overnight, food and water depths needed to initiate nesting, and nest desertion.

The nestling growth component of this model predicted recruitment success as a function of the other three submodels and the energetic requirements of the nestlings. Here is where observed mortality data was compared with stimulated data to validate model structure and where uncertainty arises from natural variability and model complexity. Rules for nestlings were similar in structure as those for adults and included the following:

1. Each nest will have 3 eggs, laid 2 days apart, and incubated for 30 days.
2. Nestlings fledge in 60 days.
3. Each nestlings needs a total of 15.5 kg of food.
4. Nestlings have thresholds for starvation, competition, and growth.

This description of the wood stork model illustrates the use of probability, process, and individual-based approaches within a model that uses spatial attributes to control animal behavior. This complexity and concurrent uncertainty are reduced by the fact that the model does not attempt to validate exactly where birds go in the landscape at any one time. Flight is a process, not a state variable. Nevertheless, results of this model (Table 2.2) produced significant landscape-level conclusions (Fleming et al. 1994). It was found that nesting for large colonies was delayed by decreasing the extent of shallow, short-hydroperiod (SSH) wetlands by only 10%. If reduced by 15% this led to colony failure. Decreasing the extent of deep, long-hydroperiod (DLH) wetlands by 10% had little impact on nesting or number of fledglings. Fleming et al. (1994) concluded that the SSH wetlands supply most of the energy for colony formation early in the nesting season, whereas later the DLH wetlands supply the energy needed to successfully fledge the offspring.

2.3 Dealing with Uncertainty

By focusing our attention upon the complex nature of uncertainty in landscape models and data, it may appear that all is uncertain, and efforts to reduce it have either been fruitless or nonexistent. Neither is true. Landscape ecologists and modelers are aware of the inherent errors in model structure, parameter estimates, and forcing functions. It is understood that ignorance, variability, chaos, and human errors can affect the reliability and predictability of landscape models. Efforts to measure and minimize these uncertainties have been discussed by Colwell (1974), O'Neill and Gardner (1979), Turner et al. (1989), Ludwig et al. (1993), Lemons (1996), Anderson (1998), and Breckling and Dong (2000).

There are a number of commonly practiced procedures for reducing uncertainties in landscape data and models. To reduce ignorance as a source of uncertainty, it is common to develop an experimental supplement for data collection and modeling. These usually involve sensor and model sensitivity tests. Field experiments to evaluate detection limits of sensors, laboratory experiments to test biogeochemical hypotheses, and computer experiments to identify the most important parameters, reduces uncertainty surrounding every aspect of model development (Fig. 2.1). To reduce model uncertainty associated with variability, it is common to develop a stochastic modeling approach for simulating likely variation in system states. To reduce data uncertainty associated with variability, it is common to develop a stochastic sampling approach to capture likely variation in system states. For simple models, dealing with the uncertainty of chaotic events may be as knowable as stochastic events (Gleick 1987). Human error is best controlled by developing a rigorous quality control and assurance program.

In general, dealing with uncertainty is best accomplished by making sure that model assumptions, data limitations, and conflicting hypotheses are clearly presented to the end user. For the ELM this was accomplished by ranking the quality and degree of confidence associated with each

TABLE 2.2. Results of the individual-based wood stork model based on 50 runs.

Percent habitat reduction	Day of colony formation when SSH wetlands are reduced	Day of colony formation when DLH wetlands are reduced	Percentage of nesting pairs when SSH wetlands are reduced	Percentage of nesting pairs when DLH wetlands are reduced	Number of fledglings when SSH wetlands are reduced	Number of fledglings when DLH wetlands are reduced
0	10	10	100	100	742	742
5	13	13	95	97	724	721
10	63	14	62	96	162	711
15	95	15	39	95	21	687
20		12		97		159
25		13		94		102
30		16		95		32

Impacts of habitat removal for shallow, short-hydroperiod (SSH) wetlands and deep, long-hydroperiod (DLH) wetlands in the Everglades, for a colony with 250 pairs. Maximum possible number of fledglings was 750.

input map and model parameter from 1 to 5 (Fitz 1994). When a parameter was based upon experiments on similar species or distant locations, Fitz (1994) would rank it low and put it on a list of highly uncertainty variables. If this variable was also found to be important according to the sensitivity analysis, then an experimental approach was recommended to reduce the uncertainty.

These days spatial data are often inputs to ecological models. Calculations or results from such models often are assumed to produced an exact result because most GIS provide no means to examine the effects of errors in the input data. The quality of the quantitative model depends on three factors: (1) the quality of the data, (2) the quality of the model, and (3) the way data and model interact. To analyze error propagation (i.e., how uncertainties accumulate and affect the end-result) we need to know: (1) sources of error estimates, (2) error propagation theory, and (3) error propagation tools.

Sources of statistical uncertainty in data include measurement errors and spatially correlated variation that cannot be explained by physical models. Mismatch in spatial and temporal correlation structures may be one of the greatest problems when combining data from different sources or from different phenomena or attributes as is common for ecological models (Burrough and McDonnell 1998). They recommend analysis of indexes of spatial covariation such as the variogram can be useful to ensure that different data are spatially compatible. The following are some of the ways mismatch can occur.

- Each phenomenon being measured on a different support (area/volume).
- Each phenomenon has a different intrinsic spatial variability.
- Some phenomena being sampled directly whereas others are collected or classified using externally imposed spatial aggregation blocks that are inappropriate.
- The spatial variation of different phenomena is governed by processes that operate at different scales.

Burrough and McDonnell (1998) discuss two approaches to "error propagation theory"—Monte Carlo simulation and analytical. If a new attribute Y is defined as a function of input $A_1, A_2, \ldots A_n$, we want to know what error is associated with Y, and what contributions are from each A_n to that error? With such information one can compare the model results from different scenarios with confidence and perform risk assessments. The Monte Carlo approach is named for its reliance on chance. One can assume that each attribute has a Gaussian (normal) probability distribution function (PDF) with a known mean and variance of each entity or cell. The simple option is to use a single PDF for all cells and assume stationarity. Conditional simulation can estimate cell-specific PDFs that reflect the location of known data points and the spatial correlation structure of the attributes. The Monte Carlo approach estimates the new attribute by repeatedly drawing (at least 100 times) a value from the PDF for each cell so a standard deviation is developed (see Chap. 11). For actual landscapes, this is a computer-intensive approach even today. Experience is demonstrating that the methods of conditional simulation aided by relevant information from controlling attributes (e.g., geology, soil, or land use) provide the best inputs for Monte Carlo simulation with environmental models (Burrough and McDonnell 1998). Many numerical models used in GIS to compute new attributes from existing properties of entities or cells can be achieved using the standard statistical theory of error propagation. The increase in error is not especially great when attributes are not correlated and addition ($Y = A_1 + A_2$) is the model operation. Subtraction can lead to explosive increases in relative errors, particularly when A_1 and A_2 are similar in value. Burrough and McDonnell (1998) provide rules to reduce error propagation using arithmetic algorithms:

- Avoid intercorrelated variables.
- Add where possible.
- If you cannot add, multiply, or divide.
- Avoid taking differences or raising variables to powers.

Errors in inputs to models can be reduced in several ways:

- Use optimal interpolation techniques.
- Use the appropriate sampling density.
- Remove or check for outliers, subgroups, systematic bias, and so on.
- Adopt appropriate classifications.
- Use sensible models.
- Improve model calibration by reducing errors in model parameters.

Heuvelink (1993) and Heuvelink et al. (1989) developed a computer program, ADAM, which can trace errors through complex numerical models in "point mode" that operate on the attributes of entities or on multiple raster overlays. ADAM (Wesseling and Heuvelink 1991) displays model output and associated errors as maps. It also makes maps of the spatial variation of the error contribution from all coefficients and input variables.

2.4 The Precautionary Principle

Even with high quality assurance, a host of sophisticated equipment and computers, and a conceptually realistic landscape model, environmental scientists and regulators are faced with contradictory data, conflicting hypotheses, and irreproducible historical trends. This occurs because ecological systems are inherently variable with an unmeasurable degree of uncertainty (Ludwig et al. 1993). Biotic systems defy the rules of inner constancy that allow mechanical systems to be reproducible and predictable (Breckling and Dong 2000). Landscape models can often uncover more uncertainty than they reduce. For example, global change models with biotic inputs of CO_2 can say something about the probabilities of different rates of sea level rise (Titus and Narayanan 1995) but they cannot say which of the possible outcomes will occur. In fact, one scenario calls for an increase in snowfall with rising temperatures, thus decreasing sea levels with more CO_2 (Titus and Narayanan 1995).

True uncertainty, where interactions are unknown and events unknowable, lies beyond the scope of scientific measures; instead, it resides within the realm of social policy. The cognition of the uncertainties associated with landscape data and models by environmental managers and policymakers can lead to regulatory paralysis (Costanza and Cornwell 1992). Their complaint is often that science has failed to provide unambiguous and certain answers needed for sound policymaking. Scientists can continue to gather data when models or experiments fail, but managers and policymakers must choose a course of action. Delaying action in the hope that new information will resolve the problem has become the "regulator's dilemma"(Weinberg 1985). The problem with delay, of course, is the risk of rapid environmental deterioration and irreversible damage.

To avoid extreme consternation about uncertainty, many regulators and managers have adopted the "precautionary principle" (Bodansky 1991). This principle states that regulators should act in anticipation of environmental harm to ensure that harm does not occur rather than await certainty. This principle has become the norm in many international environmental resolutions of the United Nations (Cameron and Abouchar 1991). The precautionary principle has mostly been applied to marine pollution where a scientific understanding of assimilative capacity and long-term damage is considered insufficient (Bodansky 1991). It can be applied to many landscape problems because all involve uncertainty. The precautionary principle was the basis for the moratorium on commercial whaling in the 1970s. It also continues to influence the "no-discharge" clause of the Federal Water Pollution Control Act.

This book, with its focus on landscape data uncertainty, will point to issues that may increase the regulator's dilemma. We mention the precautionary principle as solace for the policymaker and as hope for better communication between regulators and scientists on the risks (i.e., events with known probabilities) and uncertainties (i.e., events with unknown probabilities) associated with landscape data and models. The precautionary principle allows the policymaker to establish evidentiary presumptions on environmental hazards from similar situations as a way to overcome uncertainties about their actual environmental effects. According to Bodansky (1991), this allowed the third international North Sea conference in 1990 to call for the application of the precautionary principle to all emissions of substances that are persistent, toxic, and liable to accumulate in living tissues, even when there is no scientific evidence to prove a causal link between emissions of such substances and adverse environmental effects.

The precautionary principle, unfortunately, is not without problems. Criticisms include its lack of legal content and its vagueness (Bodansky 1991). Costanza and Cornwell (1992) suggest the *precautionary polluter pays principle* for the issuance of environmental assurance bonds that are paid by the polluter in the present, provides for worst-case environmental damage, and extends the polluter pays principle to make the polluter pay for uncertainty.

2.5 Summary

One can debate whether people who make decisions really want to know about the uncertainty associated with the data and information provided to them. It might make their task more complicated or difficult, or they might even refuse to make a decision, but this is really not a debate for ecologists; as scientists we should seek to present all of the data to the best of our ability and this includes estimates of and information on uncertainty. The challenge is how to do this and how to make it understandable. We gave an overview in this chapter of what needs to be considered when working with spatial ecological models; Eastman (Chap. 18) discusses decision support tools for effective use of spatial data resources.

Earlier we mentioned the sources of uncertainty in spatial data and models that can be traced to problems associated with: (1) data collection, (2) data processing, (3) model structure, (4) human intervention, and (5) natural variability. Stine and Hunsaker (Chap. 5) elaborate further on many of these, and other chapters give indepth case studies of these issues. Data collection issues are addressed in Chapters 3 and 4, and Chapters 12–14 address collection for remotely sensed data. Data processing is addressed in Chapters 8–11. Human intervention is the focus of Chapters 6, 7, and 17. Natural variability is a fact, and Chapters 8 and 15 examine some approaches for this topic.

We reviewed ecological model structure issues in this chapter, including (1) inputs, (2) initial conditions, (3) calibration, and (4) verification. This chapter reviewed the structure, function, and uncertainties associated with four major types of dynamic landscape models. In general, the inherent weaknesses of landscape models must be recognized and compensated by minimizing the uncertainties associated with inputs, initial conditions, calibration, and verification of the most sensitive components. If you use gradient models, then be aware that they are more susceptible to input errors than most other models. As basic input–output models, their uncertainty is reduced by implementing well-designed sampling networks at air, soil, and water boundaries. If you use transition probability models, then be aware that they are sensitive to initialization errors. If habitat change is calculated from improperly classified satellite images, then results from these models can be very misleading. Errors in these models tend to come from a combination of spatial interpretation and interpolation errors and their associated problems of spatial autocorrelation. If you use dynamic distributional mosaic models, then realize that they are very sensitive to calibration errors, whereas static distributional mosaic models are sensitive to verification errors. Finally, if you use individual-based models such as those based on biological rules of behavior, then be aware that they are most sensitive to calibration and initialization errors.

Spatial models can vary from a simple algebraic equation executed in a GIS to a complex spatial representation of organism or community locations or dynamic movements. The general sources of uncertainty stay the same no matter what the level of model complexity; their influence on outcomes may vary. The chapters of this book illustrate uncertainty issues for models in many different ways.

Acknowledgments. This work was supported by the South Florida Water Management District, the USDA Forest Service, Oak Ridge National Laboratory, and the National Center for Ecological Analysis and Synthesis. We wish to thank T. Fontaine, Y. Wu, C. Fitz, T. Case, and M. Goodchild for their thoughtful reviews and comments.

References

Anderson, J.L. 1998. Embracing uncertainty: the interface of Bayesian statistics and cognitive psychology. Conservation Ecology [online] 2(1):2. Available from the Internet. URL:http://www.consecol.org/vol2/iss1/art2.

Bodansky, D. 1991. Scientific uncertainty and the precautionary principle. Environment 33(7):4.

Botkin, D.B., J.F. Janak, and J.R. Wallis. 1972. Some ecological consequences of a computer model of forest growth. Journal of Ecology 60:849–72.

Boumans, R., and F.H. Sklar. 1990. A polygon-based spatial (PBS) model for simulating landscape change. Landscape Ecology 4(2/3):83–97.

Breckling, B., and Q. Dong. 2000. Uncertainty in ecology and handling strategies in ecological modeling. In: S.E. Joergensen and F. Muller, eds. Handbook of Ecosystem Theories and Management, CRC Press, Boca Raton, FL.

Brown, D., and P. Rothery. 1993. Models in Biology: Mathematics, Statistics and Computing. John Wiley & Sons, New York.

Burrough, P.A., and R.A. McDonnell. 1998. Principles of geographical information systems. Oxford University Press, New York.

Cameron, J., and J. Abouchar. 1991. The precautionary principle: a fundamental principle of law and policy for the protection of the global environment. Boston College International and Comparative Law Review 14:1–27.

Cerco, C.F., and T. Cole. 1993. Three-dimensional eutrophication estuarine models. Pages 1–15 in M.L. Spaulding, ed. Estuarine and Coastal Modeling, Proceedings of the First International Conference, American Society of Civil Engineers.

Christiansen, J.E. 1968. Pan evaporation and evapotranspiration from climate data. J Irrigation Drain. Div. 94:243–65.

Colwell, R.K. 1974. Predictability, constancy, and contingency of periodic phenomena. Ecology 55:1148–53.

Cosby, B.J., G.M. Hornberger, R.B. Clapp, and T.R. Ginn. 1984. A statistical exploration of the relationships of soil moisture characteristics to the physical properties of soils. Water Resources Research 20:682–90.

Costanza, R., F.H. Sklar, and M. White. 1990. Modeling coastal landscape dynamics. Bioscience 40(2):91–107.

Costanza, R., and L. Cornwell. 1992. The 4P approach to dealing with scientific uncertainty. Environment 34(9):12–42.

Costanza, R., and F.H. Sklar. 1985. Articulation, accuracy, and effectiveness of mathematical models: a review of freshwater wetland applications. Ecological Modeling 27:45–68.

DeAngelis, D.L., and L.J. Gross, eds. 1992. Individual-based models and approaches in ecology. Chapman and Hall, New York.

Fitz, H.C. 1994. Summary of task 3.2: ELM calibration analysis report. Contract Deliverable for C-1168 of the South Florida Water Management District, West Palm Beach, FL.

Fitz, H.C., E.B. DeBellevue, R. Costanza, R. Boumans, T. Maxwell, L. Wainger, et al. 1996. Development of a general ecosystem model for a range of scales and ecosystems. Ecological Modeling 88:263–95.

Fitz, H.C., and F.H. Sklar. 1999. Ecosystem analysis of phosphorus impacts and altered hydrology in the Everglades: a landscape modeling approach. Pages 585–620. In: K.R. Reddy et al., eds. Phosphorus Biogeochemistry in Subtropical Ecosystems. Lewis Publishers, Boca Raton, FL.

Fleming, D.M., W.F. Wolff, and D.L. DeAngelis. 1994. Importance of landscape heterogeneity to wood storks in Florida Everglades. Environmental Management 18(5):743–57.

Gardner, R.H., and R.V. O'Neill. 1991. Pattern, process, and predictability: the use of neutral models for landscape analysis. Pages 289–308 in M.G. Turner, and R.H. Gardner, eds. Quantitative Methods in Landscape Ecology. Springer-Verlag, New York.

Gleick, J. 1987. Chaos: making a new science. Penguin, New York.

Heuvelink, G.B.M. 1993. Error propagation in quantitative spatial modeling. KNAG, University of Utrecht Pub. 163.

Heuvelink, G.B.M., P.A. Burrough, and A. Stein. 1989. Propagation of errors in spatial modelling with GIS. International Journal of Geographical Information Systems 3:303–22.

Hornbeger, G.M., K.J. Beven, B.J. Cosby, and D.E. Sappington. 1985. Shenandoah watershed study: calibration of a topography-based, variable contributing area hydrological model to a small forested catchment. Water Resources Research 21(12):1841–50.

Hunsaker, C.T., R.A. Nisbet, D.C.L. Lam, J.A. Browder, W.L. Baker, M.G. Turner, et al. 1993. Spatial models of ecological systems and processes: the role of GIS. Pages 248–264 in M.F. Goodchild, B.Q. Parks, and L.T. Steyaertm, eds. Environmental modeling with GIS. Oxford University Press, New York.

Kadlec, R.H., and S. Newman. 1992. Phosphorus removal in wetland treatment areas. Report to the South Florida Water Management District. No. DRE 321, West Palm Beach, FL.

Kamp-Nielsen, L. 1983. Sediment-water exchange models. In: S.E. Jorgensen, ed. Application of ecological modelling in environmental management, Part A. Elsevier Scientific, Amsterdam.

Kasahara, A., and W.M. Washington. 1967. NCAR global general circulation model of the atmosphere. Monthly Weather Review 95:389–402.

Kaufmann, W.J., and L.L. Smarr. 1993. Supercomputing and the transformation of science. Scientific American Library, New York.

Lemons, J., ed. 1996. Scientific uncertainty and environmental problem solving. Blackwell. Cambridge, MA.

Lotka, A.J. 1925. Elements of physical biology. Williams and Wilkins, Baltimore. (reissued in 1956 as Elements of mathematical biology. Dover, New York).

Ludwig, D., R. Hilborn, and C. Walters. 1993. Uncertainty, resource exploitation, and conservation: lessons from history. Science 260:17–36.

MacMahon, J.A., D.L. Phillips, J.V. Robinson, and D.J. Schimpf. 1978. Levels of biological organization: an organism-centered approach. Bioscience 28:700–4.

Milne, B.T. 1991. Lessons from applying fractal models to landscape models. Pages 199–238 in M.G. Turner and R.H. Gardner, eds. Quantitative methods in landscape ecology. Springer-Verlag, New York.

Moustafa, M.Z. 1998. Long-term equilibrium phosphorus concentrations in the Everglades as predicted by a Vollenweider-type model. Journal American Water Resources Association 34:135–47.

O'Neill, R.V., and R.H. Gardner. 1979. Sources of uncertainty in ecological models. Pages 447–63 in Zeigler, B.P., M.S. Elizas, G.J. Klir, and H.I. Oren, eds. Methodology in systems modeling and simulation. North-Holland Publishing Co., Amsterdam.

O'Neill, R.V., D.L. DeAngelis, J.B. Waide, and T.F.H. Allen. 1986. A hierarchial concept of ecosystems. Princeton University Press, Princeton, NJ.

Odum, H.T. 1983. Systems ecology. Wiley Interscience, New York.

Okubo, A. 1980. Diffusion and ecological problems: mathematical models. Springer-Verlag, New York.

Parton, W.J., D.S. Schimel, C.V. Cole, and D.S. Ojima. 1987. Analysis of factors controlling soil organic matter levels in Great Plains grasslands. Soil Science Society American Journal 51:1173–79.

Plotnick, R.E., R.H. Gardner, and R.V. O'Neill. 1993. Lacunarity indices as measures of landscape texture. Landscape Ecology 8:201–11.

Press, W.H., B.P. Flannery, S.A. Teulosky, and W.T. Vetterling. 1986. Numerical Recipes. Cambridge University Press, Cambridge.

Prigogine, I., G. Nicholis, and A. Babloyantz. 1972a. Thermodynamics of evolution I. Physics Today 25(11):23–8.

Prigogine, I., G. Nicholis, and A. Babloyantz. 1972b. Thermodynamics of evolution II. Physics Today 25(12):38–44.

Reddy, K.R., G.A. O'Connor, and C.L. Schelske. 1999. Phosphorus Biogeochemistry in Subtropical Ecosystems. Lewis Publishers, Boca Raton, FL.

Reyes, E., J.W. Day, and F.H. Sklar. 1994. Ecosystem models of aquatic primary production and fish migration in Laguna de Terminos, Mexico. Pages 519–36 in W.J. Mitsch, ed. Global wetlands: old and new. Elsevier Science B.V., Amsterdam.

Riitters, K.H., R.V. O'Neill, C.T. Hunsaker, J.D. Wickham, D.H. Yankee, S.P. Timmins, et al. 1995. A factor analysis of landscape pattern and structure metrics. Landscape Ecology 10:23–39.

Ripley, B.D. 1981. Spatial Statistics. John Wiley & Sons, New York.

Risser, P.G., J.R. Karr, and R.T.T. Forman. 1984. Landscape ecology: directions and approaches. Special Publication. 2, Illinois Natural History Survey, Champaign, IL.

Rothermel, R.C. 1972. A mathematical model for predicting fire spread in wildland fuels. USDA Forest Service, Res. Paper INT-115. 40 pp.

Rutchey, K., and L. Vilchek. 1998. Air photointerpretaion and satellite imagery analysis techniques for mapping cattail coverage in a northern Everglades impoundment. Photogrammetric Engineering and Remote Sensing 64:1–7.

Saarenmaa, H., N.D.L. Stone, L.J. Folse, J.M. Packard, W.E. Grant, M.E. Makela, et al. 1988. An artificial intelligence modelling approach to simulating animal/habitat interactions. Ecological Modelling 44:125–41.

Schlesinger, M.E., and Z.C. Zhao. 1989. Seasonal climate changes induced by doubled CO_2 as simulated by the OSU atmospheric GCM/mixed-layer ocean model. Journal of Climate 2:463–99.

Show, I.T., Jr. 1979. Plankton community and physical environment simulation for the Gulf of Mexico region. Pages 432–39 in Proceedings of the 1979 Summer Computer Simulation Conference. Society for Computer Simulation.

Shugart, H.H. 1984. A theory of forest dynamics: an investigation of the ecological implications of several computer model of forest succession. Springer-Verlag, New York.

Sklar, F.H., R. Costanza, and J.W. Day, Jr. 1985. Dynamic spatial simulation modeling of coastal wetland habitat succession. Ecological Modelling 29:261–81.

Sklar, F.H., R. Costanza, and J.W. Day, Jr. 1990. Model conceptualization. Pages 625–59 in B.C. Patten, et al., eds. Wetlands and shallow continental water bodies, Vol. 1, SPB Academic Publishing, The Hague.

Sklar, F.H., and R. Costanza. 1991. The development of dynamic spatial models for landscape ecology: a review and prognosis. Pages 239–88 in M.G. Turner, and R.H. Gardner, eds. Quantitative methods in landscape ecology. Springer-Verlag, New York.

Sklar, F.H., K.K. Gopu, T. Maxwell, and R. Costanza. 1994. Spatially explicit and implicit dynamic simulations of wetland processes. Pages 537–54 in W.J. Mitsch, ed. Global wetlands: old and new. Elsevier Science B.V., Amsterdam.

Sklar, F.H., C. McVoy, R. Van Zee, D. Gawlik, D. Swift, H.C. Fitz, et al. 1999. Hydrologic needs: the effects of altered hydrology on the Everglades. Pages 2.1–2.70 in G. Redfield, ed. Everglades interim report. South Florida Water Management District, West Palm Beach, FL.

Snedecor, G.W., and W.G. Cochran. 1967. Statistical methods (sixth ed.). Iowa University Press, Ames, IA.

Sokal, R., and N.L. Oden, 1978. Spatial autocorrelation in biology. Biological Journal of the Limnean Society 10:199–249.

Thornthwaite, C.W., and J.R. Mather. 1955. The water balance. Drexel Inst. Tech., Lab. of Climatology, Climatology 8(1):1–86.

Titus, J.G., and V.K. Narayanan. 1995. The probability of sea level rise. Office of Policy, Planning, and Evaluation. EPA 230-R-95-008 (second printing). U.S. Environmental Protection Agency.

Turner, M.G. 1987. Spatial simulation of landscape changes in Georgia: a comparison of 3 transition models. Landscape Ecology 1:29–36.

Turner, M.G., R. Costanza, and F.H. Sklar. 1989. Methods to evaluate the performance of spatial simulation models. Ecological Modelling 48:1–18.

Turner, S.J., R.V. O'Neill, W. Conley, M.R. Conley, and H.C. Humphries. 1991. Pattern and scale: statistics for landscape ecology. Pages 17–49 in M.G. Turner, and R.H. Gardner, eds. Quantitative methods in landscape ecology. Springer-Verlag, New York.

Turner, M.G., Y. Wu, L.L. Wallace, W.H. Romme, and A. Brenkert. 1994. Simulating interactions among ungulates, vegetation and fire in northern Yellowstone National Park during winter. Ecological Applications 4:472–96.

Volterra, V. 1926. Varizioni e fluttuazioni del numero d'individui in specie animali conviventi. Mem. R. Acad. Naz. Dei Lincei, 2:31–113. Translated in R.N. Chapman, 1931. Animal ecology. McGraw-Hill, New York.

Walker, W.W. 1995. Design basis for Everglades stormwater treatment areas. Water Resources Bulletin 31:671–85.

Washington, W.M., and G.A. Meehl. 1984. Seasonal cycle experiments on the climate sensitivity due to a doubling of CO_2 with an atmospheric general circulation model coupled to a simple mixed-layer ocean model. Journal of Geophysical Research 89:9475–503.

Weinberg, A.M. 1985. Science and its limits: the regulator's dilemma. Issues in Science and Technology 2:59–72.

Wesseling, C.G., and G.B.M. Heuvelink. 1991. Semiautomatic evaluation of error propagation in GIS operations. Pages 1128–37 in J. Harts, ed. Proceedings EGIS '91. EGIS Foundation, Utrecht.

Wolff, W.F. 1994. An individual-oriented model of a wading bird nesting colony. Ecological Modelling 72:75–114.

Wu, Y., F.H. Sklar, K. Gopu, and K. Rutchey. 1996. Fire simulations in the Everglades landscape using parallel programming. Ecological Modelling 93: 113–24.

Wu, Yegang, F.H. Sklar, and K. Rutchey. 1997. Analysis and simulations of fragmentation patterns in the Everglades. Ecological Applications 7(1):268–76.

3
Measuring and Predicting Species Presence: Coastal Sage Scrub Case Study

Ted J. Case and Robert N. Fisher

In ecological applications of large-scale spatial data to management decisions concerning land planning and conservation, errors and biases may creep into the analysis and decision making at several steps (see Chaps. 1, 2, and 3), including:

- Uncertainty in positions of spatial locations of relevant ecological and physiographic features of the landscape.
- Uncertainty of the type and attributes of land cover at a particular location.
- Uncertainty in how different land covers at a position in space and the geometric arrangement of land covers nearby might influence an animal species occurrence or distribution, or the magnitude of some ecological process.
- Uncertainty about the relative importance of each spatial location to the overall success or persistence of a population or ecological process.
- Uncertainty about how to weight each species or ecological process in determining the overall biodiversity and functioning of ecosystems, local and national resource priorities, and consistency with legislative mandates.

We would like to be able to quantify the errors at each step, identify biases, and pass these along to the next analysis step so that our degree of uncertainty regarding potential outcomes is evident at each level (e.g., Stoms et al. 1992). This is what a quantitative ecological risk assessment should require (Chaps. 3 and 19). Uncertainties in each step in this progression, however, may operate at different spatial scales and therefore not be easily propagated to the next. Moreover, the degree of contingency of outcomes increases at each step of this progression. The error in assigning x, y coordinates to points across space is relatively invariant to where in space those points are located; however, the assignment of attributes (e.g., vegetative type) to a position is influenced heavily by attributes nearby because vegetative cover typically occurs at intervals wider than a single map unit. In addition, the type of vegetative cover in a particular point or polygon may be less revealing about an animal species occurrence in that point or polygon than the context of the position, its neighbors' types, their isolation from other such patches, and the like. A wealth of empirical evidence shows that local animal populations persist longer in large patches of suitable habitat than in small patches, and persist longer in patches close to other suitable patches than in patches more isolated from neighbors (Lande 1987; Bolger et al. 1991; Stacy and Taper 1991; Robinson et al. 1995; Luiselli and Capizzi 1997). A single patch of a particular cover type's importance in sustaining a species population may thus not be invariant to where that patch occurs on a map, particularly with respect to the configuration of neighboring patches.

One goal in ecology is to be able to understand and predict species occurrences and abundances. The National Gap Program (1997) (Scott et al. 1993) operates on the premise that at least for common species and at broad geographic scales a knowledge of land cover can be used to predict the occurrence of vertebrate species distributions, and that these predictions will be useful to land managers who must make decisions on where and how to protect these species. Many (perhaps most) land use decisions are made at the local government level through the planning efforts of cities and counties. These municipalities may band together into large cooperative programs to guide more regional decisions; however, the degree to which the gap approach (i.e., a habitat-based method for predicting species occurrences) will be successful when applied to smaller geographic scales with finer resolution, to highly fragmented landscapes with much edge between cover types, and with species that have already become so rare (as to trigger the endangered species act) that their present-day retracted and patchy geographic ranges reflect a large stochastic component is an open question.

An increasing trend for dealing with the Federal Endangered Species Act is the creation of multiple species Habitat Conservation Plans usually put together at the scale of one or a few collaborating counties (Dobson et al. 1997; Noss et al. 1997; Pulliam and Babbit 1997; Hood 1998; Kareiva et al. 1999). One such plan is California's Natural Community Conservation Program effort (NCCP) (Jasney 1997; Noss et al. 1997). The NCCP encompasses portions of five southern California counties and is in excess of 5000 square miles of landscape. This planning effort was begun in response to the listing of the California gnatcatcher and focused on coastal sage scrub habitats. This program now incorporates many habitat types in the planning process, and many land owners and jurisdictions are interested in getting coverage for a variety of species. To date several of the subregional plans have progressed and been approved. These plans need to provide assurances that sensitive species are protected. This is generally based on evaluations of habitat suitability from ground cover. Even though most of the chapters in this book deal with elements of uncertainty in the first two steps in the preceding series, this chapter focuses more on the next two steps. We will address issues of uncertainty in how different land covers influence vertebrate distribution so that habitat-affinity models based on ground cover may be tested, their errors quantified, and efficient indicators developed.

3.1 Study Objectives

We began a study of the spatial and temporal variation of reptiles and amphibians within the fragmented landscape of coastal Southern California to support the regional planning efforts of the NCCP. As a group these species are less conspicuous than most land birds, and they are not as easily censused; thus, their habitat affinities are not well known. Few studies have looked at community structure in reptiles and amphibians in California in general, and most attention has been on the desert communities (Glazer 1970; Pianka 1986). Notable nondesert exceptions are the work of Fitch (1949), Klauber (1939), Fuentes (1976), Drost and Fellers (1996), Fisher and Shaffer (1996), Morafka and Banta (1976), and Block and Morrison (1998).

We are presently operating 30 sites from the southern end of the Los Angeles Basin, south to the Mexican Border, and an additional four sites at higher elevations. A map of these sites is shown in Figure 3.1. We have been

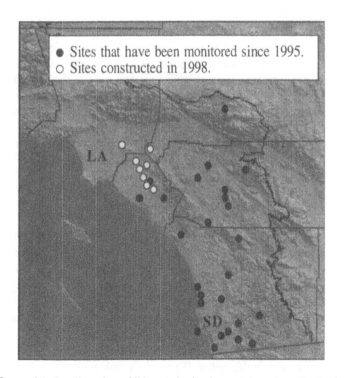

FIGURE 3.1. Reptile and amphibian study sites in coastal southern California.

monitoring 22 of these coastal sage scrub sites since 1995 (an additional eight were added in 1998, but these will not be discussed further as sampling is still incomplete). At each of these study sites we collect data at multiple sampling stations. For the purposes of measuring relevant habitat variables that influence species occurrences and abundances, we hope to characterize two spatial levels—within-patch and between patch. Individuals will respond in their local movements, birth, and death rates to local variables at the scale of a few home range sizes—the within-patch descriptors. At the between-patch or regional spatial scale, patches can be characterized by their area, the isolation of the patch from others, and other features of patch geometry. We wish to know:

- How do local variables like slope angle and exposure, substrate, local plant species composition measured at the site compare with more regional landscape variables available as geographic information system (GIS) data layers like location on the coast/interior axis, latitude, fragment size, fragment isolation and urbanization at the edge, proximity to wetlands, and fire history in predicting the abundance and distribution of these species?
- Are more easily detected species like certain conspicuous birds and reptiles reliable indicators of the presence or abundance of less conspicuous species of amphibians and reptiles?
- How reliable are transects and timed visual surveys in detecting the presence of these various reptile and amphibian species?
- Is the distribution and abundance of different reptile and amphibian species unrelated to each other, with each species individually varying across the landscape in its own idiosyncratic way? If this hypothesis is correct, then their abundances are unlikely to be predicted by gross vegetation communities mapped at a scale of 1–10 hectare resolution. If this hypothesis is incorrect, then the concept of "umbrella" species as a predictor for a suite of other species may be valid, assuming the correct umbrella species are chosen (see later).
- What species may require additional regulatory attention, and which species are adequately protected in the current planning of permanently preserved habitats?

Our study is still ongoing; however, in this chapter we can address some preliminary conclusions. To be specific we will first describe the sampling design and technique we are utilizing. We will then discuss three validation studies we have completed to test our methods for sampling these species. We will next describe the difficulties in determining species diversity using these techniques. We then analyze which species could serve as umbrella or indicator species of biodiversity, and we describe some clear results of this sampling approach and mechanisms that are becoming apparent in the decline of reptiles in this system. We conclude with implications for habitat conservation plans.

3.2 Sampling Regime

The herpetofauna of coastal southern California is very diverse due to a variety of factors including topography, geological and biogeographic history, and a diversity of climates over a relatively small area (Peabody and Savage 1958; Savage 1960; Axelrod 1989). This herpetofauna consists of more than 70 species in 19 families, of which 24 are considered sensitive at the state or federal levels (Stebbins 1985; Jennings and Hayes 1994; Fisher and Case 1997).

In order to study such a diversity of species, we needed to identify a sampling technology that would objectively sample scrub habitats with limited biases across species. From the literature it was clear that a pit-fall trapping design with the use of drift fences was the most likely technique that would minimize sampling biases across species, observers, and habitats.

The study sites range from sea-level to elevations of more than 2,000 m, and incorporate the diversity of habitat types present on the coastal slope of southern California. These habitats include coastal sage scrub, coastal sand dunes, maritime succulent scrub, chaparral, oak woodland, and coniferous forest types. Each of these study sites contains from 3 to 30 pit-fall drift-fence arrays.

Each array consists of seven 5-gallon buckets as pit fall traps, connected by shade cloth drift-fences (15-m arms), in the shape of a Y (Fig. 3.2), except the arms are allowed to make gentle bends around trees, shrubs, and boulders to

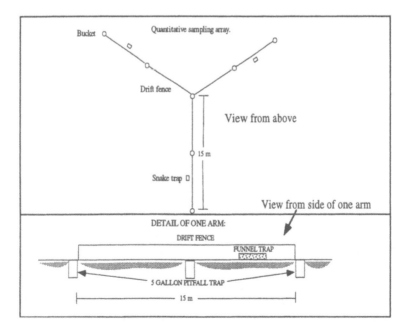

Figure 3.2. Terrestrial survey protocol and designs for arrangement of pitfall and funnel traps with drift fences.

minimize habitat alteration. A hardware cloth funnel trap is placed at each of the three arms for capturing large snakes and lizards. We currently monitor a total of 386 arrays, including the new sites that have been added more recently within Los Angeles, San Bernardino, and Orange Counties.

Figure 3.3 (see color insert) shows a photo of one such array in the ground. Sampling is conducted at each study site for 10 consecutive days every 6 weeks, for a total of 50–60 days a year, spread evenly across all seasons. The traps are kept closed between the sampling periods. All of the 22 study sites started in 1995 have been sampled for almost 200 calendar days over the span of at least three years. Each site will ultimately be studied for 3–5 consecutive years.

We have captured 45 native species of reptiles and amphibians from 17 different families, representing more than 27,000 captures from more than 20,000 individuals. The animals captured are individually marked (except for a few species that lack adequate marking techniques) either by toe-clipping or scale-clipping (snakes) and then released. Any incidental deaths are preserved as vouchers, and are deposited in the California Academy of Sciences herpetological collection. Our protocols ensure that the vast majority of the animals captured remain alive when the traps are open; we check traps once every 24 hours, in the morning. We process the reptiles and amphibians in the field and release any other animals. Processing includes marking, weighing, and measuring the body length; we keep the toe-clips and tail tips from snakes in ethanol for future molecular systematic work. We capture and release species of mammals representing eight different families (more than 8000 records). Although the most commonly collected species are in four families, two of these families (i.e., Soricidae and Geomyidae) are very difficult to collect by any other means (Wilson et al. 1996). We have recorded more than 600 individuals from these families at our sites. A wide diversity of ground-dwelling invertebrates are captured, including arachnids, centipedes, millipedes, orthopterans, coleopterans, and hymenopterans.

3.2.1 Spatial Sampling

Sites. These are simply geographic locations for clusters of arrays. We expect differences in overall coastal sage scrub composition to be associated with a broad-scale north–south axis (latitudinal climate gradient), an east–west axis (elevational and rainfall gradient), and a natural land–urbanized axis associated with differing degrees of habitat fragmentation of natural lands (Westman 1981; Axelrod 1989; Minnich 1989; Keeley and Swift 1995; Bolger, Scott and Rotenberry 1997). Sites were consequently selected to span with replication these three prime dimensions; however, because arrays are potentially attractive to vandals and poaching, it was also necessary to ensure that chosen sites had restricted access from the public.

TABLE 3.1. Array-specific variables.

Location
  Elevation
  Slope
  Aspect
Vegetation
  Mean canopy height
  Density of shrubs
  Density of grass and herbs
Flora
  % Coastal sage scrub flora
  % Chaparral flora
  % Tress
  % Grass and herbs
  % Other
  Total % Cover
Soils
  Frequency of leaf litter
  Frequency of sandy soils
  Frequency of cryptogramic soils
  Frequency of bare rock
  Frequency of organic soils
  Frequency of moss
Ants
  Argentine ants
  Harvester ants
  Crematogaster ants
  Carpenter ants

Arrays Within Sites. Small habitat fragments have fewer arrays than do large natural lands. Although we have a few arrays in habitat types bordering scrub (i.e., grassland, riparian habitats, chaparral, oak woodland) these habitat types are not as effectively sampled as scrub habitat types, which comprise the main focus of the study. We attempted to cover the range of slope and aspect variance offered by the site, with replication where possible. Array placement is haphazard within these constraints.

Independent Variables. The flora and vegetation has been recorded in the vicinity of each array following established protocols of the California Native Plant Society, and various local landscape features have also been recorded (including slope, aspect, soil type) and presence of different types of ant colonies. A list of these is given in Table 3.1.

3.3 Results and Discussion

Because one goal of our study is to provide a data set useful for testing and calibrating habitat affinity prediction models for vertebrates, a first step is to establish the efficacy of the array trapping method to other herpetological survey techniques. We summarize results from three different validation studies we undertook to determine the accuracy of our technique. Then we present several analyses and descriptions of the pitfall trap data to illustrate the uncertainty in getting accurate diversities at sites, and the geographic patterns in diversity. Finally, we address mechanisms for additional uncertainty resulting from declines associated with fragmentation, and end with an analysis of possible umbrella species for attention in reserve planning.

3.3.1 Fifteen-Minute Transects and Timed Visual Surveys

We completed a higher intensity modified form of what has been considered the "Brattstrom Transect" to determine how successful that technique is at determining the diversity of reptiles and amphibians at a study site (described in Brattstrom 1992). The Brattstrom transect is basically a 100-m transect walked through a particular habitat, and the observed lizards are recorded. Other habitat features are recorded after the transect is completed while returning to the starting point. We tried these transects and they take from under 1 minute to 2–3 minutes to complete. You typically do not see any reptiles or amphibians. We have modified this protocol to make it a 15-minute time-constrained search. Within the 15-minute period, if you were to walk the entire time you would typically do the equivalent of 8–12 Brattstrom surveys. The surveyor's goal is to find a maximal number of species and individuals during the 15 minutes; therefore, several Brattstrom surveys are usually interspersed with effort spent turning rocks and looking under other cover objects. The searcher again only spends 15 minutes. Habitat features are recorded for each survey location, and two transects are completed at each study site. These two transects are then replicated three times within 1 month; thus, the weather may be different for each survey. In Figure 3.4 we present results for two of the four study sites where we have conducted transects.

These two study sites represent the extremes of the most observations and least observations during the transects. For the Wild Animal Park, which is the study site with the most recorded species, we have completed 3 transect days in fall and 3 days in spring, for a total of 12 transects. We observed between zero and seven individual reptiles during these surveys, representing between zero and four species per survey. Seven species were detected in total during these surveys, compared with the 34 species that are known from the site. Results for Little Cedar Ridge is even more extreme. Here again we have conducted a total of 12 transects between fall and spring. Reptiles were seen for this site only on two transects, and it was just one common fence lizard, *Sceloporus occidentalis*, in both cases. For this site, therefore, there was only one species detected during the 12 transects,

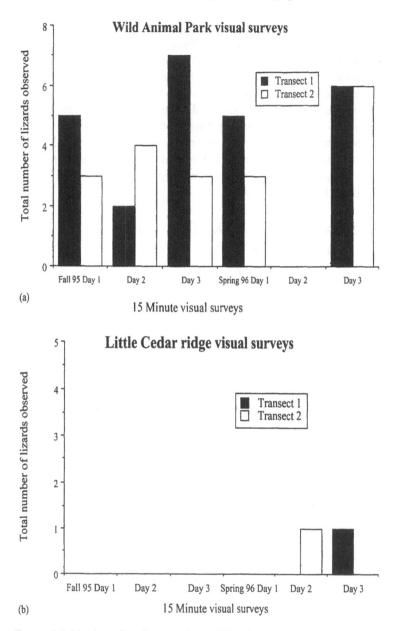

FIGURE 3.4. Number of reptiles seen during fall and spring 15-minute transects at two study sites.

compared with the 20 species that are known from the study site. Similar data was collected for Lake Perris and Starr Ranch, and again the total number of species observed at both of these sites was a fraction of the number of species we have detected in our arrays.

We find that the visual transects, even in combination with cover object searches, are not very successful at detecting the actual number of species that occur at a study site. In addition, the transect data cannot be utilized to determine the abundance or density of reptile species without capturing and marking individuals. The variance in detection of species between and within seasons was so great as to get entirely different reptile species recorded from one day to the next. In summary, these transects failed to detect some of the focal lizard species even where we knew from array results that they were present. The implications of this lack of detection of even common species puts in doubt the ability of surveyors walking transects to be certain that a species is absent from a site.

3.3.2 Comparison of 11-Day Observations to 10 Days Trapping

We compared the diversity of reptile and amphibian species observed while opening (1 day) and sampling (10 days) a study site over the 11-day sampling period, to the diversity of species that were captured in arrays during the 10 days of sampling. The point of this comparison was to determine how effective expert herpetologists are at passively observing reptiles and amphibians while sampling the arrays. They spend between 4 and 6 hours at a study site on an average sample day, and between 5 and 12 hours on opening day; therefore, this would add up to approximately between 45–72 person hours of field exposure during a sample period. This comparison of observations to trapped reptiles and amphibians was done 24 times and included several different sites, and different seasons. From Figure 3.5 we see that more species were captured 22 times out of 24 in traps than observed during the 11 days of site visits during that sample period.

On average we observed only 44% percent of the species captured during a sample period. Thus, passive site visits, even when they include 11 consecutive days (between 45 and 72 person hours), are ineffective in these scrub habitats at detecting the reptile and amphibian diversity of a location. This result is consistent with other reptile and amphibian trapping studies (Gibbons and Semlitsch 1981; Campbell and Christman 1982; Vogt and Hine 1982; Corn and Bury 1991).

3.3.3 Comparison of High-Intensity Searching to Pitfall Traps

We conducted a more exhaustive comparison of techniques in the spring of 1996 at the East Side Reservoir Site in Riverside County with Jens Vindum (Herpetology Department, California Academy of Sciences). This location is unique because there is a huge reservoir construction project underway. We have study sites both directly to the north and south of the site, and we received permission from the Metropolitan Water District and California

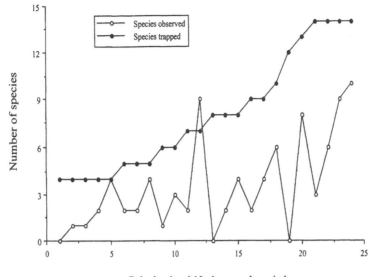

FIGURE 3.5. Comparison of visual observations to trapping data. On 22 out of 24 days, more species were captured in traps than were observed during the 11 days of site visits that sample period.

Department of Fish and Game to collect reptiles and amphibians within the construction area. We were therefore able to compare the success of pitfall traps directly with other traditional reptile and amphibian collecting techniques at successfully capturing reptiles and amphibians. We conducted high intensity herpetological searching (HIHT) by having expert herpetologists from the California Academy of Sciences and UCSD spend a considerable amount of time utilizing multiple techniques to capture reptiles and amphibians in this location. In total seven different people worked on this project, for a total of 20 person days, both in teams and individually, to try to capture as many species and individual reptiles and amphibians as possible. The data presented in Table 3.2 show that although HIHT collected more individuals in total than the pitfall traps, the species diversity represented was lower and it was biased against the sensitive species.

Thus, twice as many sensitive species were detected using the pitfall traps, and five times the number of individuals of these species were captured than by the HIHT techniques. Sixty-five percent of the total pitfall trap specimens were sensitive species, but only 9% of the HIHT specimens represented sensitive species.

3.3.4 Cumulative Species Lists

Despite the relative efficacy of the arrays at detecting reptiles and amphibians, it takes surprisingly long and intensive sampling to get an accurate

TABLE 3.2. Comparison of standard herpetological collecting to pitfall trapping.

	March pitfall sampling (10 days)	May pitfall sampling (10 days)	March HIHT (4 days)	May HIHT (8 days)	Total pitfall trap sampling (20 days)	Total HIHT (12 days)
Batrachoseps pacificus	1		6			6
Hyla regilla	1		21		1	21
Bufo boreas		3	17	2	4	19
Rana catesbeiana			12			12
Scaphiopus hammondii	4	1			5	
Coleonyx variegatus		15			15	
Elgaria multicarinatus		2			2	
Eumeces skiltonianus	21	4	10		25	10
Cnemidophorus hyperythrus	13	29			42	
Cnemidophorus tigris		8		3	8	3
Sceloporus occidentalis	12	10	2	34	22	36
Sceloporus orcutti			2	25		27
Uta stansburiana	6	10	29	44	16	73
Phrynosoma coronatum	2	8	5		10	5
Lichanura trivirgata	1		2		1	2
Lampropeltis getulus		5	1		5	1
Masticophis flagellum		4	2	1	4	3
Masticophis lateralis	1	1		1	2	1
Pituophis melanoleucas	1			1	1	1
Salvadora hexalepis		1			1	
Tantilla planiceps	1	1			1	
Crotalus viridis	1		2			2
Total number of captures:	63	102	111	111	165	222
Total number of native species:	11	15	13	8	18	15

Total number of sensitive species:	5	7	3	1	8	4
Total captures of sensitive species:					107	20
Percentage of total captures that are sensitive species captures:					0.65	0.09

HIHT: High-intensity herpetological team. This group included a contingent of people from the California Academy of Sciences and UCSD. This team used many standard methods for herpetological collecting and attempted to collect as many species/individuals as possible.

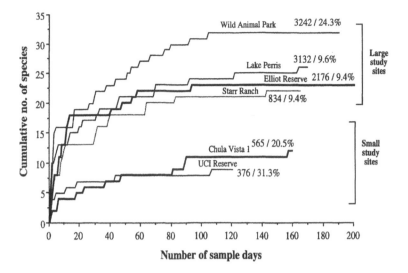

FIGURE 3.6. The cumulative number of reptile plus amphibian species recorded from a site as a function of trapping days. The numbers at the end of each curve are: total captures and the percentage of these captures that represent recaptured individuals. Four large sites and two small sites are illustrated. Only species actually caught in traps are tallied.

description of even species presence and absence at sites. This is seen by comparing the cumulative species lists over time or over capture rates. After 3 years of trapping at some larger sites, we are still occasionally recording new species. Asymptotes are approached more quickly for smaller habitat fragments (Fig. 3.6). Still, at this spatial scale, and for this relatively cryptic fauna, intense long-term sampling is required before one is certain even of species presence or absence. Notice also the relatively low ratio of recaptures to new animals even after 3 years of census. This reflects relatively high turnover of individuals in the populations, due to movements, and birth and death, at this spatial scale. The number of captures and the recapture ratio is large enough to provide estimates of population size, with reasonable confidence intervals (e.g., in *Cnemidophorus hyperythrus*) only for a few species.

3.3.5 Species Incidence Matrixes

Figure 3.7 shows tables that give the capture of a species at a study site as a black-filled square and the observation of a species (but not captured in a pitfall trap) as a gray square.

The filled boxes represent the presence of a species on the study site, or immediately adjacent to it. Species in each taxonomic group are ordered from the most occurrences (left) at sites to the least occurrences (right). Sites are ordered from those with the most species on the top to the least species on the bottom. With this arrangement, a perfectly nested pattern would yield a filled upper left triangular pattern. Such a nested pattern indicates a predictable pattern to species co-occurrences [i.e., species are "added" to sites in a common sequence (Wright et al. 1998)]. The lizards tend to show a nested pattern, the snakes less so, and the amphibians not at all; however, our sampling regime is not sufficient for most amphibians because we do not adequately sample aquatic and wetland habitat types.

The order of sites along the side also closely parallels the amount of contiguous natural habitat remaining in the area. Because all these sites were selected to have similar ground and scrub vegetation types, the absence of species from the smaller sites can be attributed to their small fragment area and possible detrimental edge effects from urbanization. For lizards, the legless lizard, *Anniella pulchra*, causes the most deviation from perfect nestedness of the lizard incidence matrix. This species is a substrate specialist, preferring loose sandy soils, yet it can occur even in suburban backyards with the appropriate soil type. However, other absences are more easily ascribed to our focus on upland scrub habitat types. Pond turtles (*Clemmys marmorata*) and the garter snakes, *Thamnophis hammondii* and *T. sirtalis*, are primarily in aquatic habitats. The California mountain kingsnake (*Lampropeltis zonata*) similarly resides primarily in high elevations in San Diego and Riverside County, but it occurs in low elevation riparian forest in Orange County; thus, it is expected only in traps from this county. The leopard lizard, *Gambelia wislizenii*, has its primary distribution east of our study area in more arid desert habitat types and no doubt would have been rare in the coastal plains even in the absence of habitat fragmentation.

3.3.6 Local Extinctions

Fragmentation is often associated with species loss (Wilcove et al. 1986; Soulé et al. 1988) although the precise mechanism responsible is often not clear. Fisher and Case (in preparation) have compared the snake species lists for our sites with those reported from the same or nearby areas by Klauber (1939, and unpublished maps at San Diego Natural History Museum) in an effort to determine whether the species absent have recently declined or have always been rare or absent. We have found that 10–50% of the snake species historically known from some of these sites may now be absent.

Several snakes commonly observed by Klauber that are now apparently extirpated at these sites include red diamond rattlesnake, California glossy snake, rosy boa, and red racer. There are also some species that we find as common, but were not reported by Klauber from some of these sites, including the western blind snake and the black-headed snake. These could be recent colonists, but it is more likely that they reflect the greater sampling effort of our study at these sites, and our ability to detect secretive species

A

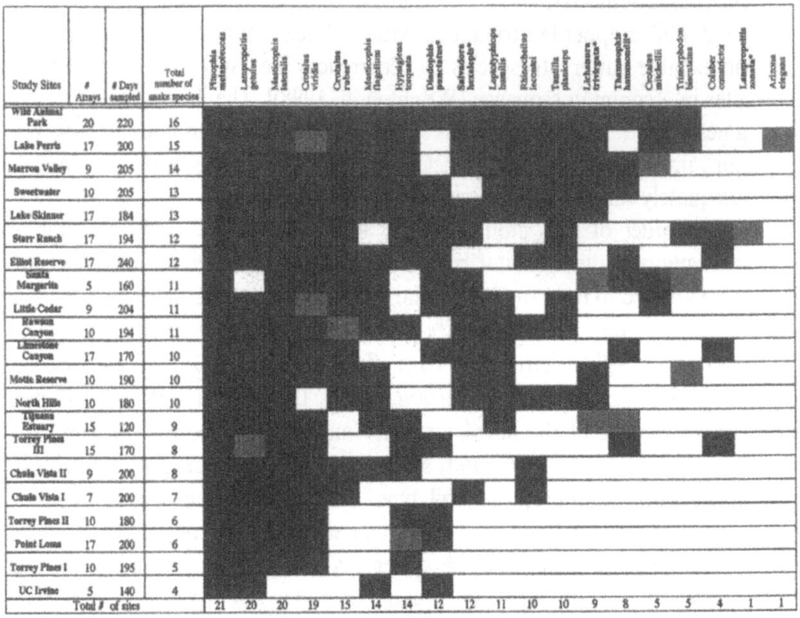
Study Sites
Arrays
Days sampled
Total number of snake species
Pituophis melanoleucus
Lampropeltis getulus
Masticophis lateralis
Crotalus viridis
Crotalus ruber*
Masticophis flagellum
Hypsiglena torquata
Diadophis punctatus*
Salvadora hexalepis*
Leptotyphlops humilis
Rhinocheilus lecontei
Tantilla planiceps
Lichanura trivirgata*
Thamnophis hammondii*
Crotalus mitchellii
Trimorphodon biscutatus
Coluber constrictor
Lampropeltis zonata*
Arizona elegans
Wild Animal Park 20 220 16
Lake Perris 17 200 15
Marron Valley 9 205 14
Sweetwater 10 205 13
Lake Skinner 17 184 13
Starr Ranch 17 194 12
Elliot Reserve 17 240 12
Santa Margarita 5 160 11
Little Cedar 9 204 11
Rawson Canyon 10 194 11
Limestone Canyon 17 170 10
Motte Reserve 10 190 10
North Hills 10 180 10
Tijuana Estuary 15 120 9
Torrey Pines III 15 170 8
Chula Vista II 9 200 8
Chula Vista I 7 200 7
Torrey Pines II 10 180 6
Point Loma 17 200 6
Torrey Pines I 10 195 5
UC Irvine 5 140 4
Total # of sites
21 20 20 19 15 14 14 12 12 11 10 10 9 8 5 5 4 1 1

B

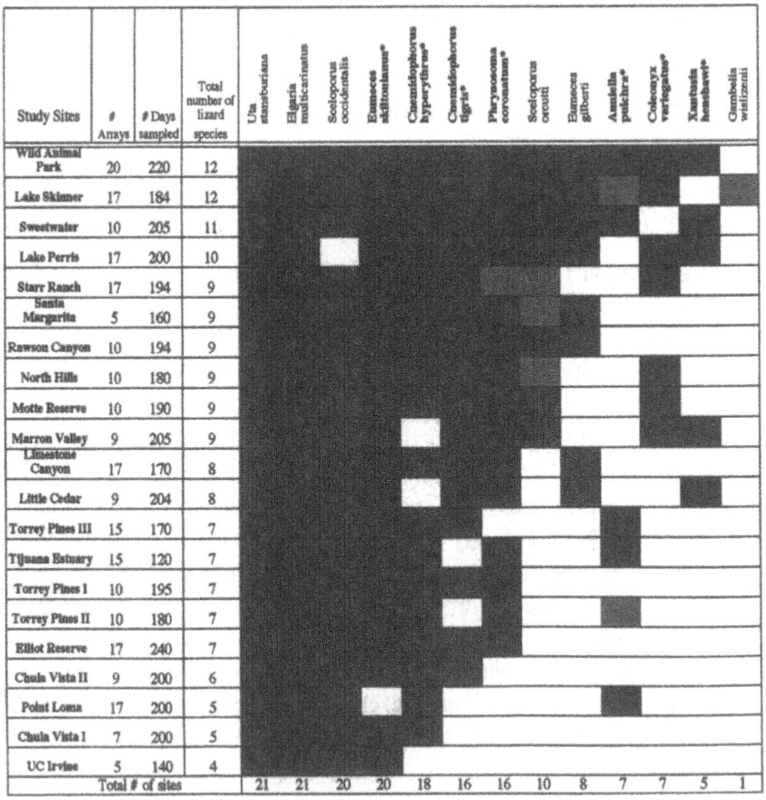
Study Sites
Arrays
Days sampled
Total number of lizard species
Uta stansburiana
Elgaria multicarinatus
Sceloporus occidentalis
Eumeces skiltonianus*
Cnemidophorus hyperythrus*
Cnemidophorus tigris*
Phrynosoma coronatum*
Sceloporus orcutti
Eumeces gilberti
Anniella pulchra*
Coleonyx variegatus*
Xantusia henshawi*
Gambelia wislizenii
Wild Animal Park 20 220 12
Lake Skinner 17 184 12
Sweetwater 10 205 11
Lake Perris 17 200 10
Starr Ranch 17 194 9
Santa Margarita 5 160 9
Rawson Canyon 10 194 9
North Hills 10 180 9
Motte Reserve 10 190 9
Marron Valley 9 205 9
Limestone Canyon 17 170 8
Little Cedar 9 204 8
Torrey Pines III 15 170 7
Tijuana Estuary 15 120 7
Torrey Pines I 10 195 7
Torrey Pines II 10 180 7
Elliot Reserve 17 240 7
Chula Vista II 9 200 6
Point Loma 17 200 5
Chula Vista I 7 200 5
UC Irvine 5 140 4
Total # of sites
21 21 20 20 18 16 16 10 8 7 7 5 1

C

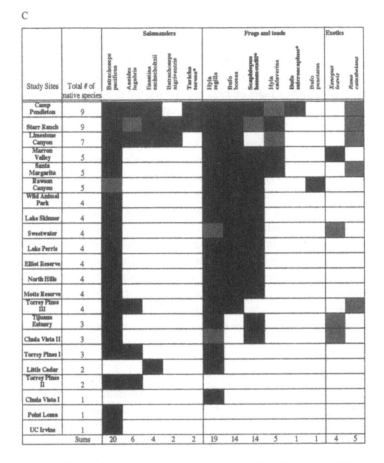

FIGURE 3.7. Distribution of species across study sites. (A) snake species; (B) lizard species; (C) amphibian species. * = Considered sensitive or protected by the U.S. Fish and Wildlife Service (USFWS) or the California Department of Fish and Game (CDFG). Solid boxes indicate captures in buckets, shaded ones indicate observation at site but not capture. The number of arrays and sample days at each site are given in the previous tables for snakes and lizards.

with pitfall traps. If the species declines are real, then it will be important to identify the contributing causes. In addition to habitat loss, we suggest two factors that seem to affect several of the reptile and amphibian species. These are the effects of roads in and around habitat patches and edge effects from surrounding urbanization.

Roads and Automobile Traffic as a Source of Fragmentation and Direct Mortality. Klauber (1939) estimated that approximately 10,000 snakes were killed in San Diego County by automobiles on roads each year. Snakes were singled out for study partly because of Klauber's personal interests, but also because roads represent an attractive nuisance for them: Several nocturnal species use the asphalt to stay warm after the sun sets, and diurnal species use this heat to extend their activity times into the early evening. The road traffic figures at the time of Klauber's estimate were about 1 million miles driven on San Diego County roads per day, whereas this number increased to about 63 million miles by 1995 and is projected to be about 92 million by 2015. Of course, we cannot linearly extrapolate these traffic figures to produce an index of present mortality and we simply have no data on current road deaths. Snakes tend to have large home ranges (Macartney et al. 1989) and mortality estimates on roads elsewhere can be high. For example, Campbell (1953) estimated 26,000 snakes killed per year on roads in New Mexico in 1953 and Rosen and Lowe (1994) estimated 13.5–22.5 snakes killed per kilometer per year along southern Arizona in the 1990s. Thus, it would be surprising if direct road mortality was not substantial for snake populations in southern California (see also Fahrig et al. 1995 for amphibian road kills). In addition to direct mortality, roads can isolate subpopulations leading to reduced genetic diversity. Hitchings and Beebee (1998), for example, found reduced genetic diversity, survival, and developmental homeostasis were found to be significantly lower in small, urban populations of the common toad, *Bufo bufo*, compared with larger rural populations.

Edge Effects from Urbanized Neighborhoods. Among other detrimental effects that result from habitat loss and subdivision, fragmentation can promote the penetration of exotic species into natural areas. We have found that fragmentation of coastal southern California scrub habitats facilitates the invasion of an exotic species, the Argentine ant (*Linepithema humile*), into remaining natural areas, subsequently reducing local native ant diversity (Suarez et al. 1998). Ant communities were surveyed in 40 isolated fragments of coastal sage scrub in San Diego County, California. The geographic scale of these fragments is much smaller than that of the reptile and amphibian study. Large fragments are only about 100 hectares, whereas large sites in the reptile and amphibian study may be several square miles of natural habitat. Across these urban fragments, native ant diversity was positively correlated with the size of the fragment, and negatively correlated with the density of Argentine ants and the number of years since the fragment was isolated from continuous scrub land. In addition, fragments had fewer native ant species than did similar-sized plots within large unfragmented areas. Native ant species vary in their vulnerability to habitat fragmentation and the subsequent presence of the Argentine ant; the most sensitive include army ants (*Neivamyrmex*) and harvester ants (genera *Messor* and *Pogonomyrmex*). In addition, other groups of arthropods respond negatively to Argentine ants, even after statistically accounting for differences due to edge effects alone (Bolger et al. 2000). Previous work on this system of fragments (Soulé et al. 1988; Bolger et al. 1991, 1997) found indirect evidence for local extinctions of birds and small mammals, but the causes of extinction in these studies are not clear.

Does this turnover in the arthropod fauna effect vertebrates? Analysis of coastal horned lizard fecal pellets revealed that the diet of wild horned lizards consisted primarily of ants (more than 95% of prey intake), particularly harvester ants. In areas where Argentine ants had invaded, however, horned lizard diets' change significantly, incorporating more nonant arthropods, particularly beetles.

At a larger geographic scale, in a preliminary analysis, we have found that after accounting for spatial autocorrelation in our data, the dependent variables (from Table 3.1) that best explain relative abundance of horned lizards is the absence of Argentine ants and the presence of ants in the genus *Camponotus* (carpenter ants). In a study elsewhere in this volume (Ver Hoef et al., Chap. 10) utilizing this data set, we similarly found that the best predictors of the abundance of orange-throated whiptail lizards are sandy soils and *Crematogaster* ants, which are also negatively affected by Argentine ants.

Crooks (2000) also found a dramatic change in the common carnivores associated with surrounding urbanization. Opossum, feral cats, striped skunk, gray fox, and raccoon are more common in smaller older fragments, whereas mountain lions, spotted skunk, coyotes, and bobcats show just the opposite correlations. The former set of species seems to thrive as partly human commensals. Thus, at the small scale of reticulating patches of urbanized and natural lands, impacts like a change in the arthropod fauna or a loss of large carnivores which are not easily categorized by ground cover types as determined from remote sensing techniques (e.g., Chaps. 4, 8, 12, and 19), can nevertheless be very important in determining differing species compositions of higher vertebrates.

Indicator Species. Using the reptile distributions available in the incidence matrixes and the species accumulation curves for each site, we can seek to identify species that (1) tend to be restricted to high-diversity communities and, (2) when present are found relatively early on with relatively little sample effort. Such species will be efficient indicators of the presence of reptile diversity. For each reptile we determined the minimum number of other species that it occurs with across our sample sites. For each site we next determined the effort taken first to detect this species at each site, where effort is defined as days of sampling at a site times the number of arrays placed at the site. These values were averaged across all sites where the species occurs. In the few cases where we know that a species is present at a site but has not yet been captured in an array, we took one more than the number of days of total sampling at that site (given in column 3 of Figure 3.7a) as the number of days. The results are plotted in Figure 3.8. Each point is a species and is labeled according to the first two letters of the genus and species names. Note that there is a significant ($p < 0.001$)

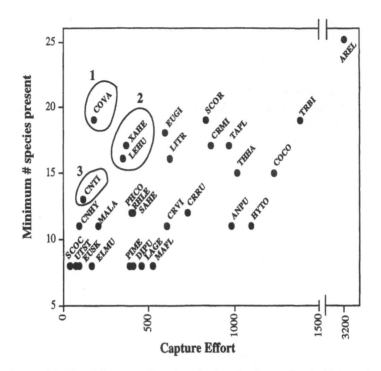

FIGURE 3.8. The minimum number of species found to be associated with a species across all study sites plotted against the average capture effort to detect that species in Array * days per site. Each point is a species and the labels indicate the first two letters of the genus and species names. The circles represent three tiers of relatively efficient indicator species (for other reptiles) indicating different degrees of reptile diversity.

correlation to these points, indicating a tendency for those species with the largest "umbrella" of other species under them also to require more average sample effort to detect, when they are present. Three sets of relatively efficient indicators are circled. The first tier, and primarily the gecko, *Coleonyx variegatus*, is indicative of the richest reptile communities. The second tier includes the granite night lizard, *Xantusia henshawi* (which is indicative of expanses of granite boulders) and the blind snake, *Leptotyphlops humilis*. Finally, the western whiptail lizard, *Cnemidophorus tigris*, is an efficient indicator of only moderately rich reptile communities.

3.4 Summary

It is impractical and prohibitively expensive to ground-truth all or even most points, so for coverage of animal species of concern, large-scale reserve design schemes are usually based on inferences about presence or absence from habitat characteristics that can be determined by remote sensing. These images can then be categorized into a set of habitats based on conspicuous vegetative and edaphic characters that can be scored from the images. On a very large spatial region and at coarse resolution (one mapping unit = 1000 hectares), and broad habitat types such as boreal forest, tundra, and desert, as in the GAP program, this method is useful for predicting sets of animal species that may occur. Most reserve design schemes in the United States, however, are implemented at the scale of counties because this is where most authority lies for land planning and zoning. At this smaller scale, where habitat types must be distinguished more finely and are typically more reticulated with one another and human-modified urban and agricultural habitats, it is an open question whether remotely sensed data on a pixel-by-pixel level will serve as useful predictors of occurrence, let alone more quantitative estimates of numerical abundance.

Our study seeks to find habitat affinity models that might be more predictive at this smaller geographic scale and to identify useful indicators of species presence and relative abundances of reptiles and amphibians. Even though this project is still in progress, we have provisionally determined that some of the best predictors (e.g., ant species) are not easily approached from a remote sensing venue. The relative abundance of different functional groups of ants has been used as an indicator of environmental change and disturbance regimes in Australia (Andersen 1990). Our work suggests that they may be equally successful in the United States. In addition, there is good evidence that some reptile species have become locally extinct within habitat patches that historically supported populations, thus adding to the uncertainty in diversity. This further obscures a tight relationship between broadly defined cover types and species occurrence at a small geographic coverage of high spatial resolution.

Finally, we find that there is not always a close correspondence between species that presently receive legal protection and those that seem to be rare based on our sampling scheme. One such example is the glossy snake *Arizona elegans occidentalis*, which presently does not have any protective status. It has been captured at only one of our sites, despite frequent collections by Klauber in the 1930s and 1940s (Klauber 1939, 1946) in the same areas as many of our study sites. This subspecies ranges from northern Baja California to the San Francisco Bay area. It is unknown whether it is equally rare in Mexico or the northern parts of its range. At the opposite extreme is the whiptail lizard, *Cnemidophorus hyperythrus*, which was considered a candidate for listing by the U.S. Fish and Wildlife Service and was one of three species originally invoked as a focal species for NCCP reserves (Hollander, Davis, and Stoms 1994). It occurs at nearly all of our study sites south of the Santa Ana River in Orange County, and is one of the three most abundant species, where it occurs, based on our capture success.

Acknowledgments. These surveys would not be possible without the day-to-day sampling by Chris Brown, Ed Ervin, Bob Haase, Adam Backlin, and the many students and volunteers for the last few years. Many land owners and agencies have granted use of their properties for our study. This work is supported by funding from the California Department of Fish and Game, Biological Resources Division of the USGS, Metropolitan Water District, California State Parks, United States Forest Service, and United States Fish and Wildlife Service. We are particularly indebted to Pete Stine, who helped design and implement this study. The manuscript benefited from helpful suggestions by Mark Friedl, Mike Goodchild, and Carolyn Hunsaker.

References

Andersen, A.N. 1990. The use of ant communities to evaluate change in Australian terrestrial ecosystems: a review and a recipe. Proceedings of the Ecological Society of Australia 16:347–57.

Axelrod, D.I. 1989. Age and origin of chaparral. Pages 7–19 in S.C. Keeley, ed. The California chaparral: paradigms reexamined. Natural History Museum of Los Angeles County, Los Angeles.

Block, W.M., and M.L. Morrison. 1998. Habitat relationships of amphibians and reptiles in California oak woodlands. Journal of Herpetology 32:51–60.

Bolger, D.T., A.C. Alberts, and M.E. Soulé. 1991. Occurrence patterns of bird species in habitat fragments: sampling, extinction, and nested species subsets. American Naturalist 137:155–66.

Bolger, D.T., T.A. Scott, and J.T. Rotenberry. 1997. Breeding bird abundance in an urbanizing landscape in coastal southern California. Conservation Biology 11: 1–16.

Bolger, D.T., A.V. Suarez, K. Crooks, S. Morrison, and T.J. Case. 2000. Arthropod diversity in coastal sage scrub fragments: contrasting effects of area, edge, and nonnative species. Ecological Applications 10:1230–48.

Brattstrom, B.H. 1992. Status survey of the Orange-throated whiptail *Cnemidophorus hyperythrus beldingi.* (Contract FG 8597.) Unpublished final report to California Department of Fish and Game.

Campbell, H.W. 1953. Observations on snakes DOR in New Mexico. Herpetologica 9:157–60.

Campbell, H.W., and S.P. Christman. 1982. Field techniques for herpetofaunal community analysis. Pages 193–200 in N. Scott, ed. Herpetological Communities, USFWS, Wildlife Research Report 13.

Corn, P.S., and R.B. Bury. 1990. Sampling methods for terrestrial amphibians and reptiles (PNW-GTR-256.) U.S. Department of Agriculture, Portland, OR.

Crooks, K. 2000. Mammalian carnivores as target species for conservation in Southern California. Pages 137–43 in J.E. Keeley, M. Baer-Keeley, and C.J. Fotheringham, eds. Second Conference on the interface between ecology and land development in California. U.S. Geological Survey, Sacramento.

Dobson, A.P., J.P. Rodriguez, W.M. Roberts, and D.S. Wilcove. 1997. Geographic distribution of endangered species in the United States. Science 275:550–53.

Drost, C.A., and G.M. Fellers. 1996. Collapse of a regional frog fauna in the Yosemite area of the California Sierra Nevada, USA. Conservation Biology 10:414–25.

Fahrig, L., J.H. Pedlar, S.E. Pope, P.D. Taylor, and J.F. Wegner. 1995. Effect of road traffic on amphibian density. Biological Conservation 73:177–82.

Fisher, R.N., and H.B. Shaffer. 1996. The decline of amphibians in California's great Central Valley. Conservation Biology 10:1387–97.

Fisher, R.N., and T.J. Case. 1997. A field guide to the reptiles and amphibians of coastal Southern California. Science Center, USGS, Sacramento.

Fisher, R.N., and T.J. Case. 2000. Distribution of the herpetofauna of coastal Southern California with reference to elevation effects. Pages 137–43 in J.E. Keeley, M. Baer-Keeley, and C.J. Fotheringham, eds. Second Conference on the interface between ecology and land development in California. U.S. Geological Survey, Sacramento.

Fitch, H.S. 1949. Study of snake populations in central California. American Midland Naturalist 41:513–79.

Fuentes, E.R. 1976. Ecological convergence of lizard communities in Chile and California. Ecology 57:3–17.

Gap Analysis Program. 1997. A handbook for Gap analysis. USGS/BRD Gap Analysis Program, Moscow, ID. http://www.gap.uidaho.edu/gap/handbook.

Gibbons, J.W., and R.D. Semlitsch. 1981. Terrestrial drift fences with pitfall traps: an effective technique for quantitative sampling of animal populations. Brimleyana 7:1–16.

Glaser, H.S.R. 1970. The distribution of amphibians and reptiles in Riverside County, California Riverside Museum Press, Riverside.

Hitchings, S.P., and T.J.C. Beebee. 1998. Loss of genetic diversity and fitness in common toad (*Bufo bufo*) populations isolated by inimical habitat. Journal of Evolutionary Biology 11:269–83.

Hollander, A.D., F.W. Davis, and D.M. Stoms. 1994. Hierarchical representations of species distributions using maps, images, and sighting data. Chapter 5 in R.I. Miller, ed. Mapping the diversity of nature, Chapman and Hall, London.

Hood, L.C., M. Senatore, W.J. Snappe, and D.A. Hosack. 1998. Frayed safety nets: conservation planning under the endangered species act. Defenders of Wildlife, Washington, D.C.

Jasney, M. 1997. Leap of faith, Southern California's experiment in natural community conservation planning. Natural Resource Defense Council, New York.

Jennings, M.R., and M.P. Hayes. 1994. Amphibian and reptile species of special concern in California. California Department of Fish and Game, Inland Fisheries Division.

Kareiva, P.L., et al. 1999. Using science in habitat conservation plans. Report from NCEAS and AIBS. http://www.nceas.ucsb.edu.

Keeley, J.E., and C.C. Swift. 1995. Biodiversity and ecosystem functioning in Mediterranean-climate California. Pages 121–83 in G.W. Davis, and D.M. Richardson, eds. Mediterranean-type ecosystems: the function of biodiversity. Ecological Studies. Vol 109, Springer-Verlag, Berlin.

Klauber, L.M. 1939. Studies of reptile life in the arid southwest. Bulletin of the Zoological Society of San Diego 14:4–100.

Klauber, L.M. 1946. The glossy snake, *Arizona*, with descriptions of new subspecies. Transactions of the San Diego Society of Natural History 17:311–98.

Lande, R. 1987. Extinction thresholds in demographic models of territorial populations. American Naturalist 130:624–35.

Luiselli, L., and D. Capizzi. 1997. Influences of area, isolation and habitat features on distribution of snakes in Mediterranean fragmented woodlands. Biodiversity Conservation 6:1339–51.

Macartney, J.M., P.T. Gregory, and K.W. Larsen. 1988. A tabular survey of data on movements and home ranges of snakes. Journal of Herpetology 22:61–73.

Minnich, R.A. 1989. Chaparral fire history in San Diego County and adjacent northern Baja California: an evaluation of natural fire regimes and the effects of suppression management. Pages 37–48 in S.C. Keeley, eds. The California chaparral: paradigms reexamined. Natural History Museum of Los Angeles County, Los Angeles.

Morafka, D.J., and B.H. Banta. 1976. Ecological relationships of the recent herpetofauna of Pinnacles National Monument, Monterey and San Benito Counties, California. Wasmann Journal of Biology 34:304–24.

Noss, R.F., M.A. O'Connell, and D.D. Murphy. 1997. The science of conservation planning: Habitat conservation under the endangered species act. Island Press, Covelo, CA.

Peabody, F.E., and J.M. Savage. 1958. Evolution of a coast range corridor in California and its effects on the origin and dispersion of living amphibians and reptiles. American Association Advancement Science, Publication 51:159–86.

Pianka, E.R., 1986. Ecology and natural history of desert lizards: analyses of the ecological niche and community structure. Princeton University Press, Princeton, NJ.

Pulliam, H.R., and B. Babbitt. 1997. Science and the protection of endangered species. Science 275:499–500.

Robinson, S.K., F.R. Thompson III, T.M. Donovan, D.R. Whitehead, J. Faaborg. 1995. Regional forest fragmentation and the nesting success of migratory birds. Science 267:1987–90.

Rosen, P.C., and C.H. Lowe. 1994. Highway mortality of snakes in the Sonoran Desert of southern Arizona. Biological Conservation 68:143–48.

Savage, J.M. 1960. Evolution of a peninsular herpetofauna. Systematic Zoology 9:184–212.

Scott, J.M., F. Davis, B. Csuti, R. Noss, B. Butterfield, et al. 1993. Gap analysis: a geographic approach to protection of biological diversity. Wildlife Monographs 123:1–41.

Soulé, M.E., D.T. Bolger, A.C. Alberts, J. Wright, M. Sourice, and S. Hill. 1988. Reconstructed dynamics of rapid extinctions of chaparral-requiring birds in urban habitat islands. Conservation Biology 2:75–92.

Stacy, P.B., and M. Taper. 1991. Environmental variation and the persistence of small populations. Ecological Applications 2:38–42.

Stebbins, R.C. 1985. A field guide to western reptiles and amphibians, second ed. Houghton Mifflin, Boston.

Stoms, D.M., F.W. Davis, and C.B. Cogan. 1992. Sensitivity of wildlife habitat models to uncertainties in GIS data. Photogrammetric Engineering and Remote Sensing 58:843–50.

Suarez, A.V., D.T. Bolger, and T.J. Case. 1998. The effects of fragmentation and invasion on the native ant community in coastal Southern California. Ecology 79:2041–56.

Vogt, R.C., and R.L. Hine. 1982. Evaluation of techniques for assessment of amphibian and reptile populations in Wisconsin. Pages 201–18 in N. Scott, ed. Herpetological communities, USFWS, Wildlife Research Report 13.

Westman, W.E. 1981. Factors influencing the distribution of species of Californian coastal sage scrub. Ecology 62:439–55.

Wilcove, D.S., C.H. McLellan, and A.P. Dobson. 1986. Habitat fragmentation in the temperate zone. Pages 237–56 in M.E. Soulé, ed. Conservation biology: the science of scarcity and diversity, Sinauer Associates, Sunderland, MA.

Wilson, D.E., et al. 1996. Measuring and monitoring biological diversity: standard methods for mammals. Smithsonian Institution Press, Washington, DC.

Wright, D.H., B.D. Patterson, G.M. Mikkelson, A. Cutler, and W. Atmar. 1998. A comparative analysis of nested subset patterns of species composition. Oecologia 113:1–20.

4
Incorporating Uncertainties in Animal Location and Map Classification into Habitat Relationships Modeling

Kevin S. McKelvey and Barry R. Noon

Our understanding of animal habitat requirements are generally based on location data. We assume that those types of vegetation in which an organism is consistently found represent important habitat. In many cases, location data are collected independently from vegetation data, and the two are combined by overlaying the points representing animal locations onto a vegetation map. The number of points that occur in each type are counted, and these proportions are compared with random expectations to infer habitat selection (Neu et al. 1974; Alldredge and Ratti 1986). This process would be fairly straightforward if the uncertainties were limited to sampling error; however, there is always additional error associated with this process: Neither the coordinates of the animal locations nor the classification of vegetation mapped at those locations are error free.

That location error exists has been recognized for at least 30 years (Heesen and Tester 1967), and various methods for measuring and minimizing error associated with radiotelemetry-based locations have been developed (Heesen and Tester 1967; Springer 1979; Lenth 1981; Lee et al. 1985; White 1985; Pace and Weeks 1990; Zimmerman and Powell 1995). The impacts of these errors on our derived habitat understandings, however, have received much less attention. Location error is seldom reported (Hupp and Ratti 1983; Saltz 1994) and formal analyses of the impacts of these errors on derived habitat relationships are almost nonexistent. This lack of reporting occurs in spite of clear understandings that error can lead to weak (White and Garrott 1986; Nams 1989) and biased (Nams 1989; Samuel and Kenow 1992; Stoms et al. 1992) results.

The potential for assigning inappropriate habitat relationships to animals due to location error is, in fact, potentially large. To illustrate this point, we arbitrarily chose a common terrestrial vegetation type (Coniferous Forest) in a classified remotely sensed image of the landscape around Westborough Massachusetts (Fig. 4.1; see color insert). We generated approximately 4000 random locations within the Coniferous Forest type and displaced them using two circular bivariate normal distributions that displaced the points on average 1.5 and 3.0 rasters (approximately 45 m and 90 m average location errors, respectively). As can be seen in Table 4.1, error both weakens the actual relationship and generates spurious relationships. With a 45 m average error, about 56% of the points displaced into alternate types leaving 44% in Coniferous Forest. The distribution that produced a 90 m average error displaced 67% of the points into alternate types. Of equal interest are a series of incorrect associations with other types produced by these displacements. The Newer Residential, Roads, Industrial/Commercial, and Water types are used much less than expectation, Golf Courses/Grass and Older Residential show less strong aversion, and types like Wetland and Deciduous Forest are used approximately at expectation levels. It is easy to make up explanations for these types of patterns: Even though we demonstrated a preference for Coniferous Forest types, the organism shows strong aversions to those types associated with human activity while using other wild areas at levels close to expectation! As Nams (1990) and Samuel and Kenow (1992) note, error displacement into other vegetation types is not random. Certain types tend to be adjacent. In this case, Deciduous Forest is almost always immediately adjacent to Coniferous Forest, whereas Newer Residential seldom is (Fig. 4.2).

Both Nams (1989) and Samuel and Kenow (1992) provide methods to correct bias. Samuel and Kenow (1992), for instance, use random sampling within the error distribution around each animal location to compute the likelihood that the location lands in an alternative type. To our knowledge, neither the Nams' (1989) nor Samuel and Kenows' (1992) methods have been widely used. The prevalent attitude is that by collecting large samples, and through careful radiotelemetry methods, spurious conclusions due to locations displacing into the wrong vegetation types can be minimized and ignored.

There are, however, strong reasons to believe that these errors will be present in most datasets, and that attempts to eliminate them through careful field methodologies will meet with limited success. Obtaining location data for many animals is difficult and expensive. In many cases, where small

TABLE 4.1. Proportion of locations in various types and random expectations.

Type	1.5 raster average displacement (%)	3.0 raster average displacement (%)	Random expectation (%)
Older Residential	7.70	9.57	12.13
Newer Residential	1.26	2.63	7.81
Industrial/Commercial	1.29	2.16	4.09
Roads	0.29	0.97	3.53
Water	0.66	0.92	2.36
Agriculture/Pasture	0.55	0.97	2.11
Deciduous Forest	24.84	27.99	30.64
Wetland	15.56	16.43	14.65
Golf courses/Grass	3.99	6.18	9.49
Coniferous Forest	43.81	32.09	12.64
Shallow Water	0.05	0.11	0.58

All locations, prior to displacement, were located randomly in the Coniferous Forest type.

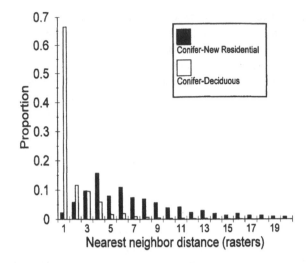

FIGURE 4.2. Nearest neighbor distances from rasters typed as Coniferous Forest to rasters typed as Deciduous Forest and Newer Residential. Newer Residential rasters are more distant from Coniferous Forest than are Deciduous Forest rasters. Locations of an organism using the Coniferous Forest type will be much more likely to be displaced by error into Deciduous Forest than into Newer Residential.

radiotransmitters are used, the sample size is constrained by transmitter life and sampling constraints associated with spatial autocorrelation (Swihart and Slade 1985). Displacing points into different vegetation types will similarly continue to occur even if location errors are low. The reasons for this are based in geometry. For any polygon, a large proportion of the area will be close to the edge: For a circle with a radius of 100 m, for instance, about 20% of the total area lies within 10 m of the perimeter. In vegetation maps, polygons are often small and convoluted, having high edge–area ratios. In these surfaces, randomly located points will often lie adjacent to edges (Zabel et al. 1994; Fig. 4.3).

4.1 Vegetation Typing Errors

Discussions of map error historically revolved around line placement: A primary cartographic problem is to draw lines that represent landscape features accurately. One concern is associated with the process of digitization. Because digitization only samples a small number of the potential points on a line, digitized lines are a series of linear approximations of the actual line. If the approximation is a best fit, then the true line will exist with an "epsilon band" surrounding each line segment (Blakemore 1984; Dunn et al. 1990) representing a zone of uncertainty around the line. Vegetation

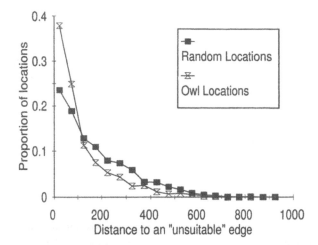

FIGURE 4.3. Distance to the nearest edge for random and northern spotted owl locations in northern California (Zabel et al. 1994). Twenty-four percent of the random points and 38% of the owl locations fell within 50 m of an edge of a vegetation polygon. This was in spite of the fact that the map was simplified into only two vegetation types: types considered suitable for owl use based on rules developed by the United States Forest Service, and those considered unsuitable.

maps have been particularly prone to errors associated with polygon delineation because (1) there tend to be many small and irregular polygons or (2) lines have historically been delineated on air-photos, transferred to maps, and finally digitized or scanned—with unavoidable errors at each stage of the process. In addition, vegetation often occurs in gradients, making polygon delineation arbitrary, and, because typing is generally based on the interpretation rather than direct measurement, vegetation typing tends to be arbitrary as well.

4.2 Uncertainties Associated with Vegetation Maps Derived from Remotely Sensed Information

Vegetation maps have largely been based on remotely sensed data. Although many of the error sources associated with air photogrammetry are not present in remotely sensed data, misclassification error remains a major problem. Remotely sensed imagery consists of rasters containing light intensity data associated with a number of frequency bands. These rasters can be clustered based on their spectral similarities to one another (unsupervised classification) or can be grouped based on their similarities to specified areas referred to as a training set (supervised classification). Supervised classification is commonly used to generate vegetation maps, and we concentrate on uncertainties associated with supervised classification.

When a raster is associated with a specific component of a training set, there is generally some uncertainty in the association. If P_x is the probability that a raster is of type x, then $1 - P_x$ is the probability that it in fact belongs to some other type and is a measure of the uncertainty in classification. $1 - P_x$ can be large for a variety of reasons, but two major causes are (1) the spectral signature of a raster is not close to any of the potential training areas and (2) the raster contains a spectral mixture (often associated with a mixture of vegetation within the raster) and could, with similar probability, type to several of the training areas. If $1 - P_x$ is large, then large errors will be associated with the classification of the raster (Zhu 1997).

4.2.1 Fuzzy Set Classification

To avoid these errors, "fuzzy" classification methods were developed (Wang 1990; this volume, Chap. 15). In fuzzy classification, similarity indexes are developed associating each raster with the spectral signatures of a training set, but rasters are not formally typed to the most similar signature. These indexes are instead kept as attributes of the raster and can be utilized in subsequent analyses. Fuzzy classification approaches have the advantages both of avoiding errors and of retaining far more information. In addition, fuzzy classification methods encourage examination of classification uncertainty—once a fuzzy system has been "hardened" (i.e., classified based on the maximum P_x) raster-level classification uncertainty is lost.

An examination of $1 - P_x$ suggests a number of interesting features. First, the classification uncertainty is far from uniform. Certain areas and types are classed with much greater certainty than others (Zhu 1997; Fig. 4.4). Edges in particular have high-classification uncertainties. For developing wildlife habitat associations with vegetation maps generated by remote sensing, classification success should be evaluated locally for the area and the vegetation types of interest. Global misclassification rates will often be unreliable metrics of the error associated with a specific area. In particular, if the organism in question uses a rare type or one that tends to have a high edge–area ratio (e.g., riparian areas), then the typing error associated with these areas will probably be high.

For purposes of determining habitat utilization, fuzzy classification methods should allow for greater separation between used and random points than hardened classifications that only retain partial information. This should be particularly true if the utilized types are rare and/or have high edge–area ratios. In these cases, the signature of the rare type will in many cases be overwhelmed by the signatures of adjacent types leading to erratic classification. Because the proportion of a raster of a rare type (and hence its likelihood of being classified as that type) is a function of the size of the rasters, the number of rasters classified as the rare type and the consequent association of animal locations to that type will all be scale dependent. These problems should be at least partially negated by using all of the

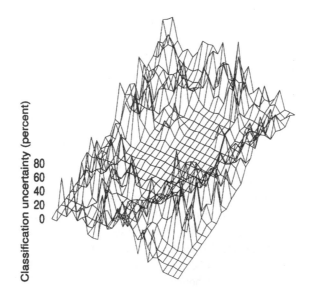

FIGURE 4.4. Classification uncertainty $(1 - P_x)$ after Baysian maximum-likelihood classification (Eastman 1997, pp. 11–13) for an area adjacent to Westborough, Massachusetts. The area of high uncertainty in the upper left-hand corner is associated with a Newer Residential area, and the area of high uncertainty in the lower right-hand corner is associated with a highway. The areas that classify with low uncertainty are deciduous forest.

similarity indices to statistically test for group separation between animal and random locations. If this is done, then the similarities to the rare signature will contribute to group separation even if they are not dominant. In essence, retaining all of the similarity indexes allows the retention of subraster information.

4.3 Methods

We argue that for determining animal use patterns, an analysis that retains more data should provide a truer and more robust model. We extend fuzzy logic concepts to incorporate uncertainties in both location and classification error. Rather than treating a location as being in a raster of a particular type, we approach the problem by stating that a location could potentially be in one of many rasters, and that each of those rasters could be one of many types. A formal statement of the probability that a location is in a type is therefore the probability that an organism location is in a raster and that raster is of a particular type summed across all rasters:

$$P[l, t_x] = \sum_{j=1}^{R} \sum_{k=1}^{C} P[l_{j,k}] P[t_{x,j,k}] \qquad (4.1)$$

where $P[l, t_x]$ is the probability that an animal (l)ocation is in a particular (t)ype x, R is the number of rows in the coverage and C is the number of columns. The full description of a location is therefore given by a vector **L**:

$$\mathbf{L} = P[l, t_1], P[l, t_2], P[l, t_3], \ldots, P[l, t_n] \qquad (4.2)$$

where n is the number of potential types.

We propose that an appropriate method to determine habitat preference is to generate probability vectors for each organism location and for a random set of locations. The elements of the vector are descriptors that can be used to separate the organism and random locations through the use of logistic regression. The relative contribution of each vector element (representing the probability of being in a type) to group separation is used to determine those habitats that are used or avoided (See Manley et al. 1993, Chap. 5). This method can be used where uncertainties concerning either location or classification or both are estimable. For a map that has been classified, and for which the original spectral data are not available (or for maps produced using other methods such as aerial photogrammetry) $P[t_{x,j,k}]$ will be set to either 0.0 or 1.0 for all rasters. If location error is unknown, $P[l_{j,k}]$ will be set to 1.0 for the raster enclosing the estimated location and 0.0 for all other rasters.

4.4 A Simulation

To test whether this method provides a reasonable approach to determining habitat selection, we utilized the Westborough site (Fig. 4.1). We chose three potential habitat associations. In the first, the organism used only the Roads type, in the second it used only the Coniferous Forest type, and in the third it used the Water type. We chose these types because they have different properties. The Roads type is poorly classified, highly fragmented, and rare. The Coniferous Forest type is more common, and it is classified

TABLE 4.2. General descriptions of three land cover types chosen for analysis.

	Percent area in type	Percent classification success	Average number of adjacent cells having the same type
Roads	3.2	69.0	3.5
Conifer	12.3	82.7	3.9
Water	2.5	99.9	7.3

Classification success is the proportion of cells typed similarly in the maximum likelihood classification and in the random instance.

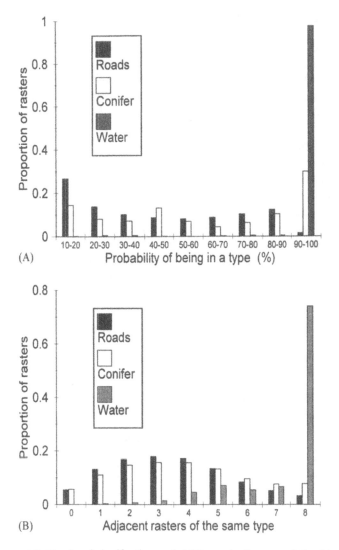

FIGURE 4.5. Graphs of classification probabilities and adjacency relationships for the three chosen types. (A) The probability that a raster was typed to either Roads, Coniferous Forest, or Water ignoring those rasters that were classified to these types with less than 10% probability. For both Roads and Coniferous Forest types, many of the rasters had intermediate probabilities of being classified to these types, leading to uncertain typing. (B) is a histogram, given that a raster is of type Roads, Coniferous Forest, or Water, of the number of adjacent rasters which are of the same type.

more successfully, but it is still highly fragmented. The water type, though rare, is typed with almost absolute certainty and is unfragmented (Table 4.2, Fig. 4.5).

4.4.1 Generating the Classification Probabilities

We generated classification probability surfaces by using IDRISI's BAY-CLASS procedure with no prior probabilities (Eastman 1997; Chap. 18, this volume). This procedure produces separate probability maps for each type and a classified map. We used IDRISI's RECLASS procedure to break the classified map into separate surfaces for each type where the values were 1.0 if the raster was classified to the specified type and 0.0 if the raster was typed to an alternate type.

To test the models' abilities to determine habitat selection with both location and map classification error, we needed to define "reality." That is, for the simulation purposes, we needed a map which was considered to be correct, and simulated locations that occurred exclusively in the selected types as defined by this correct map. To create this map we used the probability vectors generated by the BAYCLASS procedure to produce a valid instance—a map that could be correct based these probabilities. We produced this instance using transformation methods (Press et al. 1986), creating a cumulative probability distribution for each raster and assigning the raster to the type associated with the value of a uniform random deviate (Fig. 4.6). We considered this instance to be the "real" distribution

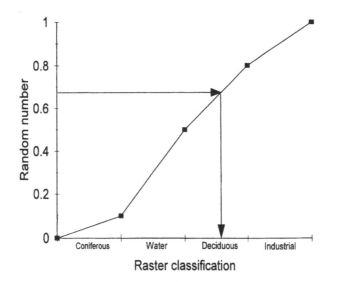

FIGURE 4.6. An example of the transformation method used to classify the map representing "reality" for purposes of simulation. The example raster had a 10% chance of being Coniferous Forest, a 40% chance of being Water, a 30% chance of being Deciduous Forest and a 20% chance of being Industrial/Commercial. These probabilities are summed together to form a cumulative probability distribution, and the value of a uniform random number between 0 and 1 defines the type. For instance, if the value of the random number is 0.67, the raster will be typed as Deciduous Forest.

of types on the landscape. We then generated random sets of coordinates and intersected these with our "real" landscape. Simulated locations were created by eliminating all points which fell outside the specified types (Roads, Coniferous Forest, or Water). These points were considered to be the actual locations of the organism at the times it was censussed. In addition, we generated random sets of coordinates to serve as the random expectation data sets. For each simulation we generated a separate group of between 3000 and 4000 locations to represent the organism's locations and random points.

4.4.2 Displacing Organism Locations

The organism locations and random locations were displaced using bivariate normal distributions. We displaced the locations using two distributions, one which displaced the points on average 1.5 rasters (displacement 1), and one that displaced them on average 3 rasters (displacement 2).

Generation of the "correct" map, coordinate location and masking, and bivariate normal weighting were accomplished using computer programs written in Turbo Pascal 7.0 (Borland International Inc.).

4.4.3 The Models

Our primary focus was to test a model with fuzzy classification and known location error against a classified map using the measured point locations; however, there are a variety of other models of interest. For instance, we may know the error distribution of the location data, but be working with a classified vegetation map. For each selected type and displacement we therefore developed three models:

1. Both the locations and the maximum likelihood classifications were assumed to be accurate.
2. The classification was assumed accurate, but the location error was simulated by using bivariate normal weighting around each location.
3. The classification was fuzzy and the location error was simulated by using bivariate normal weighting around each location.

For Models 2 and 3 the bivariate normal distribution used to weight the probability of the location being in a specific raster was the same distribution that was used to displace the points, simulating a known error distribution. For Model 3 and with displacement 1, we generated an additional model in which displacement 2 was the assumed error distribution. By generating a mismatch between the distribution used to displace and evaluate the points, we simulated a case where the location error distribution was poorly understood (Model 4). In all cases, because the "real" organism use patterns were confined exclusively to a type, better models would demonstrate this pattern more strongly, would minimize spurious alternative explanations, and would achieve greater group separation between the used and random points.

4.4.4 Statistical Analyses

We used logistic regression to discriminate between random points and organism locations. We used SAS's logistic procedure (SAS Institute Inc.), and used forward stepwise regression to generate the models. To compare models we used the χ^2 scores associated with the utilized type and of other spurious types and the concordance.

4.5 Application to Spotted Owl Data

In addition to these pure simulations, we wanted to test the method on measured location data for a species. The spotted owl (*Strix occidentalis caurnia*) has been heavily studied through radiotelemetry-based location data. Spotted owls forage at night, and their habitat use patterns, other than nest and daytime roost sites, are only available through radiotelemetry. For most habitat use studies of spotted owls tracking movement has been accomplished by attaching a small transmitter to the bird and using two directional receivers simultaneously to locate the bird at night. Both directional antennas are pointed in the direction of the perceived maximum signal strength, an azimuth is taken, and lines are drawn on a map from the antenna locations in directions indicated. The intersection of the lines is assumed to be the location of the bird.

There are obviously errors in this process associated with antenna location and direction, and signal bounce. These errors are unfortunately difficult to correct in the field. Creating fewer but more precise antenna locations necessarily increases the distance to the bird and therefore increases displacement associated with directional error (Heezen and Tester 1967). For these reasons, those spotted owl telemetry studies that reported error generally had average displacement errors of approximately 100 m (Cary et al. 1992). In owl telemetry studies in northern California, the average telemetry error was 110.6 m (Zabel et al. 1995). Combined with these error levels, owls in northern California used habitat edges more frequently than expectation (Zabel et al. 1995). For this reason, even in a highly simplified map with only two defined types (i.e., suitable habitat as defined by the Forest Service and unsuitable) 55% of the locations were less than 100 m from an edge (Zabel et al. 1995; Fig. 4.3).

We reanalyzed these data using a combined data set for all owl locations within the Mad River study area (Zabel et al. 1995). The data were 1700 nighttime foraging locations for 15 owls. Transmitters were placed on all known territorial owls within the study area, and the area was exhaustively surveyed to ensure that no territorial owls were missed. We can therefore state with reasonable certainty that unused areas within the study area were not used by noncensussed territorial spotted owls.

The map, which was originally composed of polygons based on aerial photogrammetry was rasterized to 30 m rasters using ARC/INFO's (ESRI, Inc.) POLYGRID function. For these analyses we generated two sets of simulated locations to compare with the owl locations: a random sample and a "habitat specialist" similar to the habitat specialists used in the simulations (see earlier). Both of these sets were confined to an area defined by a minimum convex polygon (MCP) enclosing the owl locations. The locations representing the habitat specialist were randomly distributed in "suitable" habitat and displaced using a bivariate normal distribution with an average displacement of 110 m. We used Models 1 and 2 to compare the owl locations to random locations within the MCP. Error was assumed to be normally distributed for Model 2, with an average displacement distance of 110 m and centered on the 30 m raster containing the owl location.

4.6 Results

4.6.1 Simulation Results

The instance of the map, which represented reality in the simulation, was not very divergent from the map generated through maximum likelihood classification. For the entire map, 83.6% of the rasters had the same classification in both coverages. Of the three types chosen to represent preferred habitat, 69.0, 82.7, and 99.9 were the same in both coverages for Roads, Coniferous Forest, and Water, respectively.

As hypothesized, the models in which all of the data were retained (Models 3 and 4) produced the most satisfactory results: The χ^2 score associated with the utilized type and the concordance were always maximized. Model 2, which utilized the hardened, classified map, but with weighted location error, produced models that were nearly as good. In all cases, Model 1, which did not compensate for either classification or displacement errors produced the poorest model (Table 4.3). For all three vegetation types, in Models 2 and 3, the χ^2 score associated with the utilized type was at least three times as large as the largest spurious χ^2 score. For the Roads type, using Model 1, the strongest association was an "aversion" to Deciduous forest. Doubling the displacement distance (displacement 2) decreased the power of the models in all cases, but did not affect the qualitative understandings. It is of interest that using the wrong error distribution for habitat evaluation (Model 4) produced models that were nearly as good (in the case of Roads, better) than the results associated with Model 3.

TABLE 4.3. Model results for simulations.

Roads
Avg displacement = 1.5 rasters

Classification/location	χ^2 roads	Max nonroads χ^2	Concordance (%)
Model 1	542	978	76*
Model 2	1828	474	90
Model 3	2163	295	90
Model 4	2249	267	90

Avg displacement = 3.0 rasters

Model 1	272	716	72*
Model 2	1664	360	87
Model 3	1886	227	87

Coniferous Forest
Avg displacement = 1.5 rasters

Classification/location	χ^2 Conifer	Max nonconifer χ^2	Concordance (%)
Model 1	791	151	64
Model 2	1585	68	81
Model 3	1683	55	82
Model 4	1545	37	81

Avg displacement = 3.0 rasters

Model 1	341	72	56
Model 2	1117	39	76
Model 3	1168	30	77

Water
Avg displacement = 1.5 rasters

Classification/location	χ^2 Water	Max nonwater χ^2	Concordance (%)
Model 1	4076	1124	96
Model 2	4427	397	97
Model 3	4427	370	98
Model 4	4375	264	98

Avg displacement = 3.0 rasters

Model 1	3265	769	91
Model 2	3905	288	98
Model 3	3906	274	98

*In these models, the strongest relationship was a negative association with deciduous forest. Model 1 assumes that locations and types are accurate. Model 2 represents locations as bivariate normal distributions, but assumes that map typing is accurate. Model 3 represents locations as bivariate normal distributions, and map rasters as vectors of classification probabilities. Model 4 uses a 1.5 raster average displacement, but uses a bivariate normal distribution with twice the variance for analysis.

4.6.2 Spotted Owl Results

When analyzed in aggregate, Models 1 and 2 both demonstrated that suitable habitat, as defined by Region 5 of the USDA Forest Service, was used significantly more than would be expected by random chance ($p < 0.0001$). Model 2 was arguably a better model (65% vs. 36% concordance; $\chi^2 = 287$ vs. 194), but at this level of analysis the two models provided the same understandings. The advantages of applying the error distributions to the locations lay primarily in the greater understandings allowed due to retained information. For instance, a histogram comparing the probability of owl and random locations being in suitable habitat showed that there were more owl locations than random expectations for locations with greater than 50% chance of being suitable, but that most of the difference between these two distributions lay in having far fewer owl locations distant from suitable habitat (P[Suitable] < 10%; Fig. 4.7). There are a fair number of owl locations (about 16%), however, that have low probabilities of being in suitable habitat (Fig. 4.7). To determine whether this proportion of points was likely to have been generated by error displacement from suitable habitat, we compared the owl locations with those of the simulated "habitat specialist." The locations for the habitat specialist differed from the owl locations in having far more locations in areas distant from unsuitable habitat (P[suitable] > 0.9), and far fewer locations that displaced into areas with low likelihoods of being suitable (Fig. 4.8). These two comparisons allow several statements to be made concerning the habitat use by owls. We reject the hypothesis that the owls are using habitat randomly: They are definitely located more frequently in suitable than in unsuitable habitat, but we also reject the hypotheses that they use suitable habitat randomly and exclusively. Their habitat associations are not absolute: The proportion of

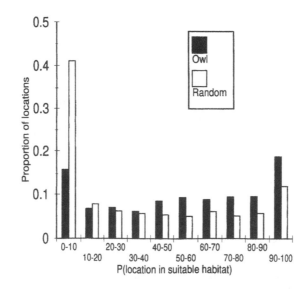

FIGURE 4.7. Proportions of owl and random locations as a function of their probability of being in suitable owl habitat. The largest differences between these two samples lies in the proportion of points that have less than a 10% probability of being in suitable habitat. These points that are distant from patches of suitable habitat.

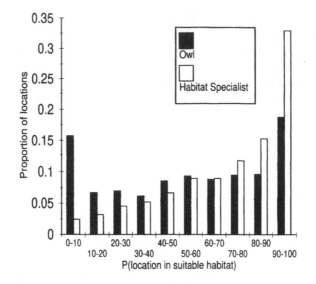

FIGURE 4.8. Proportions of owl and "habitat specialist" locations as a function of their probability of being in suitable owl habitat. The habitat specialist locations are randomly located within suitable habitat. The largest differences between these two samples lies in the proportion of points having less than 10% and greater than 90% probability of being in suitable habitat. Compared to the habitat specialist, there is a higher proportion of owl locations that are unlikely to be suitable and a lower proportion that are almost certainly suitable.

points that are distant from suitable habitat is greater than is likely based on error displacement; however, they are equally unlikely to utilize habitat that is distant from unsuitable patches. They appear to forage close to the edges between suitable and unsuitable habitat, with more of their time spent on the suitable side of the edge (Fig. 4.9). In Zabel et al. 1995, we were able to demonstrate edge utilization by spotted owls in the Mad River drainage by utilizing a separate analysis based on ARC/INFO's (ESRI Inc.) NEAR statistic, but we had no vehicle to ascertain the impact of error displacement on our habitat understandings.

4.7 Discussion

Deriving habitat associations from point-location data has traditionally been problematic. Errors in map classification combined with location error can lead to weak and biased interpretations of the data, and these problems were clearly evident in our simulations. Despite using very large samples, relatively modest location errors, and absolute selection for a type, we were unable to demonstrate clearly selection for the Roads type if we accepted the

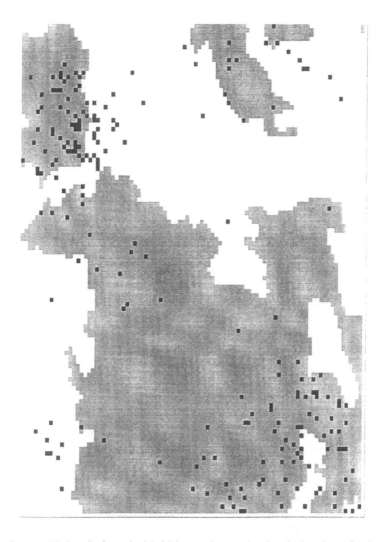

FIGURE 4.9. Details from the Mad River study area showing the locations of owls. Raster size is 30 m. The edge utilization of the owls, avoiding interior areas in both suitable and unsuitable habitat is evident.

map typing and organism locations as accurate (Model 1). By including location error, however, we were able to demonstrate strong selection. Incorporating estimates of location and classification error, although not entirely removing spurious relationships, greatly reduced their magnitude. Incorporating uncertainties of both location and classification produced the best models in all cases (Models 3, 4), but most of the improvements were associated with incorporating location error, while including fuzzy map classification produced only marginally stronger models. We have no reason to believe that this result is general—other maps and location-error structures could lead to different understandings. Still, we were encouraged by the model improvement associated with Model 2 because in many cases vegetation classification uncertainty cannot be estimated.

We were equally encouraged by the quality of Model 4, where we used an incorrect location-error distribution. If location-error distributions are not precisely known, then it may be advantageous to estimate a reasonable error distribution and use it for subsequent analyses. This is particularly true if the utilized type is rare and/or fragmented. If this is the case, location error will frequently displace animal locations into adjacent types. If we estimate an error, we will enclose the rare type within the error distribution. Its signal will not be as strong as it would be if the distribution were correct, but the probability of the location being within the utilized type will probably be greater than random expectation.

The preliminary results of these analyses show that the evaluation methods we describe here can, in many cases, lead to improved habitat relationship modeling. For organisms that have strong associations with fairly uncommon types (what we tested) we feel that the technique has much to offer. We stress, however, that we have not fully tested the approach under all circumstances. There may be combinations of habitat selection patterns and landscape attributes for which this method will lead to greater confusion than a simple comparison of use with availability. One potential source of confusion is seen in the analysis of spotted owl use patterns. Because a location that is farther from a type has lower likelihood of being displaced into that type, the probability of a location being in a type is a function both of error and of proximity. If an organism's behavior includes the use of edges or other proximal relationships, then incorporating location error into the analysis will improve the model partially because it accounts for location error and partially because it provides a metric that describes the organism's behavior.

The odd result that Model 4 produced the best model for Roads is probably also explained by the confusion between the measurement of error probabilities and neighborhood properties. In Model 4, different distributions were used to displace locations and evaluate habitat relationships. Because uncertainty due to location error ($P[l_{j,k}]$ in Eq. 1) is properly accounted for using the displacement distribution, the difference between the heights of these distributions: $P[l_{j,k}]_D - P[l_{j,k}]_E$, where D is the displacement distribution and E is the evaluation distribution, represented an unspecified neighborhood function, which in this case increased the influence of more distant rasters. For Roads, the combination of having the displaced locations reasonably close to the road and evaluating a broader neighborhood apparently increased the relative number of rasters that were strongly typed to Roads.

In the case of the simulations in this chapter, the influence of these neighborhood effects was relatively benign, but this is probably not always the case. We believe that in complex landscapes where one is attempting to infer behavior from the observed patterns, incorrect inference remains a strong possibility.

Acknowledgments. We would like to thank Carolyn Hunsaker and Ted Case for reviews of this chapter and J. Ronald Eastman for providing us with a complimentary copy of IDRISI GIS and for advice as concerning its use in these analyses. In addition, we would like to thank Cynthia Zabel for providing the telemetry locations and vegetation data that allowed evaluation of spotted owl habitat use.

References

Alldredge, R.J., and J.T. Ratti. 1986. Comparison of some statistical techniques for analysis of resource selection. Journal of Wildlife Management 50:157–65.

Blakemore, M. 1984. Generalization and error in spatial databases. Cartographica 21:131–39.

Cary, A.B., S.P. Horton, and B.L. Biswell. 1992. Northern spotted owls: influence of prey base and landscape character. Ecological Monographs 62:223–50.

Dunn, R., A.R. Harrison, and J.C. White. 1990. Positional accuracy in digital databases of land use: an empirical study. International Journal of Geographical Information Systems 4:385–98.

Eastman, J.R. 1997. Idrisi for Windows, users guide version 2.0. Clark University, Worcester, MA.

Heezen, K.L., and J.R. Tester. 1967. Evaluation of radiotracking by triangulation with special reference to deer movements. Journal of Wildlife Management 31:124–41.

Hupp, J.W., and J.T. Ratti. 1983. A test of radiotelemetry triangulation accuracy in heterogeneous environments. Proceedings for International Wildlife Biotelemetry 4:31–46.

Lee, J.E., C.G. White, R.A. Garrott, R.M. Bartmann, and A.W. Alldredge. 1985. Accessing accuracy of radiotelemetry system for estimating animal locations. Journal of Wildlife Management 49:658–63.

Lenth, R.V. 1981. Robust measures of location for directional data. Technometrics 23:77–81.

Manley, B., L. McDonald, and D. Thomas. 1993. Resource selection by animals. Chapman and Hall, London.

Nams, V.O. 1989. Effects of radiotelemetry error on sample size and bias when testing for habitat selection. Canadian Journal of Zoology 67:1631–36.

Neu, C.W., C.R. Byers, and J.M. Peek. 1974. A technique for analysis of utilization-availability data. Journal of Wildlife Management 38:541–45.

Pace, R.M., and H.P. Weeks, Jr. 1990. A nonlinear weighted least-squares estimator for radiotracking via triangulation. Journal of Wildlife Management 54:304–10.

Press, W.H., B.P. Flannery, S.A. Teukolsky, and W.T. Vetterling. 1986. Numerical recipes. Cambridge University Press, London.

Saltz, D. 1994. Reporting error measures in radio location by triangulation: a review. Journal of Wildlife Management 58:181–84.

Samuel, M.D., and K.P. Kenow. 1992. Evaluating habitat selection with radio-telemetry triangulation error. Journal of Wildlife Management 56:725–34.

Springer, J.T. 1979. Some sources of bias and sampling error in radio triangulation. Journal of Wildlife Management 43:926–35.

Stoms, D.M., F.W. Davis, and C.B. Cogan. 1992. Sensitivity of wildlife habitat models to uncertainties in GIS data. Photogrammetric Engineering and Remote Sensing 58:843–50.

Swihart, R.K., and N.A. Slade. 1985. Testing for independence of observations in animal movements. Ecology 66:1176–84.

Wang, F. 1990. Improving remote sensing image analysis through fuzzy information. Photogrammetric Engineering and Remote Sensing 56:1163–69.

White, G.C. 1985. Optimal locations of towers for triangulation studies using biotelemetry. Journal of Wildlife Management 49:190–96.

White, G.C., and R.A. Garrott. 1986. Effects of biotelemetry triangulation error on detecting habitat selection. Journal of Wildlife Management 50:509–13.

Zabel, C.J., K. McKelvey, and J.P. Ward, Jr. 1994. Influence of primary prey on home-range size and habitat use patterns of northern spotted owls (*Strix occidentalis caurina*). Canadian Journal of Zoology 73:433–39.

Zhu, A-Xing. 1997. Measuring uncertainty in class assignment for natural resource maps under fuzzy logic. Photogrammetric Engineering and Remote Sensing 63:1195–202.

Zimmerman, J.W., and R.A. Powell. 1995. Radiotelemetry error: location error method compared with error polygons and confidence ellipses. Canadian Journal of Zoology 73:1123–33.

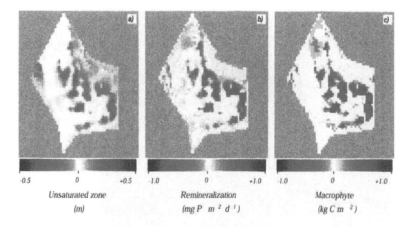

FIGURE 2.5. Simulated annual mean difference between two 17-year ELM runs for Water Conservation Area 2A. A test run with increased evapotranspiration was compared with the base run by subtracting the nominal from the test run values (1989). (a) Depth of the zone of unsaturated water; (b) remineralization rate of P in the soil; and (c) macrophyte total biomass. Reprinted from Fitz and Sklar, 1999, with permission from CRC Press.

FIGURE 3.3. An example of an array in the ground (at the Point Loma site).

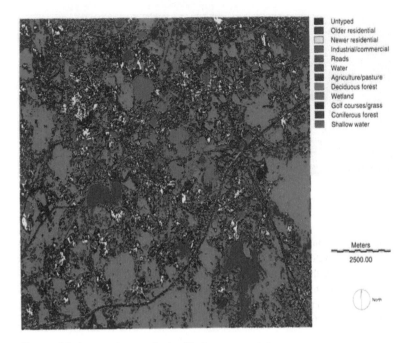

Untyped
Older residential
Newer residential
Industrial/commercial
Roads
Water
Agriculture/pasture
Deciduous forest
Wetland
Golf courses/grass
Coniferous forest
Shallow water

Meters
2500.00

North

FIGURE 4.1. A remotely sensed, classified coverage of the area around Westborough, Massachusetts. Spectral data associated with this coverage are used for all simulations in this chapter. This coverage is provided as part of the IDRISI tutorial. Raster size is 30 m, and the map covers 309 km².

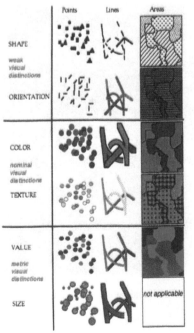

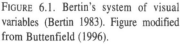

FIGURE 6.1. Bertin's system of visual variables (Bertin 1983). Figure modified from Buttenfield (1996).

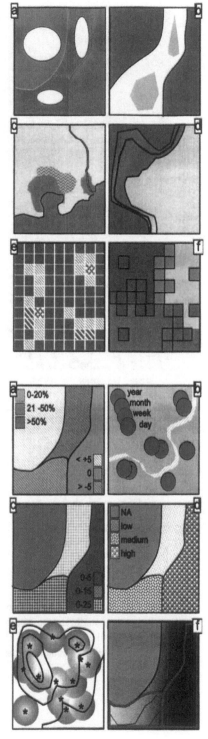

FIGURE 6.2. Examples of mapping ecological uncertainty. (a) Positional accuracy, discrete data: small raptor habitats; (b) positional accuracy, discrete data: stream deposition; (c) positional accuracy, categorical overlay: wetlands location; (d) positional accuracy, categorical overlay: eutrophication; (e) attribute accuracy, discrete data: bird sightings; (f) attribute accuracy, categorical overlay: misclassed vegetation types.

FIGURE 6.3. Examples of mapping ecological uncertainty (continued). (a) Attribute accuracy, categorical partitions: crown heights; (b) currentness, discrete data: wolf sightings; (c) currentness, categorical overlay: duration of parasite infestation; (d) currentness, categorical overlay: nitrogen depletion by landcover type; (e) positional accuracy, continuous data: interpolated nitrogen levels; (f) currentness, continuous data: rate of nitrogen depletion.

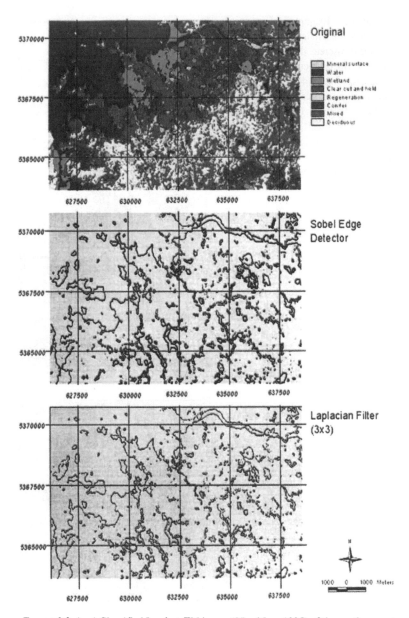

FIGURE 8.3. (top) Classified Landsat-TM image (25 × 25 m, 1995) of the southern part of Lac Duparquet (Abitibi region, Quèbec). (middle) Sobel edge detector and (bottom) Laplacian filter (3 × 3) on the classified image. Both algorithms find the main boundaries (in black), but there is some variability for the secondary boundaries (in red).

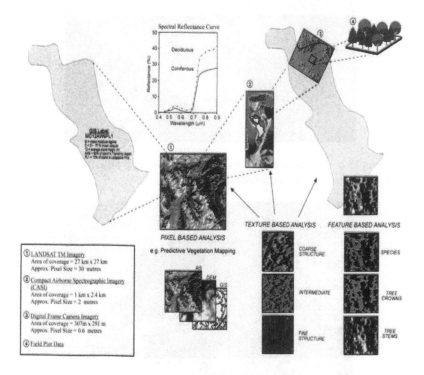

FIGURE 13.2. Uncertainty and image complexity. Depending on the image processing approach, there will be increasing uncertainty associated with decreasing spatial resolution, and, paradoxically, increasing uncertainty associated with increasing spatial resolution in the assignment of forest cover-type labels to forest stands.

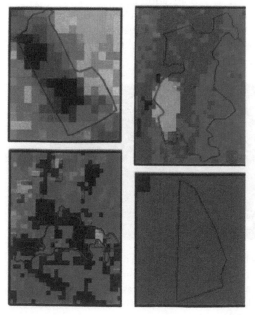

FIGURE 13.3. Illustrative examples of TM pixel variability within GIS polygons that are assumed to be homogeneous forest stands/cover type; aerial photointerpretation overlaid on the satellite image can show the degree to which variability was included within polygons and that boundaries were arbitrarily placed on the landscape. (a) Softwood, >50% conifer (dark brown); (b) mixed wood, 60% mixed (light brown); (c) mixed wood, 50% hardwood (orange); and (d) hardwood, 50% conifer arbitrary boundary. From Franklin (1997a), used by permission.

FIGURE 18.2. The Cua-Lo estuary near Vinh in north-central Vietnam: an area susceptible to the impact of accelerated sea-level rise.

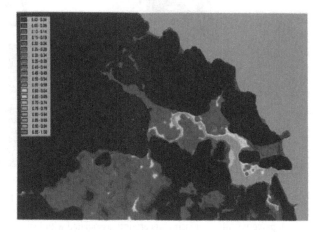

FIGURE 18.3. A mapping of the probability of land surfaces being below the projected new level of the sea. The uncertainty arises as a result of measurement error in the elevation model (i.e., the evidence). The probability estimate, created with a *soft* thresholding GIS software procedure, can also be used as a direct estimate of Type II Decision Risk—the probability that an area would be inundated if we were to assume that it would not.

FIGURE 18.4. A *hard* decision achieved by thresholding the decision risk surface from Figure 18.2 at the 5% decision risk level. Although the result is portrayed as a flood level, the area depicted is actually a statement of all areas that one would have to consider as potentially being flooded at that risk level. The actual result would be expected to cover only a subset of this region.

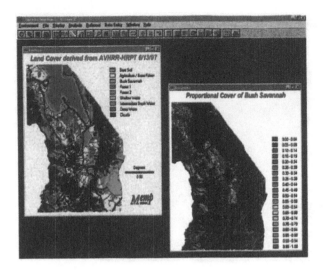

FIGURE 18.5. A classification of land cover based on coarse-resolution (1.1 km) AVHRR HRPT imagery (left). At this resolution, the evidence (measure refectance in various areas of the electromagnetic spectrum) at each location will show some degree of support for more than one land-cover class (because each pixel contains more than one cover type). This uncertainty in decision set membership is normally ignored in favor of a hard decision in favor of the land cover class with the greatest degree of support; however, this uncertainty can also be used to make statements of the proportional cover of specific land-cover types (e.g., the Bush Savannah mapping on the right).

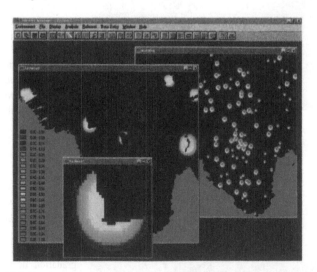

FIGURE 18.6. Expressions of uncertainty in the relationship between evidence and the decision set. These images show maps of *Dempster-Shafer* belief in the presence of land degradation as a consequence of fuelwood gathering and grazing around bore holes (wells) in southern Mauritania. This expression of the relationship between evidence and the decision set is based partly on local knowledge (which established the shape of the distance decay function) and expert opinion (which scaled the function into units of belief).

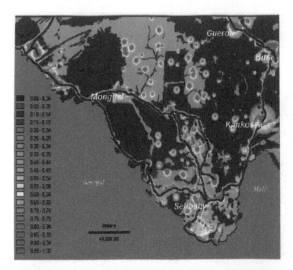

FIGURE 18.7. Using a Demp-ster-Shafer aggregation pro-cedure in the GIS, belief maps from 10 indicator vari-ables (e.g., those shown in Fig. 18.6) are combined to produce an aggregate state-ment of belief in land degradation.

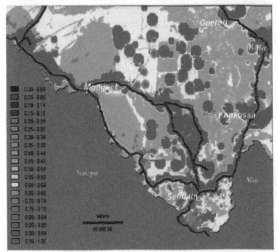

FIGURE 18.8. The Dempster-Shafer procedure can also be used to produce state-ments of the *plausibility* of a hypothesis. In this instance, the map shows the plausibil-ity of land degradation as a counterpart to the belief map in Figure 18.7. Plausi-bility is a statement of the degree to which the evidence does not refute the possibil-ity of the hypothesis, where-as belief measures the degree to which the evidence sup-ports the hypothesis.

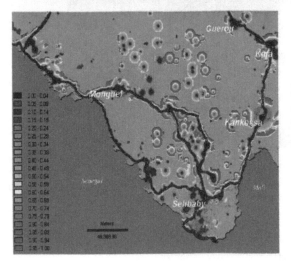

FIGURE 18.9. A Dempster-Shafer *belief interval* map expressing the uncertainty in the model of land de-gradation developed. The spatial arrangement of ar-eas of high uncertainty im-mediately suggested the nature of the source of that uncertainty, and thus how the model could be im-proved.

5
An Introduction to Uncertainty Issues for Spatial Data Used in Ecological Applications

PETER A. STINE and CAROLYN T. HUNSAKER

Ecologists study the dynamics of populations of living organisms, including the interactions that occur both within populations, and between populations and their environment. These phenomenon are therefore inherently dynamic in both space and time. Dynamic processes are driven by interaction between an organism and its congeners, competitors, prey, predators, and the physical environmental properties that it may encounter sometime during its life. These interactions occur upon a spatially explicit landscape. Physical separation is a matter of degree, depending on the organism and how it has responded to selection pressures through its evolutionary history. There is virtually a continuous spectrum of spatial domains that may be relevant to the viability of a given individual organism, population, community, or ecosystem.

Spatial relationship of organisms with their environment has become the subject of numerous lines of research and investigations, particularly with the advent of computer tools that allow for simulation and modeling of the spatially explicit environment. Geographic information systems (GIS) data and digital remotely sensed data have opened entire new areas of assessment and research in ecology and biogeography. New disciplines of ecology, such as landscape ecology and conservation biology, have developed and gained favor. Simulation models developed in these fields now can define the vitally important spatial relationship of any given environmental feature with countless others, enabling a quantitative measure of the ecologically significant nature of spatial relationships. For example, any organism can be precisely located with geographic coordinates, defining separation (distance and direction) with important environmental features (e.g., a nest site) directly through a metric, and this separation can be evaluated as a parameter in models concerning the ecology of the organism. Global positioning systems (GPS) have also become an almost indispensable piece of equipment for the field ecologist. The speed and accuracy with which an ecologist can obtain location data for any variety of field measurements has enabled research and management applications not possible just a few years ago.

The variety of environmental features to which spatial relationships are relevant is enormous. There are many methods by which such data are collected and represented in a GIS. This range of features and methods of data collection introduces many sources of uncertainty in both precision and accuracy. Such uncertainty can seriously confound the results of analysis. In this chapter we will attempt to describe how spatial uncertainty in ecological data can introduce errors to analysis.

The relevancy of uncertain spatially explicit data in this discussion is directed at applied ecological analyses. Many studies of biogeographical relationships of organisms depend on or are enhanced by analyses that quantify the spatial relationships of organisms and populations with their environment. This chapter addresses the relevancy of spatially explicit data in ecology, why ecologists should care about uncertainty in spatial data, the availability of digital spatial data, and the sources of uncertainty in ecological data.

5.1 The Relevancy of Spatially Explicit Data in Ecology

There are two vitally important components of common definitions in ecology that concern spatial organization: the *distribution* of organisms and the *relationship* between organisms and their biotic and abiotic environment. Organisms can be found virtually everywhere on earth. Each individual species, however, has a restricted geographic range due to both intrinsic limits of environmental tolerance and interactions with other species. Both biotic and abiotic processes are significantly influenced by spatial variation of the seemingly infinite number of environmental attributes. In other words, the geographic location of a given environmental attribute is a crucial piece of information in almost any ecological investigation.

Ecological organization and the levels of organizational hierarchies complicates the matter. Ecological processes operate over a continuum of domains in space and time, and ecologists have an intuitive appreciation for the existence of ecological hierarchies. It is clear, however, that a simple means of describing these relationships, such as the commonly used hierarchy of cell, organism, population, community, and ecosystem (Noss 1990), is unlikely to be useful across the range of observation sets and spatio-temporal scales involved in ecosystem analysis (O'Neill et al. 1986). This range of observations typical in ecological investigations results in a continuum rather than a discrete hierarchy.

A growing number of ecological research activities are directed toward applied problems. Today, all facets of the scientific community, including academia, government, and the private sector, are heavily involved in ecological research that has a direct application to conservation problems. Community ecology, population biology, and biogeography, as well as the applied correlate—conservation biology—all have inherently spatial characteristics and require some kind of analysis of spatially explicit data. The application of ecological risk assessment to large geographic areas (regional risk) was reported in 1990 by Hunsaker et al. They identified unique aspects for regional assessments: (1) the extent of interaction between the source terms, endpoints, and reference environment; and (2) the degree to which boundary definition and spatial heterogeneity are significant in determining uncertainty.

Landscape ecologists seek to better understand the relationships between landscape structure and ecosystem processes at various spatial scales (Forman and Godron 1986; Risser 1987; Turner 1987, 1989). We use the word *structure* to refer to the spatial relationships of ecosystem characteristics such as vegetation, animal distributions, and soil types. *Processes* or *function* refer to the interactions (i.e., the flow of energy, materials, and organisms) between the spatial elements. Landscape ecologists have largely come from the geographical and/or biological backgrounds and represent a synthetic view of these disciplines as they are projected over large geographic domains. Traditional ecological investigations have focused on the relationships among plants, animals, air, water, and soil within a relatively homogeneous unit that was often a small geographic area. Landscape ecology is unique due to its focus on the relationship among spatial units (Forman and Godron 1986). Landscape ecology has become the integrative discipline of applied research that considers a wide variety of ecological and human land uses patterns. Applied ecological research, which addresses many issues, now can and often does take a "landscape ecology" view to provide a valuable contextual backdrop to research questions.

Because ecologists are studying the relationships between organisms and their biotic and abiotic environment, many different kinds of data can be required. Even simple studies would likely look at a basic data theme such as climate (e.g., temperature and rainfall). More complex studies would likely evaluate data on climate as well as one or more of the following: vegetation type and structure, topography (elevation, slope, and aspect), soil characteristics, biogeochemical measurements, species locations, and land use (see Chap. 3). Watershed studies or ecosystem process models are good examples of analyses that often use many of these data types (see Chap. 2 for examples). All spatial data structures or models are necessary to represent such a diversity of data: point, raster, vector, and polygon. Ecological analyses would be enhanced if GIS were more effective at integrating raster and vector data, and uncertainty from data transformations could be reduced.

The need for such a diverse set of data also means that complex ecological studies usually cannot collect all the needed data in a designed and structured way with known and adequate quality control. Thus, data from various sources and with various levels of error and uncertainty (which are usually not documented) are often brought together in a single analysis. Little guidance exists, however, to help ecologists choose among this amalgam and what the resulting degree of uncertainty might be for the overall analysis. Documentation of spatial data with meta-information that includes source and uncertainty information, original purpose, and lineage or alteration information would be a welcome start for ecological analyses (see Chap. 17 for a discussion of meta-information).

5.1.1 Spatially Explicit Data Play an Important Role in an Increasing Number of Ecological Investigations

There are numerous examples of specific ecological issues where spatial pattern is at least relevant if not of fundamental interest to the research. Many of these scientific topics have become important new specialty disciplines in applied conservation biology (i.e., the scientific and technical means for the protection, maintenance, and restoration of life on this planet) in the mid-1980s or 1990s. Much of this trend in ecological investigations is rooted in biogeography. Early pioneers in biogeography recognized the significance of geographic variation over space and time and its influence on the distribution and abundance of species.

More sophisticated and quantitative investigations have increased the significance of spatially explicit data. Even traditional research on ecosystem processes now address spatial relationships and take advantage of the availability of baseline geographic data as part of their investigations. Early ecologists viewed spatial heterogeneity as a "problem," whereas heterogeneity has more recently been viewed as an important and tractable source of spatial variation in biophysical processes (Levin 1974; Kolasa and Pickett 1991). Ecologists now typically view nature as being better characterized by dynamic change with large fluctuations in magnitude of state variables rather than the earlier concepts of steady-state or "balance of nature" (Egerton 1973; Botkin and Sobel 1975; Botkin 1990; Schneider 1994). The development of models with spatial processes is one way to examine landscape effects. Another is analyze existing landscape patterns (e.g., the pattern metric work in landscape ecology). A third is to develop an understanding of processes (e.g., seed dispersal and wind effects on forests) that create spatial patterns. Several examples will illustrate how important spatially explicit data are to ecologists.

Metapopulation Dynamics. Most species are naturally patchily distributed (Soule 1987). Under certain circumstances species that are distributed in this manner can be considered a population of populations, known as a *metapopulation* (Levin 1974). This is a vitally important concept from an ecological and a spatial perspective. Subpopulations are separated from each other by some distance and across some kind of heterogeneous landscape. Each subpopulation has some probability of becoming extirpated (inversely proportional to its area) and some probability of being colonized by individuals from other subpopulations. The spatial relationship of subpopulations help ecologists understand how certain species persist, and, by contrast, why certain species are so vulnerable to habitat fragmentation. Metapopulation dynamics is an emerging area in both ecological theory and management applications (Gilpin 1996). Investigators and resource managers now pay particular attention to the data that describes spatial patterns and the relationships of animal and plant populations to environmental features that present barriers on conversely link populations.

Immigration and Emigration. The geographic distribution of any given taxon inevitably involves fluctuations in the sizes and distribution of peripheral populations. This is especially crucial to metapopulations that are sustained by source populations where conditions for that taxon are more optimal. The status of a given population is in part a function of the immigration and emigration rates, which in turn are related to intrinsic and extrinsic factors. Intrinsic factors will dictate how an organism is able to contend with the challenge of dispersal. Some taxa are innately capable of traversing large areas of inhospitable terrain such as migratory birds or animals with large home ranges. Some taxa are restricted to very small geographic areas and never move from suitable habitat such as many amphibians, reptiles, and small mammals.

Extrinsic factors include the large variety of barriers that exist across the landscape. These barriers include physical and biological barriers that can range in size from meters (e.g., creeks) to hundreds of kilometers (e.g., oceans) and both natural barriers and human-induced barriers (e.g., highways, pipelines, or urban–rural interfaces). Examples of barriers for aquatic species include waterfalls and dams, a transition from a perennial to an intermittent stream, and temperature or salinity gradients. The size and location of these barriers to movement, relative to size and location of a given population, and its dispersal abilities will have a profound influence on the rates of immigration and emigration.

Dispersal of Propagules. Another ecological consideration of populations, similar to immigration and emigration of adults, is the dispersal of propagules. Populations persist through reproduction and the survival of new generations of individuals. Both plant and animal species exhibit dispersal strategies that have evolved in response to their environment, including interspecific competitors. The ability to reestablish populations through reproduction depends on successful dispersal. A successful dispersive strategy will necessarily be a compromise between locating areas of inherent favorability and avoiding threats of predation, competition, and parasitism. A variety of landscape features will influence this outcome, and the location of these features is a primary concern.

Environmental Gradients. The spatial structure of a community is a fundamental subject of inquiry to ecologists. The horizontal distribution of species and other environmental factors exhibit complex patterns. Boundaries between two assemblages of species can be abrupt where the ecotone is distinct and easily identified, as with an obvious change in geologic formation or open water next to emergent vegetation (see Chap. 8); however, an overlap zone of two recognizable ecological communities is often typically wide with a gradual transition from one assemblage of species to another forming a continuum. Environmental gradients, as evidenced in mosaics of natural communities such as grasslands–oak savannah–oak woodlands, will commonly involve a large number of environmental factors (e.g., soil type, climate conditions, etc.) that are unevenly or patchily distributed over space. This environmental variation, interacting with the dispersal of propagules across space, influences the distribution and abundance of species.

Intraspecific and Interspecific Competition. Competition between individuals of the same and/or different species is one of the fundamental theoretical and empirical concepts of ecology. If two different but sympatric species utilize a common resource (e.g., nutrients, cover, light, etc.) that is potentially limited and actually becomes so, they are said to be in competition for it. For sessile organisms like plants, barnacles, or ant colonies, competitive interactions occur with neighboring organisms. For more mobile organisms, interactions are still confined within some spatial context. An understanding of the spatial distribution of individuals and populations over the landscape necessarily involves how relevant features of the landscape moderate the strength of local interactions and their impacts on demographic variables like birth, death, and dispersal rates.

Biogeochemical Cycling. The fundamental ecological process of nutrient cycling continues to be the subject of basic research on ecosystems. Life is dependent on the availability of a variety of elements required in the dynamics of life processes. The process of plant photosynthesis and the uptake of more complex organic substances depends on spatial characteristics of basic nutrients found upon the earth and flows from one part of the landscape to others. Latitude, aspect, soil chemistry, moisture, and soil texture are just a few of the spatial variables that affect biogeochemical cycles. Water, nitrogen, phosphorus, and carbon budgets/cycles are the basic drivers for primary productivity.

Ecologists have long been involved in investigations that attempt to understand the cyclical patterns of nutrients and other building blocks of life. The hydrologic cycle, gaseous nutrient cycles, and sedimentary nutrient cycles are some of the fundamental ecological processes that make ecosystems operate. Much is known about how these cycles function, but we are now attempting to include the actual spatial juxtaposition of the salient variables in our understanding of how ecosystems operate.

5.2 Quantification of Landscape Spatial Pattern

There is a growing demand for the measurement and monitoring of landscape-level patterns driven by the premise that ecological processes are linked to and can be predicted by some ecological pattern exhibited at somewhat coarse spatial scales. The development of landscape ecology has been instrumental in this interest in spatial heterogeneity. Gustafson (1998) reviews techniques and issues for quantification of landscape spatial pattern, and we will review some of the key points he makes. Although he offered a discussion on the interpretation of pattern indexes, described the appropriate application of spatial analysis techniques, and suggested lines of research, he overlooked the influence of data uncertainty on pattern indexes.

Spatial heterogeneity can be defined as the complexity and variability of a system property in time and space (Li and Reynolds 1994), including composition, configuration, and temporal aspects. Some important measures of heterogeneity are clearly not spatialy explicit (e.g., number of land types and proportions of these types); rather, they have important spatial effects. Structure and configuration should be used to refer to the spatial components of heterogeneity. Heterogeneity is also a function of scale (Wiens 1989; Allen and Hoekstra 1992). Two primary scaling factors affect measures of heterogeneity: grain and extent as used by ecologists. *Grain* is the resolution of any given landscape feature as it is perceived through the source of data used (minimum mapping unit, pixel size, and time interval); *extent* refers to the size of the area mapped or analyzed, or to the time period over which observations were collected (O'Neill et al. 1986). Grain and extent are ideally selected to match the features of processes of interest to the study.

There are now hundreds of quantitative measures of landscape pattern that have been proposed to quantify various aspects of spatial heterogeneity, often called *landscape pattern metrics* or indexes (Baker and Cai 1992; Hunsaker et al. 1994; McGarigal and Marks 1995; Riitters et al. 1995). Gustafson (1998) reviews some of these. Li and Reynolds (1994) divided spatial heterogeneity into five components using theoretical considerations: (1) number of types, (2) proportion of each type, (3) spatial arrangement of patches, (4) patch shape, and (5) contrast between neighboring patches. Riitters et al. (1995) used emprical data to identify five landscape factors: (1) average patch compaction, (2) overall image texture, (3) average patch shape, (4) patch-perimeter scaling or fractal measures, and (5) number of types. McGarigal and McComb (1995) also identified three principal components: (1) patch shape and edge contrast, (2) patch density, and (3) patch size. It is critical to consider the influence of grain and extent on the calculation of these metrics, as well as the number of classes and the effect of uncertainty in the spatial data. Some of these metrics are very complicated to calculate, and, in general, the expected variation of the metric with scale and uncertainty is not known. Statistical techniques can detect significant changes in the mean of a pattern index with known variation, but ecologically significant change is much more difficult to assess (Gustafson 1998).

Gustafson (1998) makes an important point that categorical- and point-data models are complementary, even though they are not usually combined in ecological analyses. In this book we present techniques to estimate or quantify uncertainty for both data models. Categorical-data models (the-

matic or chloropleth) map the system property of interest (i.e., soil, vegetation, land use) by identifying patches that are relatively homogeneous with
respect to that property at a particular scale and that exhibit a relatively
abrupt transition (boundary) to adjacent areas that have a different intensity
(or quality) of the system property of interest. The criteria for defining a
patch may be somewhat arbitrary depending on how much variation will be
allowed within a patch, on the minimum size of patches that will be mapped
(minimum mapping unit), and the components of the system that are
ecologically relevant to the organism or process of interest. A landscape may
exhibit many different patch structures (realizations), depending on the
system properties measured and the mapping decisions made. Patch-based
analysis uses a simplified description of the system that is perhaps more
amenable to interpretation and is more directly relevant to much of ecological theory than is point-data analysis. Point-data analysis assumes that the
system property is spatially continuous. The landscape is sampled to
generate spatially referenced information about system variables and is
analyzed using geostatistical techniques such as kriging. This data model is
most commonly used for geologic, soil, and water attributes.

The appropriate spatial and temporal scales at which pattern metrics are
applied should be determined by geostatistical analysis. Point-data analysis
can provide insight into the scale of patchiness, whether there are hierarchies
of scale, and whether the spatial distribution is random, aggregated, or uniform (Gustafson 1998). This is seldom done because it usually involves
collecting extra (point) data and conducting spatial statistical analyses with
unfamiliar techniques. If maps are not available at the scale identified as
appropriate, compilation of a new landscape map may be required.

5.3 Availability of Digital Spatial Data

Although the concern for the spatial location of environmental features has
always been important to ecologists there are certain developments in
science and technology that have made the uncertainty of spatial data an
issue. We have established the fact that many of the trends in ecology have
tended toward a more applied purpose, and that these investigations often
require a landscape or regional perspective; however, the explosion in the
availability of spatially explicit digital data has had a profound effect on
ecological investigations. It has opened up many avenues for new investigations in the various ecological disciplines. Some examples of these new
sources of ecologically relevant data or methods are discussed in this
section: GIS data, GPS data, remotely sensed digital data, radiotelemetry
data, and analytical procedures.

• *GIS data of physical environmental attributes (climate, soils, elevation, slope,
aspect, landform, etc.).* Ecological data are notoriously variable. In fact,

ecologists spend much of their field and analytical efforts attempting to explain variation observed in the natural world. Nevertheless, there is a rapidly growing body of digital spatial data available that can be analyzed with a GIS. Some of these data are biological in nature (e.g., vegetation maps that characterize the spatial extent of vegetation communities across a landscape). These data are necessarily subject to interpretation and their real values change over time. Such data may have somewhat limited value to some users other than those who developed the data; however, fully developed meta-information can improve the long-term value of data (see Chap. 17). Physical environmental data (e.g., climate, soil, elevation, landform, geology) and derived data (i.e., slope and aspect from elevation) are useful to a diverse set of disciplines. Some of these data change more slowly than biological data.

Although applications of digital geospatial data vary greatly, users have a recurring need for a few common themes. Many organizations are active in creating these "framework" data (a few commonly needed data themes, developed, maintained, and integrated by public and private organizations within a given geographic area) that will provide the fundamental building blocks of many scientific investigations. The Federal Geographic Data Committee (FGDC) is currently involved in promoting the development of framework data through a coordinated effort of various federal agencies. The framework should provide a base on which to collect, register, or integrate information accurately. To be successful, the framework data must be dependable and trustworthy, be created from the "best" data available, and be easy to access and use.

Both data contributors and users will enjoy benefits from the framework. These benefits include reduced expenditures for data, increased ease of obtaining and using data collected by others, accelerated development of mission-critical applications, and increased number of customers for data products linked to the framework (FGDC 1997). A more consistent use of a standardized set of basic data for a geographic location can also help quantify and, we hope, reduce uncertainty.

The information content for the framework will include the data themes of geodetic control, digital orthoimagery, elevation, transportation, hydrography, governmental units, and cadastre (i.e., public record survey maps of land ownership). The features encoded will include a minimum set of information needed to classify, name, and uniquely identify a feature. These features are generally stable within time frames relevant to investigations of current ecological conditions and are generally free of human interpretation variability. Such variables are also directly relevant to many ecological processes and can provide important insights into the understanding and management of ecosystems.

• *GPS data.* Global Positioning System (GPS) satellites have been fully operational since the mid-1990s, providing an incredible new tool for ecologists. A handheld ground receiver must be able to receive signals from at least three NAVSTAR GPS satellites that provide details of their orbits and an atomic clock correction. Virtually any ecological field investigation can now provide geo-referenced coordinates, accurate to within about 100 m of the true position on the surface of the earth. Handheld GPS devices for as little as $100 can provide these data within minutes. The main limitations are the accuracy with which the geographical position can be derived because the precise time code is dithered by the U.S. Department of Defense, the number of satellites in view, and the quality of the GPS receiver. Kennedy (1996) states that differential GPS measurements, using a local base station on a well-located object, can improve spatial resolutions to within 1 m accuracy. The value of having accurate locational data for sightings of rare species or for other field observations is enormous.

• *Remotely sensed digital data.* Landscape ecologists need a wide variety of information about the characteristics and condition of the earth's surface. Remote sensing, in the form of aerial photography, has been used for over 50 years to gather these kinds of information. There are now a number of satellites that have been operational for as long as two decades that are providing digital "pictures" of the planet with varying degrees of resolution and geographic extent. Newer sensors (e.g., laser altimetry) are just starting to be explored for uses such as quantifying vertical vegetation structure. A multitude of quantitative computer-aided analysis techniques for processing such date also exist. Significant developments in the interrelationships between remote sensing and GIS include the ability to fully integrate digital image processing software with geographical analyses using a GIS. These developments have created additional dimensions of complexity as well as opportunities to use various data sources and analysis techniques to provide valuable information for ecological investigations. The integration of several kinds of remotely sensed data (i.e., data fusion) to define biological attributes also has a high potential for ecologists.

• *Radiotelemetry data.* Field zoologists have been using radiotelemetry as an investigational technique for more than 20 years. The daily, seasonal, and annual movement patterns of many different kinds of animals has been established using sophisticated electronics. The geographic coordinates of each recorded location is established using different methods with varying degrees of accuracy depending on terrain, distance, equipment, and effort. The accuracy of these data is crucial when analyzing the results, especially when the analysis involves other spatially referenced data (e.g., GIS vegetation data) (see Chap. 4).

• *Diversity and computational power of analytical procedures.* GIS and other computer technologies have exploded since 1990. Access to these technologies is both much easier and less expensive; thus, many applications have emerged in the academic, governmental, and private sector arenas. Full-capability GIS enables virtually endless spatial analyses and can now be performed by students or professionals with only limited training. Although this can create potential problems that can add sources of error (i.e., inappropriate use of map projections, mislabeling, etc.), that topic is not addressed in this book.

Many analytical approaches now combine various tools. Some GIS analyses can be linked with simulation models that attempt to reproduce an ecological process (e.g., plant succession or sediment transport within a watershed). The linkage allows for incorporation of the spatially explicit variables within the simulation. Modeling has become a popular scientific endeavor both to understand how ecosystems work and to examine alternative future scenarios through sensitivity analyses where individual variables are varied and the resulting effect evaluated (see Chaps. 1 and 2 for more on this topic).

5.4 What Are the Sources of Uncertainty in Ecological Data?

In order for ecologists to minimize the potential error due to data uncertainty they must be able to determine the sources and magnitude of uncertainty. There will always be some error in data; however, some sources of error can be reduced or eliminated if recognized and managed. This subject will be addressed in detail in Chapters 11 and 14.

5.4.1 Types of Measurement Uncertainty

Spatial data may be compiled to comply with one level of accuracy in the vertical component and another in the horizontal components. A data set may contain themes or geographic areas that have different accuracies (FGDC 1997). Much of the uncertainty in spatially explicit data can be traced directly to the original capture of the data. The manner in which the data are collected, how it was collected, and by whom, can result in significant sources of uncontrolled and unbalanced uncertainty.

• *Geometric error and/or uncertainty.* Most data collected are from a spherical surface (i.e., the earth) and there is often significant error in the process of transferring the georeferenced point(s) from a sphere to a plane (i.e., two-dimensional map) with which we commonly work.

• *Attribute error and/or uncertainty.* All kinds of variables, from nominal (categorical) to ratio (numeric) variables, are measured with a variety of tools and methods. Inherent in these tools and methods is some degree of error that leads to uncertainty. This error is sometimes biased because of a flaw within the construction of the tool or the design of the method.

Human error, through unstable or biased interpretation of observations or the simple variation among people, is also an important potential source of error in the value of an attribute. For example, the slope of a site could be measured by a technician to be 15% when it is actually 13%.

A variation on this problem is the issue of classification error. Ecologists are quite familiar with the troubles of classification schemes. Soils, vegetation, geologic substrate, and other landscape features are a function of human interpretation. Classification schemes for any of these features are all fundamentally flawed because they exist for the convenience of human description of the feature. Nevertheless, we need these schemes to reduce complex features into units that we can understand and communicate to others. Categorical classed data will always be plagued with this problem, especially when the data are subsequently manipulated and transformed.

• *Locational and boundary uncertainty.* Both sources of error cited earlier can combine to produce errors of location. Objects of interest on the surface of the earth, identified through various means, may be located incorrectly relative to their true position. This may be especially problematic when the object is an abstraction such as a vegetation type boundary, which has no true location on the earth's surface.

• *Physical changes of the attributes over time and space.* Many physical and certainly most biological variables measured at any given time are likely to change at a subsequent time step. Many variables change virtually continuously (e.g., temperature or soil moisture). This is nothing new to ecologists; however, spatially referenced data sets must often represent a variable as a static value. Capturing change in values of a landscape feature over time within a georeferenced dataset is a difficult matter; however, most geographically referenced data only represent a single occurrence or state in time. The implications of this uncertainty to a particular ecological analysis could be significant.

• *Data compatibility, combining data of different qualities.* Many analytical activities that use spatially explicit data will derive new data using combinations of existing data. The FGDC framework data, mentioned earlier, are intended, in part, to be used as a means for deriving new data. Digital elevation models (DEMs), for example, can be used to derive many other kinds of data that are useful to ecological investigations (e.g., aspect, slope). If combined with other spatially explicit data to "synthesize new data," what kinds of errors are potentially introduced? For example, DEMs can be used to delineate watershed boundaries, and if combined with a drainage network derived from a digital line graph (DLG) at a finer scale, a line representing a stream course could fall outside the boundaries of its correct watershed as depicted by the DEM. Errors can be generated or propagated by combining data from different sources.

• *Interpretation and manipulation of existing data.* Many ecological observations are scale dependent, as are the results of many data manipulations that create new composite data or interpretations. A common practice of ecologists who use spatial data is map generalization where an analyst will simplify or degrade the detail of an existing vector or raster dataset to suit some new purpose (e.g., aggregate 20 vegetation classes into eight superclasses). Ecologists usually have no, or at best, an inadequate understanding of the consequences of changing data (e.g., locational or classification values) through some method of computer manipulation. Another example is the conversion of data from raster to vector (or the opposite); many potential sources of uncertainty can be introduced in this process. Map projections are another potential source of error introduction. If an original data source, obtained using a UTM projection, is reprojected into another map projection and then back into UTM, the two UTM projections cannot be expected to match exactly. The magnitude and significance of these errors should be known to the investigator.

• *Inability to detect attribute of interest accurately (inappropriate resolution).* A primary consideration, particularly when using remotely sensed data, is a suitable data collection protocol to ensure the ability to detect the attribute of concern. Data collection techniques vary widely, and some are more effective than others for the variable of interest. It is possible that the method (including the tools) will have unreliable or inconsistent detection, resulting in data that have undetermined levels of error. The inability to interpret the attribute of interest accurately, and insufficient sensitivity of data collection devices (limitations in data collection abilities) can produce significant, sometimes unknown uncertainty. For example, classifying vegetation using remotely sensed digital imagery can often fail due to inadequate spectral and spatial resolution of the raw data. Many key species of plants or stands of plants that characterize vegetation types are not detectable using seven discrete bands (spectral resolution) or 30-meter pixels (spatial resolution) that are available from conventional satellite imagery.

5.4.2 Analysis Uncertainty

Problems during the analysis phase can result from assumptions concerning the exactness of spatial entities. Burrough and McDonnel (1998) state that most procedures commonly used in geographic information processing assume implicitly that (1) the source data are uniform, (2) digitizing procedures are infallible, (3) map overlay is merely a question of intersecting boundaries and reconnecting a line network, (4) boundaries can be sharply defined and drawn, (5) all algorithms can be assumed to operate in a fully deterministic way, and (6) class intervals defined for one or another "natural" reason are necessarily the best for all mapped attributes. They also list the main factors governing the errors that may be associated with geographic information processing. Sources of errors in spatial data include accuracy of content, measurement errors, field data, laboratory errors, locational accuracy, and natural spatial variation. Factors affecting the reliability of spatial data include age of data, areal coverage, map scale and resolution, density of observations, relevance, data format exchange and interoperability, accessibility, costs and copyrighting, and numerical errors in the computer.

5.4.3 Describing Uncertainty

The FGDC (1994) has developed standard procedures for describing positional accuracy of spatial data, in both digital and graphic forms, derived from sources such as aerial photographs, satellite imagery, or maps. It provides a common language to report accuracy to facilitate the identification of spatial data for geographic applications; however, this standard does not define pass–fail criteria. Data and map producers must determine what accuracy exists or is achievable for their data. Users identify acceptable accuracies for their applications (FGDC 1997). This standard describes how to express "the applicability or essence of a data set or data element" and includes "data quality, assessment, accuracy, and reporting or documentation standards."

Meta-information or metadata are vitally important to informing and warning data users about the limitations of the data. The purpose of a metadata standard is to provide a common set of terminology and definitions for documentation related to metadata. The main reason to document data is to maintain an organization's investment in its geospatial data. Organizations that do not document their data often find that, over time or because of personnel changes, they no longer know the content or quality of their data (FGDC 1994).

Data users are ultimately responsible for judging the appropriate uses of data. More and more digital spatial data are being made available over the world wide web and through other means. Unless the end-user is cognizant of the inherent and known uncertainties of the data they are using, then significant error in analysis and interpretation can result.

Ecologists are very familiar with reporting on the uncertainty of their data through various statistical methods. Interpretation and description of spatial data uncertainty involves other techniques and interpretations that may not be familiar to ecologists. Chapters 9, 12, 14, and 15 all discuss issues and provide guidance about how to evaluate uncertainty in spatial data and its overall contribution to uncertainty in an ecological application.

5.5 Why Should Uncertainty Be Considered?

Error and uncertainty in spatial data are often ignored within the scientific community, but this should not happen unless one knows the effect of this uncertainty is insignificant for the application of the data. Multiple reasons probably account for this attitude, including ignorance about sources and magnitude of error in spatial data, experiences with "inaccurate" maps that still get you to your destination, the general acceptance of information when it is represented in a map or figure, and the lack of tools to make addressing uncertainty straightforward. Consider the amount of time and effort that the field biologist or toxicologist puts into experimental design and replication of sites/trials in order to estimate different forms of variability. Should we not expect a similar level of effort with digital data and maps? Scientists should know that uncertainty is not a flaw to be avoided or ignored; rather, it is an inherent attribute of data and data manipulation processes (see Chap. 18) and statistical analysis (see Chap. 10). More than other analyses, ecological risk assessment requires such information.

There are two areas where uncertainty information often accompanies the spatial data product: remotely sensed data and cartographic maps. For example, a vegetation map derived from remote sensing data will often have an estimate of overall map accuracy based on a confusion or contingency matrix (see Chaps. 12 and 14). Cartographers have standards for map production that should be documented and printed on the map. Whether scientists without geography training know how to incorporate this information into their analyses is questionable. Clarke et al. (1991) produced a map of ecoregions for Oregon where they used variable width boundaries to represent their confidence in the boundary location and its abruptness. Such approaches are uncommon, however, and how one would represent this information in a GIS and utilize it in an analysis is neither standardized nor straightforward.

Risk assessment approaches can improve the rigor and credibility of environmental impact assessment and environmental regulation/policy. Risk assessment is a rigorous form of assessment that uses formal quantitative techniques to estimate probabilities of effects on well-defined endpoints, estimates uncertainties, and partitions analysis of risks from decision making concerning significance of risks and choice of actions (Suter 1993). It is the emphasis on characterizing and quantifying uncertainty that distinguishes risk assessment from other impact assessment. Barnthouse and Suter (1986) named the study of risks to the natural environment, *ecological risk assessment*. Early work focused on the estimation of risks posed by chemicals at the project site or local scale; however, the rigor of risk assessment is needed for other hazards (e.g., thermal effluents, flow regulation, resource harvesting, habitat protection for species of concern, climate modification, and the reality of multiple stressors). Graham et al. (1991) illustrate the influence of spatial pattern on a regional ecological risk assessment, and Hunsaker et al. (1993) illustrate the importance of boundary definition or analysis units for regional assessments. A logical need to arise from applied fields such as risk assessment, conservation biology, and landscape ecology is accepted techniques to quantify uncertainty in the spatial data often used in regional assessments for ecological issues. Regional ecological risk assessment was one impetus for this book.

References

Allen, T.F.H., and T.W. Hoekstra. 1992. Toward a unified ecology. Columbia University, New York.

Baker, W.L., and Y. Cai. 1992. The r.le programs for multiscale analysis of landscape structure using the GRASS geographical information system. Landscape Ecology 7:291–302.

Barnthouse, L.W., and G.W. Suter II, eds. 1986. User's manual for ecological risk assessment. ORNL-6251. Oak Ridge National Laboratory, Oak Ridge, TN.

Botkin, D.B., and M.J. Sobel. 1975. Stability in time-varying ecosystems. American Naturalist 109:625–46.

Botkin, D.B. 1990. Discordant harmonies: a new ecology for the twenty-first century. Oxford University Press, Oxford.

Burrough, P.A., and R.A. McDonnell. 1998. Principles of geographical information systems. Oxford University Press, Oxford.

Clarke, S.E., D. White, and A.L. Schaedel. 1991. Oregon, USA, ecological regions and subregions for water quality management. Environmental Management 15:847–56.

Egerton, F.N. 1973. Changing concepts of the balance of nature. Quarterly Review of Biology 48:322–50.

Forman, R.T.T., and M. Godron. 1986. Landscape ecology. John Wiley & Sons, New York.

FGDC (Federal Geographic Data Committee). 1994. Content standards for digital geospatial metadata. Federal Geographic Data Committee, Washington, DC.

FGDC (Federal Geographic Data Committee). 1997. Content standards for digital geospatial metadata. Federal Geographic Data Committee, Washington, DC.

Gilpin, M.E. 1996. Metapopulation biology. Academic Press, San Diego.

Graham, R.L., C.T. Hunsaker, R.V. O'Neill, and B.L. Jackson. 1991. Regional ecological risk assessment. Ecological Applications 1:196–206.

Gustafson, E.J. 1998. Quantifying landscape spatial pattern: what is the state of the art? Ecosystems 1:143–56.

Hunsaker, C.T., R.L. Graham, G.W. Suter II, R.V. O'Neill, L.W. Barnthouse, R.H. Gardner. 1990. Assessing ecological risk on a regional scale. Environmental Management 14:325–32.

Hunsaker, C.T., R.L. Graham, P.L. Ringold, G.R. Holdren, R.S. Turner, and T.C. Strickland. 1993. A national critical loads framework for atmospheric deposition effects assessment: II. Defining assessment endpoints, indicators, and functional subregions. Environmental Management 17:335–41.

Hunsaker, C.T., R.V. O'Neill, B.L. Jackson, S.P. Timmins, D.A. Levine, and D.J. Norton. 1994. Sampling to characterize landscape pattern. Landscape Ecology 9:207–26.

Kennedy, M. 1996. The global positioning system and GIS. Ann Arbor Press, Inc., Ann Arbor, MI.

Kolasa, J., and S.T.A. Pickett, eds. 1991. Ecological heterogeneity. Springer-Verlag, New York.

Levin, S.A. 1974. Dispersion and population interactions. The American Naturalist 108:207–28.

Li, H., and J.F. Reynolds. 1994. A simulation experiment to quantify spatial heterogeneity in categorical maps. Ecology 75:2446–55.

McGarigal, K., and B.J. Marks. 1995. FRAGSTATS: spatial pattern analysis program for quantifying landscape structure. USDA Forest Service, Pacific Northwest Research Station, General Technical Report PNW-GTR-351.

McGarigal, K., and W.C. McComb. 1995. Relationships between landscape structure and breeding birds in the Oregon Coast Range. Ecological Monographs 65:235–60.

Noss, R.F. 1990. Indicators for monitoring biodiversity: a hierarchical approach. Conservation Biology 5:355–64.

O'Neill, R.V., D.L. DeAngelis, J.B. Waide, and T.F.H. Allen. 1986. A hierarchical concept of ecosystems. Princeton University, Princeton, NJ.

Riitters, K.H., R.V. O'Neill, C.T. Hunsaker, J.D. Wickham, D.H. Yankee, S.P. Timmins, et al. 1995. A factor analysis of landscape pattern and structure metrics. Landscape Ecology 10:23–39.

Risser, P. 1987. Landscape ecology: state of the art. Pages 3–14 in M.G. Turner, ed. Landscape heterogeneity and disturbance. Springer-Verlag, New York.

Schneider, D.C. 1994. Quantitative ecology: spatial and temporal scaling. Academic Press, San Diego.

Soule, M.E., ed. 1987. Viable populations for conservation. Cambridge University Press, Cambridge, England.

Suter, G.W., II. 1993. Ecological risk assessment. Lewis Publishers, Boca Raton, FL.

Turner, M.G., ed. 1987. Landscape heterogeneity and disturbance. Springer-Verlag, New York.

Turner, M.G. 1989. Landscape ecology: the effect of pattern on process. Annual Review of Ecology and Systematics 20:171–97.

Wiens, J.A. 1989. Spatial scaling in ecology. Functional Ecology 3:385–97.

Part III
Methods

The chapters of the previous section discussed the many ways in which uncertainty can affect spatial ecology, through uncertainty in data, in models of process, and in methods of analysis. In the best of all possible worlds, this next section would present a series of recipes, much like the flowcharts found in statistics texts, that prescribe exactly what to do in every conceivable circumstance. These methods would be implemented in software, allowing the spatial ecologist to deal with uncertainty by simply pressing a button labeled "fix". Procedures might be specified for converting "soft" data to data that are rigorously based, for exploring the effects of uncertainties on the results of modeling and analysis, and for visualizing them in ways that make the attendant message easy to communicate.

It is not surprising that our arsenal of methods for dealing with uncertainty has not yet reached that ideal state. We have not yet worked out exactly how to describe and measure uncertainty in all circumstances, or how to store that information so that it can be communicated to others or used in analysis to determine sensitivity. Much useful research has been done, and we are much further ahead now than we were, but it is still necessary for an ecologist to immerse himself or herself in the methods, to learn enough about them to make intelligent choices, and to understand something of their theoretical bases. The quantification of uncertainty in spatial data is not yet a world of simple recipes and prescriptions. This situation unfortunately usually means it is not done. It is more realistic, perhaps, that an ecologist is likely to seek out help from experts in the more technical areas rather than commit to a substantial investment of personal time and effort.

On a more fundamental level, research on uncertainty has followed a number of tracks that are inherently incompatible because they incorporate different paradigms or world views. It is impossible, for example, to merge an approach based on fuzzy sets with one based on probability theory because their conceptual frameworks and axioms are incompatible. Other incompatibilities arise because of differences of objectives: Are we trying to describe uncertainty and cope with it as an inevitable adjunct to research, or would we rather remove it? Are we more interested in describing generic uncertainty in data, or in decisions that must be made in specific application domains? Hence, the diversity of approaches discussed in the chapters in this section is both inevitable and a sign of the richness and importance of the field.

The section contains 13 chapters, each offering one perspective on the problem of uncertainty and its impacts on spatial ecology, and reviewing the current state of knowledge about methods for coping with it. We selected the authors from the full domain of research interests, drawing from disciplines as different as cognitive science, statistics, and geography. The prevailing approach in statistics is to take a problem and to define a formal version that is tractable and rigorous in the full knowledge that the problem is actually much more nebulous and messy. In return, one gets a formalization that is useful, robust, and productive. At the same time, however, one runs the risk that the formalization will not be communicable to people without a statistical background. Other approaches are more intuitive, requiring little technical background on the part of the user, but arguably being less able to meet the normal standards of scientific rigor, and less likely to lead to rigorous and productive results. This tension between intuition and formality runs through much of this research, and is readily apparent in the chapters that follow. Technologies such as GIS are by now widely accessible to people who want to make decisions about ecological systems but often have little in the way of formal training in fields like spatial statistics, forcing us to think about how the tension might be resolved.

This issue is encountered head-on in the first two chapters. In Chapter 6, Barbara Buttenfield looks at uncertainty from the perspective of a cartographer, asking how visualization of spatial data can be used to convey messages about uncertainty. Cartography itself has gone through several transitions over the past few centuries, as different perspectives (e.g., the earth's surface as a reflection of God's perfect design; uncertain areas not yet explored by Europeans; and today's comprehensive satellite-based views) fought each other for dominance in the human imagination. GIS is an intensely visual technology, and we know that "a picture is worth a thousand words," so the visual channel clearly offers great potential if it can be harnessed appropriately. Cartographic research has focused on the different dimensions of visual representation, allowing one to ask whether certain ones (e.g., hue, symbols, or depth) are more naturally effective than others at communicating the concept of uncertainty.

In Chapter 7, Geoff Edwards and Marie-Josée Fortin introduce the cognitive perspective, arguing that mapping and the production of spatial data can be understood only in a human context. Although spatial data are often input to scientific models, much of the tradition of mapping is inescapably subjective, and maps are designed to satisfy diverse human needs, even though that leads at times to divergence from the rigid principles of scientific measurement. They review some of the growing literature on cognitive approaches to GIS and the handling of spatial data, and show how a cognitive perspective can be applied at all stages of production and use.

Many of these issues come to a head in the concept of vegetation boundaries, or the longstanding tradition in mapping vegetation that partitions the world into areas separated by sharp boundaries, and assigns each area to a single class or vegetation type. The same approach is used in mapping land cover, land use, soils, surficial geology, and many other properties. Marie-Josée Fortin and Geoff Edwards provide a comprehensive review of the issues in Chapter 8. In one view, the practice arose in the era of paper mapping, when it was the only option available to a cartographer armed with a pen and with the ability to fill areas with patterns or colors; it follows that reinvention of mapping should be possible in an era of technological change, when many of these technical constraints have already disappeared. The opposing view holds that boundaries satisfy some deeper human need to see the world as simpler than it really is. To the modeler, however, these arguments may seem frivolous: If spatial data are to provide the boundary conditions for models, then they should represent the relevant variables as accurately as possible; anything less is not worthy of serious attention.

Statistics provides an immensely powerful battery of concepts and techniques for dealing with uncertainty, and several chapters show how geostatistics in particular can be used as an encompassing conceptual framework for a scientific approach to uncertainty. In Chapter 9, Phaedon Kyriakidis presents a comprehensive discussion of geostatistical principles, and shows how they can be used to deal with uncertainty. Spatial variables measured on continuous scales can be analyzed using kriging, even though indicator kriging is designed to deal with categorical data. Simulations can be conditioned to existing data through conditional simulation, and datasets can be combined using cokriging. Although the field is mathematically complex, excellent sets of tools are available in the form of standalone packages, subroutine libraries, or functions embedded in GIS.

In Chapter 10, Jay Ver Hoef and his coauthors take a model-centric approach within a similar statistical framework. Modeling in space has always been problematic because of the endemic presence of spatial auto-correlation, which defies the assumption of independence that underpins much statistical method. Drastic approaches (e.g., the rejection of sufficient samples to ensure that the remaining ones are far enough apart to show independence) seem too wasteful of precious data, and less desirable than methods that deal explicitly with spatial effects. The chapter reviews work on the linear model, as well as available methods for dealing with spatial autocorrelation. One of the greatest difficulties in modeling with explicit spatial effects is to discriminate between real spatial trends and the patterns created by spatial dependence, which can appear to persist as trends over long distances.

The series of statistical chapters ends with Chapter 11, in which Ashton Shortridge examines uncertainty in elevation data and its effects on modeling. Effects can be simulated by creating a sample of outcomes or realizations of an error model, or a set of equally likely possible maps that differ by no more than the noise represented by the model. One can then examine the variation in results after repeated analysis using different realizations. This approach has become popular in the GIS literature because it is robust and readily understood by analysts and others with only a rudimentary understanding of statistics.

Remote sensing is an increasingly important source of data for spatial ecology. Satellite observation avoids much of the subjectivity inherent in ground-based mapping, and offers the potential for direct access to the variables of interest to ecologists. It is not surprising that many sources of uncertainty exist in satellite-derived data, and Mark Friedl and colleagues review them in Chapter 12, along with commonly used metrics of uncertainty (e.g., the confusion matrix). In Chapter 13, Steve Franklin describes a typical modeling application of remotely sensed data, showing how the results are affected by the spatial resolution of the data, and by other forms of uncertainty. In Chapter 14, Ken McGwire and Peter Fisher show how the confusion matrix is essentially stationary, and inadequate in several ways as a comprehensive description of uncertainty in classifications derived from remote sensing.

Many researchers have been attracted to the formalisms of fuzzy sets as a way of escaping the uncertainties associated with boundaries, and with mapping classifications. A-Xing Zhu has pioneered the implementation of fuzzy sets in soil mapping, using a variety of techniques to determine the degree of membership of a location's soils in each of a set of classes. He reviews this work in Chapter 15, showing how it is possible to map ecological variables using fuzzy methods, to visualize the results, and to characterize statistical properties of fuzzy memberships in ways that are helpful to ecologists and others. In Chapter 16, Peter Fisher positions fuzzy sets within a broader conceptual framework that includes rough sets, Dempster-Schaeffer theory, subjective probability, and many other attempts to formalize the notion that the world is not best described by assigning each place to a single class.

The section ends with two chapters that address the communication aspects of uncertainty, or the ways in which people interested in sharing data tell each other about the data's properties. Kate Beard, in Chapter 17, addresses the essential issue of meta-information (i.e., metadata or data about data). Much work in the areas of digital libraries, data clearinghouses, and other electronic mechanisms for sharing data focuses on how one discovers the existence of data, and determines their fitness for a particular use. Meta-information is much more than the electronic equivalent of the library catalog card because it must also give the user the information needed to open and work with the data, and to handle it successfully. The purpose of sharing information is ultimately to allow decisions to be made, and Ron Eastman ends this section on methods with a review of methods for coping with spatial data uncertainty in decision making, showing how it has been possible to implement many of these methods within the powerful Idrisi GIS.

Taken together, the chapters in this section cover the range of the literature without any major gaps, though they inevitably only scratch the surface in some areas. Choosing an appropriate method for a given problem can be a daunting task, but it is made easier by the deep divisions represented by this collection: The choice between fuzzy and statistical methods, for example, will be driven by the user's background, experience, and preferences as much as by any sense that one approach is objectively right or wrong. The cognitive scientist's concern for the human mind will similarly be essential in dealing with stakeholders whose professional values include a strong respect for subjectivity, or for techniques that accommodate the innate workings of the human mind. Similar respect for stakeholder values is essential in decision making, where the role of technology must be to facilitate consensus rather than to impose it. In short, coping with uncertainty in spatial ecology requires above all a willingness to adapt to the circumstances of a problem rather than to follow prescription.

6
Mapping Ecological Uncertainty

BARBARA P. BUTTENFIELD

The application of information technology to ecological analysis is growing. Munroe and Decker (1991) assert that roughly 60% of habitat management professionals report using GIS, and nearly all find it useful. In particular, GIS software provides capabilities to map data quickly. GIS software, however, does not necessarily produce good maps by default. This is due in part to the diversity of mapping situations that arise, and the difficulty of establishing a single set of graphical defaults to accommodate all mapping needs. Many professionals in the ecological and environmental science disciplines do not traditionally have background training in information design or in visualization. Moreover, their interest in any particular ecological application will likely center on domain issues, as opposed to focusing attention on the design of graphical displays.

The need for presentation of data reliability or uncertainty needs to be coupled with a powerful mapping technology. Ecology deals with relationships between organisms and their environment, and the search for proximate and ultimate factors determining these relationships (Forman and Godron 1986). The intention to synthesize data gathered at multiple spatial levels (e.g., patch, mosaic, or landscape) and within multiple slices of time is impeded by uncertainty. Data collected in many ecological studies are difficult to ascertain as being positionally accurate, correctly categorized, complete, or logically consistent. Habitat delineation provides an excellent example of these problems. In studies monitoring daily and seasonal migration, researchers determine three criteria of uncertainty, using presence of animal tracks, presence of mating signs, and presence of offspring to establish habitat boundaries with high, medium or low degrees of uncertainty, respectively (Andrle and Carroll 1985). It is difficult to establish all three criteria for many studies. The complexity of plant processes and animal activities, and of their environments, makes it difficult to replicate findings.

This chapter presents a taxonomy to guide display of uncertainty in ecological data. Knowledge about uncertainty is important for effective use of ecological map displays and overlays. It impacts upon reliability and credibility of data representation, decision making, and model building. The

Proposed Standard for Digital Cartographic Data (FIPS 1992; FGDC 1994, 1998) includes specifications for reports of data uncertainty, including components of *positional accuracy*, *attribute accuracy*, *completeness*, (and, more recently, *currentness*), *lineage*, and *logical consistency*.

Uncertainty about horizontal or vertical location is a straightforward concept in definition. Attribute uncertainty refers to confidence that an item has been attributed correctly. Completeness refers to issues of missing data, missing attribute classes, and sampling regimes. *Currentness* refers to temporal attributes (e.g., uncertainty about age of a tree stand, the date of the most recent fire scars, the duration of a parasite infestation, or the periodicity of reproductive cycles). Lineage refers to the chronology of processing applied to a data set. Logical consistency refers to database validity, as for example in checking a triangulated irregular network (TIN) model to insure that links are not established between spot heights, that elevation values progress logically up and down slopes, and so on. These last two components of uncertainty will not be covered further in this chapter, which focuses instead on mapping ecological uncertainty.

The proposed standard states that when spatial variation in uncertainty occurs, thematic overlays may be constructed as diagrams or thematic map depictions; unfortunately, no guidelines are provided for symbolization, level of generalization, or other graphic design criteria for these reports. The chapter will not dwell directly on the modeling of error, but rather on its appropriate graphical display. The choice of data display strategy may bias comprehension of spatial variations in uncertainty, which, in turn, may affect subsequent modeling or decision making. The taxonomy presented builds upon aspects of ecological phenomena, as well as conventions of cartographic design. We acknowledge distinctions between three components of uncertainty (i.e., positional accuracy, attribute accuracy, and currentness), incorporating various data models (i.e., raster, vector and TIN), types of geographic phenomena (i.e., discrete, categorical and continuous), and specific visual variables. The framework is illustrated by graphic example.

6.1 What Is Uncertainty in a Mapping and Modeling Context?

Definitions for uncertainty abound. Many are presented throughout this volume. *Uncertainty* in the context of ecological mapping may be defined in a number of contexts. A cartographic definition is based in the community of spatial data producers, and culminates in the Metadata Content Standard (FGDC 1994). Uncertainty was historically subsumed under the label, "data quality." The assessment of quality in a map or product was paper-based, and founded in specifications for manual compilation of topographic maps.

The U.S. National Map Accuracy Standard (NMAS) specifies thresholds for horizontal and vertical positional uncertainty that are variant with respect to map scale (Thompson 1975). These thresholds were determined in part by the limits of data collection technologies available in the era of World War II, and in part by the smallest manufactured drafting pen stylus (one fiftieth of an inch).

In subsequent decades, a transition from manual (paper map) compilation to digital cartographic data caused a shift in perspective from "threshold specification" to "truth-in-advertising." That is, specification of a stated threshold of accuracy was succeeded by a request simply to include a data quality report with the dataset. A national committee (the Spatial Data Transfer Standard Committee) of mapping scientists worked for more than 12 years to define data models and formats amenable to GIS mapping and modeling. The resulting Spatial Data Transfer Standard (FIPS 1992) identified five types of uncertainty information to be reported. The five types (i.e., positional accuracy, attribute accuracy, completeness, logical consistency, and lineage) were subsequently expanded to include temporal data quality by adding temporal currentness (FGDC 1998).

As numeric modeling was incorporated into GIS software, and digital data sets grew in volume, it became more difficult to establish uncertainty reports on the basis of SDTS protocols. These protocols involve (for example) comparing two data sets of independent compilation to determine relative degrees of accuracy. Positional accuracy is assessed by comparing the mapped position of "well-defined" points (e.g., bridges, major highway intersections, landmark buildings, survey benchmarks). Attribute accuracy is also determined by paired comparisons (Fitzpatrick-Lins 1981; Chrisman 1997), but here the independent source is drawn from ground truth. Attribute classifications observed in the field are compared with mapped attribute classifications to determine areas of attribute divergence on the map.

When the comparison involves small numbers of feature records, such comparisons are manageable if tedious. For gigabyte-size data files, such comparisons become excessively expensive and slow. Sampling and inference were incorporated into SDTS to streamline the data quality reporting process. Instead of comparing every identifiable point in both datasets, a random sample of points are compared. This is similar to a selection of ecological mesh size that is sensitive to the grain size of landscape or ecological elements of interest (Forman and Godron 1986). Too large a mesh will render small-grained species invisible (Levins 1968); too small a mesh size will require redundant sampling of larger predators (Deveaux 1976). In uncertainty comparison, a large sampling net has a higher probability of missing positional errors, and the smallest nets will require extra comparisons that may not improve the estimate of positional errors. In both cases, one wants to establish a balance between characteristics of the element under observation and the number of samples taken.

With sampling and inferential statistical methods came a shift in terminology from "reporting data quality" to "establishing levels of uncertainty." Current definitions for uncertainty are not codified in the policy of data production, but in mathematical probability, and based upon an assumption of independent random samples. Subsequent refinements have occurred to relax assumptions of spatial and temporal independence. The taxonomy for uncertainty used in this chapter will be based jointly upon the cartographic specifications of different types of uncertainty, and the statistical distinctions between different types of data.

6.2 Why Map Ecological Uncertainty?

Bretherton (1993) argued the importance of environmental modeling and mapping on the basis of "a crucial need for efficient science-driven natural resource management." Model refinement and interpretation has increasingly been based on output in the form of maps. Mapping ecological data is therefore clearly important. Mapping ecological uncertainty is equally important for a number of reasons. Just as data values vary spatially, one can expect that levels of uncertainty will also vary. This does not mean that the variations will be identical (i.e., that the magnitude of uncertainty will be a function of data magnitude). The level of uncertainty is more often associated with related variables. For example, growth rates for particular vegetative species can vary with levels of insolation, which is partially constrained by terrain slope and aspect. Insolation may be difficult to interpret in the absence of terrain data.

Mapping the distribution of uncertainty about one's data communicates information "about the content, quality, condition, and other characteristics of data" (FGDC 1994). Map displays of uncertainty serve multiple roles in ecological analysis. The relative importance of these roles will vary for specific analytical situations. Uncertainty mapping can help viewers to identify patterns of dense or sparse field data observations and evaluate the fitness of a data set for particular applications. Data access and exchange can also be facilitated in this way (Miller 1993).

Maps of uncertainty can help to model ecological data. For example, Clarke and Olsen (1993) calibrated and refined wildfire modeling parameters on the basis of patterns derived from visual sources such as remotely sensed imagery. Recognition of the significance of visualization to guide model building and refinement is apparent in a National Science Foundation solicitation. The solicitation (NSF 1999) targets new tools for "visualization of program state, . . . program behaviors, . . . debugging, and performance analysis."

Mapping uncertainty takes on particular significance for ecological applications (Aspinall and Pearson 1993). Ecologists have developed a deep tradition of mapping applications from as far back as the work of

Charles Darwin (see Forman and Godron 1986 for a concentrated bibliographic chronology). As much (or more perhaps) than in any other field-based earth science, mapping creates the foundation for planning ecological fieldwork, for collecting field observations, and for developing ecological models (Veblen 1989). Evidence and status of ecological processes are often established after the fact (fire or hurricane disturbance patterns provide an excellent example). Detection of longitudinal patterns forms a backbone to many ecological studies (Veblen and Lorenz 1991). The documentation of major landscape change is often possible only through mapping and photography (Rogers et al. 1984). Temporal data models began to proliferate in the GIS literature since 1990 (Langran 1991; Wieshofer 1994), but they have not been successfully implemented in GIS and mapping packages.

Uncertainty seems unavoidable in ecological analysis. Ecological concepts (e.g., "patch," "habitat," and "mosaic") are derived empirically and are therefore difficult to formalize (Forman 1995). Quantification of ecological process (e.g., "disturbance," "fragmentation") is often indirect, and is based on mapping composition, configuration, and heterogeneity (Gustafson 1998). Uncertainty arises in part because of the complexities of formalizing definitions of pattern and process, and establishing standard units of measurement. Ecological patches (plant or animal) are dependent upon the size of the species under study, on spatial requirements for nutrition, and on the aggressiveness of territorial expansion (e.g., the spread of invasive plants such as kudzu).

Another source of uncertainty is that the measurements that identify ecological processes tend to be sensitive to resolution or grain size, and extent. The scale of an ecological study is functional (i.e., dependent on organism or process location, purpose, etc.) (Hammer 1998; Innes 1998). Functional units of measurement make comparative analysis difficult. The scale component is central to many aspects of ecological study (Wiens 1989; Levin 1992; Jelinski and Wu 1996; Bian 1997; Patterson and Parker 1998). Scale functions are well-developed in GIS software (Hunsaker et al. 1993), allowing users to input data at one scale and subsequently zoom in repeatedly, and often beyond the usable limits of data detail. Data models intended to create multiple views on a single detailed database have been a matter of scrutiny for decades, going back to original work on self-similarity in mathematics (Mandelbrot 1967) and political science (Richardson 1961). On the other hand; much basic spatial data that ecologists need is now available in standard data sets [e.g., DEMs, soil, and (increasingly) land cover]. This means that ecologists may use data at a readily available resolution, rather than developing new data at the ideal resolution for a particular function.

Ecological modelers commonly integrate data from multiple sources, data collected by differing methods, and data measured at varying levels of detail multiple scales. It is important for a researcher to keep integrated data sets within their usable limits of resolution, to preserve reliability of inference and reasoning. Introduction of uncertainty and error during data integration is not well understood (Goodchild and Gopal 1989; Goodchild 1992). In some instances, data integration can reduce systematic error (e.g., in surveying triangulation). In other instances, the impact is harder to predict, beyond the certainty that data are sensitive to being combined. An example is wetlands mapping, which traditionally combines input from soils, vegetation, and hydrology to determine extent. Wetlands maps for the same region can differ dramatically depending upon which of the three source data types the wetlands delineation uses as the base layer. One only has to inspect the wetlands maps of a single region generated by the NRCS (formerly the Soil Conservation Service), by National Wetlands Inventory (NWI), or by the U.S. Fish and Wildlife Service to document this sensitivity (Kracker 1993).

Field observation is time-consuming and labor-intensive, as is the case in many field-based earth sciences (e.g., stratigraphy and physical oceanography). When one adds in the cost of collecting ecological data to the need for multiple data sets, one begins to understand that the use of existing data may be preferable to re-collection. In the case of historical studies (e.g., fire burn histories) the temporal record may go back 50–80 years. In these situations, it would be impossible for an ecologist to collect the full data set personally. It would also diminish the model results to ignore data collected by others. Information about the levels of uncertainty in the data become essential for merging data and for establishing appropriate use.

6.3 Visualization as a Modeling Tool

Methods for visualizing uncertainty can be invoked to mediate the challenges presented by ecological data and process. The visualization clearly cannot make data more certain, but by reading, analysis, and interpretation of spatial and temporal variations in uncertainty patterns, ecological researchers can begin to identify aspects of data that may be suspect.

A large body of cartographic research has demonstrated that the design of a visual display may affect reader comprehension of data patterns (for summaries, see texts by MacEachren 1995; Robinson et al. 1995; Dent 1999). The human brain has strong acuity for visual pattern detection (McCormick et al. 1987). Moreover, humans associate characteristics of equivalence and difference on the basis of iconic stimuli (Marr 1992). Graphical displays can be designed to facilitate these associations, and this forms a basis for cartographic design. Optical illusions result when graphical displays obstruct these associations.

A graphical system acknowledging visual acuity and visual differentiation has been adopted in thematic cartography. The system is iconic, including a set of visual variables. The system was first proposed by Bertin (1983). The visual variables include size, shape, value, orientation, visual texture, color (hue and saturation), and two-dimensional position. (*Visual texture* refers to the granularity of a graphic element: coarse or fine screen, solid or dashed line, plain or fancy typeface.) Only the first six are incorporated into the taxonomy presented later. Figure 6.1 (see color insert) illustrates how the visual variables are manipulated in mapping points, lines and areas. Bertin argues that his system is comprehensive (i.e., there is nothing beyond the visual variables that can be manipulated in graphical design). For static mapping, his argument has held up under intense scrutiny by map designers.

Bertin's second argument that a hierarchy of visual distinctions exists has also held up, confirmed by empirical studies in cartography too numerous to itemize here (see MacEachren 1995 for an extensive bibliography). Weak visual variables are shown at the top of Figure 6.1. Most map viewers do not distinguish displays of symbols varying only in shape or orientation. A map showing sightings of various bird species by triangles, squares, and so forth will be seen by many viewers as a uniform scatter of dots. Those who take the time to discern shape differences will find the map hard to read. Fatigue levels will rise quickly and errors in judgment will begin to occur over a long viewing period.

Color (i.e., hue and saturation) and visual texture comprise stronger visual variables, providing nominal or categorical visual distinctions. Map viewers readily distinguish a red line and a blue line as categorically different (e.g., a rail line and a stream), but the color or texture differences are generally not associated with rank or metric progressions. That is, the rail line is not "more" or "less" than the stream; it is simply different. Progressions and data sequences are best illustrated using the strongest visual variables; namely, value (darkness) and size. Empirical studies of map symbols show that people will match specific numbers to a sequence of light-to-dark blue (for example) and associate specific shades to specific data values with good accuracy. Similar progressive associations have been empirically confirmed for graduated point symbols and for flow line maps (cf. MacEachren 1995).

6.4 The Taxonomy

The taxonomy presented here is modified from Buttenfield and Weibel (1988) and cross-tabulates cartographic measures of uncertainty with statistical data type. The cartographic measures are defined in Section 6.1, and include positional uncertainty, attribute uncertainty, completeness, and currentness.

The taxonomy utilizes three data types: discrete, continuous, and categorical. Discrete data are characterized by having finite boundaries, and point and line features. Discrete data may have an areal extent. For example, a riparian ravine may be represented by a line symbol at a coarse scale, and as an areal buffer at finer resolutions. Continuous data can be measured everywhere on the ground (e.g., elevation, temperature, barometric pressure). One may represent continuous phenomena by a field view in a raster data model. A field view is defined by Goodchild (1987) as a matrix whose values are everywhere defined (e.g., a satellite image, or a gridded map of soil type). Another common data model creates a linked network (e.g., a TIN data model of sampled elevations). In a vector data model, continuous data are often displayed as an interpolated surface (e.g., contours of snow depth).

The third data type is categorical data. This type includes tesselations such as ecological mosaics, tiled data, and areal coverages. The tiles or areal units are characterized with boundaries that are either data-derived (as in patch boundaries) or enumerated on the basis of administrative regions (counties or ownership plots), or on the basis of a related variable. For example, one could map species type (nonmetric categorical data) or species diversity (metric categorical data) within regions delineated by watersheds. Overlay operations are commonly applied to nonmetric categorical data. Metric categorical data must be standardized to the size of enumeration unit to compensate for the discrepancy in resolution between data source and data representation. GIS operations can set class breaks for metric categorical data by partition (equal interval or natural breaks) or by aggregation (quantiles or minimized deviations from mean or median).

Table 6.1 cross-tabulates each uncertainty measure with each data type. Cells of the table contain graphic syntax for mapping uncertainty for all possible combinations. In subsequent sections of the chapter, mapping techniques are described and illustrated to suggest how ecological uncertainty could be mapped. Cells with italic type indicate problem areas, where uncertainty measures are not meaningfully defined, or where the graphical syntax breaks down.

TABLE 6.1. Taxonomy for mapping ecological uncertainty.

Data type/ uncertainty type	Positional accuracy	Attribute accuracy	Currentness
Discrete	Size (Fig. 6.2a,b) Shape (Fig. 6.2a,b)	Texture (Fig. 6.2e) Color saturation	Color (de)saturation Color fade (Fig. 6.3b)
Categorical (Overlay)	Texture (Fig. 6.2c,d) Color saturation	Color (Hue) Mixing (Fig. 6.2f)	Texture overprint (Fig. 6.3c,d)
Categorical (Partition)	*Not meaningful*	Texture (Fig. 6.3a)	*Insert text into map margins*
Continuous	*No clear distinction between positional & attribute accuracy*	*No clear distinction*	
	See adjacent cell →	Color saturation Point gradients (Fig. 6.3e)	Color saturation Area gradients (Fig. 6.3f)

Visual variables are cross-tabulated with data types. Table cells contain exemplar visual variables from Figure 6.1, and refer to specific illustrations in Figures 6.2 and 6.3. Cells with italic text indicate mapping situations where the taxonomy breaks down.

6.4.1 Established Cartographic Conventions

The first row of Table 6.1 contains suggestions for mapping uncertainty associated with discrete data. Positional accuracy is customarily represented with ellipses to show both the magnitude of locational uncertainty as well as any directional bias. For example, Figure 6.2a (see color insert) illustrates a hypothetical map of vegetation cover types. Ellipses identify possible nesting habitats for small raptors (e.g., a flamulated owl). In low-lying riparian areas (blue and green), the directional bias may indicate that trees will cluster along a stream channel. In the third (brown) vegetative area, an ellipse appears on one side of the stream, but not on the other. It is possible that the habitat lies on the south-facing slope of the ravine. The larger size and more circular shape indicate a wider area of suitability within this vegetative cover type.

Figure 6.2b illustrates a different example for mapping positional uncertainty of discrete data. The symbol works upon the same (size and shape) principles to illustrate uncertainty of sediment deposition within a stream bed. The length of the symbol indicates a longer deposition path due to local stream velocities. The width may be a function of channel widening, as well as a function of the granularity of stream bottom material (e.g., silt, gravel, boulders, etc.). The central point is that one can design unique symbols for representing positional uncertainties of discrete data, tailoring the design to a particular ecological domain.

The second row of Table 6.1 cross-tabulates uncertainty mapping strategies with categorical map overlays. Figure 6.2c shows an example of one common type of map overlay, the comparison of two independently compiled wetlands databases. Where the databases coincide about wetlands attribution, one sees a solid green areal patch. Here, one can be more certain that the area is a wetland because the two databases concur. Where the wetland is mapped with a textured areal patch, one or the other (but not both) databases attribute the area as marsh, lagoon, bog, or other type of wetland. The texture provides graphic "noise" in the form of a coarse textured polygon fill, to parallel the contradictions in the two databases.

Figure 6.2d shows a different type of positional uncertainty for land–water distinctions. It is difficult to reconstruct a previous waterline in a lake, oxbow, or meander once eutrophication has occurred. Aerial photography can help if it is properly georeferenced. An ecologist will often overlay multiple sources to determine the historical waterline. Note that the uncertainty of interest here is categorical (e.g., land–water distinctions) and not currentness (e.g., the temporal sequence of shorelines). The cartographic method is to identify the highest and lowest waterlines to create a polygon. The polygon is filled with a color gradient that ranges from the land tint (in this case, green) into the water tint (in this case, blue). The current shoreline is overlaid in black. One can surmise from the map that the shoreline has migrated within this polygon during eutrophication, even if one is uncertain where the shoreline fell at a specific time.

Figure 6.2e illustrates attribute accuracy for aggregated categorical data. The focus of attention in this map is the (un)certainty that a species of bird is actually residing in any one of the quadrat samples, based on sighting. The solid gray quadrats indicate a lack of sightings. The finest texture indicates quadrats where a bird was seen either perched or flying. The coarser texture indicates quadrats where a nest was sighted. The basketweave is most prominent graphically and indicates those quadrats most certain to contain the birds. Here, the ecologist sighted a nest with a bird sitting. This example is adapted from work by Andrle and Carroll (1985).

Figure 6.2f illustrates a mapping technique common to remote sensing. When a digital image is trained and classified, image maps are subsequently generated identifying image pixels whose data do not exactly fit the pattern expected for their assigned class. A similar method can be applied to map uncertainty for any gridded attribute. The figure illustrates an enlarged portion of a vegetative cover type map, showing three classes (red, yellow, and blue). Color mixing has been applied to grid cells where either of two cover types would be equally possible. Uncertainty between the red and yellow class is identified by mixing red and yellow to make orange. Uncertainties between the red/blue and blue/yellow classes are identified by mixing these hues to make purple and green, respectively. This mapping method can be used to check a newly compiled attribute database efficiently. The human eye will pick out the color mixing quickly to determine if color mixing occurs only at class boundaries, or if it occurs in the core of classed polygons as well.

In the third row of Table 6.1 lies the cell that cross-tabulates attribute accuracy for categorical partitioned data. Partitioning is a metric classification method for metric attributes. The data are ordered from lowest to highest value, and then divided (or partitioned) into a small number of categories (usually 5–9) according to some numeric method. Methods for partitioning that are commonly available in GIS packages include Equal Steps, Quantiles, and Natural Breaks. Figure 6.3a (see color insert) shows how uncertainty about partitioned data may be mapped in an ecological context. The figure demonstrates that texture provides a good visual variable for mapping the uncertainty of partitioned attributes, in this case the density of tree canopy in a watershed. Uncertainty in this situation refers to the uncertainty of the actual value at a particular point on the classed map. The legend in the upper left corner of the figure indicates three classes: 0–20%, 21–50%, and greater than 50%. These are graphically depicted using three shades of blue (the visual variable is value). Although it is possible to know that any point in the darkest blue class has a canopy density of at least 50%, it is uncertain whether canopy density is 51%, 70%, or 100%.

A second legend in the lower right corner of the map (Fig. 6.3a) shows uncertainty information by texture. The middle class (no texture overprint) means the classification is highly accurate, and the best guess of canopy density would be the class midpoint (i.e., 10%, 35%, and 75% canopy density, respectively). The whiter texture indicates that overestimation is probable, but not to exceed 5%. The bluer texture indicates that underestimation is probable, but not to exceed 5%. There are five map categories moving diagonally from northwest to southeast. The tree canopy is in the lower class and possibly overestimated (10–15% canopy cover) in the northwest corner. The tree canopy is most dense (>50%) in the north central patch and is likely underestimated (probably 75–80%) canopy cover or higher. The middle density canopy is likely underestimated in the south, and in the east, the ecologist is pretty certain that the canopy cover is in the lowest class (no uncertainty).

The third cell in the first row of Table 6.1 suggests a way to map currentness of discrete data. In Figure 6.3b, circles identify wolf sightings within a stream basin. The size (or shape) of the point symbols could once again be utilized to show positional uncertainty (this is not shown in the figure). The point of this illustration is to show how temporal uncertainty can be embedded into a map. The brighter red (more saturated) symbols identify more recent sightings, and the faded red (de-saturated) symbols identify sightings that occurred some time ago. It is more uncertain whether a wolf may still be sighted at these places.

Figure 6.3c and 6.3d illustrate texture overprints to show temporal uncertainty. In Figure 6.3c, the ecologist has mapped parasite infestations for adjacent watersheds. Temporal uncertainty can vary with the cycle of updating field data because air photography may not be available for the entire area, or partially available from different flight dates. If field checks have been performed for all watersheds, and current evidence of infestation can be determined, but previous dates may be uncertain. The ecologist uses color to distinguish the watersheds, and overlaying textured screens to show historical evidence. Finer textures indicate that photography is current within the past 5 years. Coarser textures indicate previous infestations within the past 15 and 25 years, respectively.

Figure 6.3d illustrates currentness of categorical overlay data. The mapped variable is the rate of nitrogen outflow by landcover type. Landcover types are shown by color hue. The blue patch represents a lake, and the other patches represent a glacier (green), a talus slope (purple), and a gravel patch (yellow). Uncertainty refers to the difficulty of sampling water flowing through (or under) each landcover. The cover type with highest uncertainty is talus, which is the most difficult to sample. The glacier may be cored to sample snowmelt flowing beneath it, but again the placement of the core could hit or miss an actual stream, thus uncertainty is still present. An ecologist may simply dig out a channel in the gravel bed to direct water flow; thus, the uncertainty of a representative sample in this cover type is low.

Figure 6.3e illustrates one of the problem cells in Table 6.1 as discussed in Section 6.4.2.

In Figure 6.3f, the same lake is mapped as in Figure 6.3d, but here the variable is continuous (rate of nitrogen depletion) not categorical (land cover type). Contours of elevation are mapped to show how slope affects the rate of outflow. The last two figures (Figs. 6.3d and 6.3f) complement each other. Working with the two maps together, an ecologist sees the uncertainty of sampling nitrogen in the water flowing through various landcover types, and the slope of the landscape that will affect flow rates. A color gradient indicates the rate (darker is faster). The outline of the talus and the glacier polygons has been copied onto the slope map to help delineate portions of the map where uncertainty is highest. This final example is admittedly somewhat complicated, but it is nonetheless a realistic demonstration of ways that ecological data and uncertainty can be mapped and interpreted.

6.4.2 Problem Cells in the Taxonomy

The first problem cell in Table 6.1 cross-tabulates positional accuracy with categorical partitions. The problem in the taxonomy arises because the positional data refer to features, not to the attributes being partitioned. Partitioning is most often applied to data enumerated by administrative boundary (e.g., cadastral plot, national forest boundary). The boundaries are not data derived, and their positional accuracy does not reflect on uncertainty in the categorical partitions. Representing the positional accuracy of these boundaries is of course possible, and it fits into another cell of the table (the cell cross-tabulating positional accuracy of discrete points and lines).

The second problem cell cross-tabulates currentness for partitioned categorical data. Metric class breaks partition the entire data set uniformly; therefore, no single spatial unit will be more current than another. Instead of mapping uncertainty with a graphical syntax, the ecologist could utilize text in the map margin to annotate dates when source data were compiled or partitioned.

The third problem area involves two table cells, and cross-tabulates *positional* and *attribute accuracy* for continuous data, which are commonly mapped as interpolated surfaces. The problem is that there is no clear distinction between the positional accuracy of the z value and the attribute accuracy because the z value can also be thought of as an attribute. Moreover, interpolated data are derived and thus positional uncertainty does not apply to every point on the map, only to the original spot heights. Cartographic convention suggests mapping the positional accuracy of the posting points, or representing the spatial attenuation of the interpolating algorithm. (In kriging, attenuation is referred to as the *kernel size* and, by definition, the interpolated surface touches each original spot height. Thus, the following technique would only be appropriate for Inverse Distance Weighting or Splining Interpolation algorithms.)

First, a buffer is generated around each posting, to the attenuation scale. Each buffer is then filled with a point gradient (Fig. 6.3e). White means the interpolation accuracy is good, and colored areas indicate the probability of increasing discrepancy between the height of the interpolated surface and the original spot height. Notice that each buffer gradient has a slightly different direction of its uncertainty. This is due to proximity of other spot heights. One could also insert text in the map margin to describe the interpolation algorithm as well as posting the original spot heights.

6.5 Discussion

A search through the ecological literature does not uncover direct references to mapping uncertainty; most examples come from the cartographic and GIS literature. Notable here is the work of MacEachren (1992) on graphic defocusing to show attribute uncertainty. MacEachren proposes two additions to Bertin's system of visual variables, and applies both to mapping uncertainty of several environmental toxicity measures for the Chesapeake Bay. Fog is proposed to represent areas where uncertainty is high. The graphic depiction is to place an irregular gray patch over the map, obscuring the data beneath. Transparency is proposed to represent areas where uncertainty is low. The graphic depiction is to turn off the background color of a graphic layer, and let other layers underneath it "shine through." As MacEachren describes these two extensions, the cartographic technique is (in the case of fog) a straightforward manipulation of value and (in the case of transparency) a manipulation of color and ink effects. Most mapping packages (Macromedia FreeHand, Avenza Map Publisher, and Adobe PhotoShop) have a transparency tool. ESRI ArcView has a "brightness" option available in Spatial Analyst that drapes a transparent version of one theme over another.

An NCGIA Research Initiative on Visualizing the Quality of Spatial Information (Beard, Buttenfield, and Clapham 1991) initiated a good deal of effort to design graphical tools for showing uncertainty in environmental information. McGranaghan (1992) offers several suggestions for mapping attribute uncertainty, arguing that embedding uncertainty information is likely to increase visual complexity for the user; however, he does not test this empirically. Fisher (1987; 1994) has produced interesting uncertainty maps using point pattern animations and even sound, and Weber (1993) has established through subject testing that sonic information may be associated with progressions of spatial data magnitude and local density values. Eventhough Weber's work demonstrates that sonic presentations can amplify visual patterns, he does not test sound in the absence of visualized data.

Research has continued to explore graphical depictions of uncertainty. The dissertation by Hassen (1995) demonstrates positional uncertainty with deformation grids, similar to those first published by Thompson (1961) to illustrate evolutionary comparisons of fish and other animals. In Hassen's work, a regular (planimetric) lattice is stretched in proportion to the degree of positional uncertainty measured at lattice intersections. Working with attributes instead of features, Beard and Sharma (1996) design small glyphs to represent availability and quality of data within a very large archive. Beard and Buttenfield (1999) provide an extensive review of the statistical and related literature on detecting error by graphical means.

It has been demonstrated empirically that embedding uncertainty in a thematic map display can reduce viewer response times for some types of spatial decision making (Leitner and Buttenfield 1996; Leitner 1997; Leitner and Buttenfield 1997; Leitner and Buttenfield 2000). People respond to uncertainty displays as if they reduce rather than increase the information load in a map, which is counter to several arguments presented at the NCGIA Initiative meeting at the beginning of the 1990s.

Visualizing temporal variation in data uncertainty is a more complex issue. Temporal characteristics include both the initiation of an event as well as its duration, its periodicity (regular, irregular, chaotic), and its intensity (phased, catastrophic, or steady). Establishing temporal certainty is additionally hampered by a lack of data models that can define temporal characteristics or relations and interactions between them. It is a principle (in database creation at least) that what cannot be stored explicitly becomes difficult to represent and archive (Starr and Bowker 1994).

Spectral variation in uncertainty is not limited to particular bands of electromagnetic wavelength. In GIS, data themes and domains provide a metaphor for the spectrum. Uncertainty in some themes (e.g., terrain) varies with resolution and with processing algorithms and is usually predictable. The volume of work on interpolation error (e.g., see Lam 1983 or MacEachren 1982) comes to mind. For domains (vegetation or soils data) that may be created by merging data, the probability of uncertainty is high but quite difficult to visualize except by inference. Soil data provide a good example. One maps soil inclusions by noting the percentage of a polygonal region that is expected to be covered by several soil types. One cannot determine the precise location of any single soil type; thus, uncertainty cannot be attached to specific locations. Similar problems arise with the habitat determination for raptors (see example given earlier), if only because the animals are so mobile.

6.6 Summary and Prospects

Establishing uncertainty in ecological data is complicated by the nature of a disciplinary endeavor that searches for environmental patterns that are known to be difficult to formalize and difficult to document empirically. Customary units of measurement are defined functionally, and (in the case of fauna) they may also be mobile. Data sources are varied and often sensitive to the very kinds of data integration that make the evidence of uncertainty easy to expect but difficult to predict. Ecological mapping basically presents a very tricky situation.

Uncertainty information can help to determine the consequences of integrating a data set from an external source into an existing model. At a higher level, uncertainty information can be used to maintain an organization's internal investment in geospatial data.

Research questions arise when mapping uncertainty, and most have yet to be fully answered by cartographic or cognitive science research. First, does uncertainty vary with temporal, spatial, and spectral resolution? If so, what is the nature of this variation? Second, what computational and logical models can be established to explore the nature of uncertainty in data management, in data analysis, and in data browsing and archiving? Finally, to what extent can the collection and maintenance of uncertainty information be automated? If the collection can be accomplished without draining attention from the domain of an ecologist's research topic, the chances rise for consistent and comprehensive recording of uncertainty data. The same principle holds true for other science disciplines as well and is particularly relevant for any field-based environmental research area.

A lot of what is known about mapping uncertainty has arisen from actual practice, proceeding inductively rather than on the basis of established theory; however, ecologists need not wait until cartographic conventions are established to utilize uncertainty information to their advantage. With some common sense and the guidelines in this chapter as a starting point, incorporating uncertainty into ecology mapping is within the reach of most field scientists. The graphical tools are available; mapping uncertainty need not take a lot of extra time. Moreover, becoming familiar with patterns of uncertainty can speed up the process of learning how best to map it, even if starting with the easier methods, and accepted conventions. The more often that uncertainty is visualized, the more we will learn about the effectiveness of exploratory cartographic methods.

Acknowledgments. This research forms a portion of the The Niwot Ridge Long Term Ecological Research (LTER) Program (NSF award DEB-9810218). Funding from the National Science Foundation is gratefully acknowledged. An early version of the taxonomy was developed in 1988 with assistance from Professor Robert Weibel, University of Zurich, Switzerland.

References

Andrle, R.F., and J.R. Carroll, eds. 1985. The atlas of breeding birds in New York state. Cornell University Press, Ithaca, NY.

Aspinall, R.J., and D.M. Pearson. 1993. Data quality and spatial analysis: analytical use of GIS for ecological modeling. Pages 35–38 in Goodchild, M.F., L.T. Steyaert,

B.O. Parks, C. Johnston, D. Maidment, M. Crane, et al., eds. GIS and environmental modeling. GIS World Books, Fort Collins, CO.

Beard, M.K., B.P. Buttenfield, and S.B. Clapham. 1991. Scientific report for specialist meeting on NCGIA research initiative 7, visualizing the quality of spatial information. NCGIA Technical Report 91-26, Santa Barbara, CA. 170 pages.

Beard, K., and V. Sharma. 1996. A multidimensional ranking for data in digital spatial libraries. International Journal of Digital Libraries, Special issue on metadata.

Beard, M., and B.P. Buttenfield. 1999. Detecting errors by graphical methods. Pages 219–233 in Longley, P., M.F. Goodchild, D. Maguire, and D. Rhind, eds. Geographical information systems: principles, techniques, management and applications. John Wiley & Sons, London.

Bertin, J. 1983. Semiology of graphics: diagrams, networks, maps. University of Wisconsin Press, Madison.

Bian, L. 1997. Multiscale nature of spatial data in scaling up environmental models. In Goodchild, M.F. and D.A. Quattrochi, eds. Scale in remote sensing and GIS. Lewis Publishers, Boca Raton, FL.

Bretherton, F. 1993. Why bother? Pages 3–6 in Goodchild, M.F., L.T. Steyaert, B.O. Parks, C. Johnston, D. Maidment, M. Crane, et al., eds. GIS and environmental modeling. GIS World Books, Fort Collins, CO.

Buttenfield, B.P., and R. Weibel. 1988. Visualizing the quality of cartographic data. Presented third international geographical information systems symposium (GIS/LIS 88), San Antonio, CA, November, 1988.

Chrisman, N.C. 1997. Exploring geographic information systems. John Wiley & Sons, New York.

Clarke, K.C., and Olsen, G. 1993. Refining a cellular automata model of wildfire propagation and extinction. Pages 333–38 in Goodchild, M.F., L.T. Steyaert, B.O. Parks, C. Johnston, D. Maidment, M. Crane, et al., eds. GIS and environmental modeling. GIS World Books; Fort Collins, CO.

Dent, B.D. 1999. Cartography: thematic map design; fifth edition. WCB/McGraw-Hill, Boston.

Deveaux, D. 1976. (cited in Forman and Godron 1986) Repartition et diversite des peoplements en carabiques en zone bocagere et arasee. Pages 377–84 in Les bocages: histoire, ecologie, economie. Institut National de la Recherche Agronomique, Centre National de la Recherche Scientifique, Universite de Rennes, France.

FGDC (Federal Geographic Data Committee). 1994. Content standards for digital geospatial metadata. Federal Geographic Data Committee, Washington, DC.

FGDC (Federal Geographic Data Committee). 1998. Content standard for digital geospatial metadata (version 2.0) Federal Geographic Data Committee, Washington, D.C. FGDC-STD-001-1998. (Also on the WWW at http://fgdc.er.usgs.gov/publications/publications.html.)

FIPS (Federal Information Processing Standard). 1992. The spatial data transfer standard. Federal Information Processing Standard 173, Department of Commerce, Washington, DC.

Fisher, P. 1987. The nature of soil data in GIS: error or uncertainty. Proceedings international geographic information systems (IGIS) symposium: the research agenda III:307–18.

Fisher, P.F. 1994. Animation and sound for the visualization of uncertain spatial information. Pages 181–86 in Hearnshaw, H.M., and D.J. Unwin, eds. Visualization in geographical information systems. John Wiley & Sons, London.

Fitzpatrick-Lins, K. 1981. Comparison of sampling procedures and data analysis for a land use and land cover map. Photogrammetric Engineering and Remote Sensing 47:343–51.

Forman, R.T.T., and M. Godron. 1986. Landscape ecology. John Wiley & Sons, New York.

Forman, R.T.T. 1995. Land mosaics: the ecology of landscapes and regions. Cambridge University Press, London.

Goodchild, M.F. 1987. A spatial analytical perspective on geographical information systems. International Journal of Geographical Information Systems 1(4): 327–34.

Goodchild, M.F., and S. Gopal. 1989. The accuracy of spatial databases. Taylor and Francis, London.

Goodchild, M.F. 1992. Geographical data modeling. Computers and Geosciences 18:401–8.

Gustafson, E.J. 1998. Quantifying landscape pattern: what is the state of the art? Ecosystems 1:143–56.

Hammer, R.D. 1998. Space and time in the soil landscape: The ill-defined ecological universe. Pages 105–40 in Peterson, D.L., and V.T. Parker, eds. Ecological scale: theory and applications. Columbia University Press, New York.

Hassen, K.M. 1995. Reference grid: a visual tool for viewing positional accuracy for spatial data. Unpublished Ph.D. Dissertation, Department of Spatial Information Science and Engineering, University of Maine–Orono.

Hunsaker, C.T., R.T. Nisbet, D. Lam, J.A. Browder, W.L. Baker, M.G. Turner, t al. 1993. Spatial models of ecological systems and processes: the role of GIS. In: Goodchild, M.F., B.O. Parks and L.T. Steyaert, eds. Geographic information systems and modeling. Oxford University Press, Oxford.

Innes, J.L. 1998. Measuring environmental change. Pages 429–258 in Peterson, D.L., and V.T. Parker, eds. Ecological scale: theory and applications. Columbia University Press, New York.

Jelinski, D.E., and J. Wu. 1996. The modifiable areal unit problem and implications for landscape ecology. Landscape Ecology 11:129–40.

Kracker, L. 1993. Analyzing inconsistencies in wetlands mapping using GIS. Unpublished Master's Thesis, Department of Geography, SUNY-Buffalo.

Lam, N.S-N. 1983. Spatial interpolation methods: a review. The American Cartographer 10:129–49.

Langran, G.E. 1991. Time in geographic information systems. Taylor and Francis, London.

Leitner, M., and B.P. Buttenfield. 1996. The impact of data quality displays on spatial decision support. Proceedings, GIS/LIS '96, Denver, CO, November 1996:882–94.

Leitner, M. 1997. The impact of attribute certainty displays on spatial decision support. Unpublished Ph.D. Dissertation, Department of Geography, SUNY-Buffalo.

Leitner, M., and B.P. Buttenfield. 1997. Cartographic guidelines for visualizing attribute accuracy. Proceedings, AUTO-CARTO 13, Seattle, WA, April 1997:184–94.

Leitner, M., and B.P. Buttenfield. 2000. Guidelines for the display of attribute certainty. Cartography and GIS 27(1):3–14.

Levin, S.A. 1992. The problem of pattern and scale in ecology. Ecology 73(6):943–67.

Levins R. 1968. Evolution in changing environments. Princeton University Press, Princeton.

MacEachren, A.M. 1982. The role of complexity and symbolization method in thematic map effectiveness. Annals, Association of American Geographers 72:495–513.

MacEachren, A.M. 1992. Visualizing uncertain information. Cartographic Perspectives 13(3):10–19.

MacEachren, A.M. 1995. How maps work: representation, visualization, and design. Guilford, New York.

Mandelbrot, B.B. 1967. How long is the coast of Britain? Statistical self-similarity and fractal dimension. Science 156:636–38.

Marr, D. 1992. Vision. W.H. Freeman & Co., San Francisco.

McCormick, B.H., T.A. DeFanti, M.D. Brown, eds. 1987. Visualization in scientific computing. SIGGRAPH Computer Graphics Newsletter 21(6), October. Condensed version appears in IEEE Computer Graphics and Applications, July: 61–70.

McGranaghan, M. 1993. A cartographic view of data quality. Cartographica 30(2): 8–19.

Miller, R.B. 1993. Information technology for public policy. Pages 7–10 in Goodchild, M.F., L.T. Steyaert, B.O. Parks, C. Johnston, D. Maidment, M. Crane, et al., eds. GIS and environmental modeling. GIS World Books, Fort Collins, CO.

Munroe, L., and E. Decker. 1991. GIS use by wildlife resource agencies. GIS World 4(3).

NSF (National Science Foundation). 1999. Large scientific and software data set visualization. NSF Program Announcement 99–105, Target Date: July 6, 1999.

Patterson, D.L., and V.T. Parker, eds. 1998. Ecological scale: theory and applications. Columbia University Press, New York, Series on Complexity in Ecological Systems.

Richardson, L.F. 1961. The problem of contiguity: an appendix to the statistics of deadly quarrels. General Systems Yearbook 6:139–87.

Robinson, A.H., J.L. Morrison, P.C. Muehrcke, A.J. Kimerling, and S.C. Guptill. 1995. Elements of cartography, sixth edition. John Wiley & Sons, New York.

Rogers, G., H. Malde, and R. Turner. 1984. Repeat photography for evaluating landscape change. University of Utah Press, Salt Lake City.

Starr, S.L., and G. Bowker. 1994. Knowledge and infrastructure in international information management: problems of classification and coding. Pages 187–213 in Bud-Frierman, L. Information acumen: the understanding and use of knowledge in modern business. Routledge, New York.

Thompson, D.W. 1961. On growth and form. Cambridge University Press, London.

Thompson, M.M. 1975. Maps for America: cartographic products of the U.S. Geological Survey and others. U.S. Geological Survey, Washington, DC.

Veblen, T. 1989. Biogeography. In: Gaile, G.L. and C.J. Wilmott, eds. Geography in America. Merrill Publishing, Columbus, OH.

Veblen, T.T., and D.C. Lorenz. 1991. The Colorado front range: a century of ecological change. University of Utah Press, Salt Lake City.

Weber, C.R. 1993. Sonic enhancement of map information: experiments using harmonic intervals. Unpublished Ph.D. Dissertation, Department of Geography, SUNY–Buffalo.

Wiens, J.A. 1989. Spatial scaling in ecology. Functional Ecology 3:285–397.

Wieshofer, M. 1994. Dynamic and temporal aspects of GIS. Unpublished M.A. Thesis, Department of Geography, SUNY–Buffalo.

7
A Cognitive View of Spatial Uncertainty

Geoffrey Edwards and Marie-Josée Fortin

It seems appropriate that a book concerned with understanding spatial uncertainty in the context of ecology should contain a chapter that discusses cognitive issues. Ecology as a discipline represents the development of a particular way of thinking (i.e., holistic thinking) about the world; a way of thinking that is foreign to the mechanistic world view that dominated much of the twentieth century. Furthermore, although a great deal has been said about uncertainty, particularly about spatial uncertainty, little effort has been made seriously to address the cognitive roots of spatial uncertainty despite widespread lip service concerning their importance.

In addition, it is worth pointing out that although cognition usually refers to what goes on in our heads, it is also possible to model cognition. Cognitive modeling, whether using simulated neural networks or symbolic architectures, is being used increasingly in research in cognitive science and in artificial intelligence, as well as in other disciplines, and the term *cognition* is often used as if it embraces cognitive models. For the purposes of the discussion in this chapter, we shall use the term *cognition* to refer to what goes on in our heads, except where we explicitly refer to the other use of the term; however, neither of the authors are cognitive scientists, and hence the treatment presented explains the issues largely from a practitioner's, and to a lesser extent, a modeler's point of view, rather than the perspective one would expect from a specialist in the field of cognition.

The working definition of uncertainty we have collectively adopted for this book underlines the importance of cognition for the study of uncertainty: *Uncertainty* is defined as the difference between phenomena in the world and the description of these phenomena. Where the word *description* is to be found, cognition cannot far behind—without a cognizing entity, no description is possible. Furthermore, there is no direct way to determine the nature of a thing in order to perform the comparison.[1] Its nature can be determined only by perception, by measurement, or by theorizing. Measurement requires choosing what is to be measured, which in itself depends on perceptions or theories. *Theories* are the result of applying abstractions or generalizations to perceptions, albeit within the historical context of a discipline. Hence, perception is the source of all information concerning a thing, both its measurement and its description.

Description and measurement both require categorization. The conceptualization of a category is rooted in human cognition and perception, although it may be socially or culturally systematized or incorporated within a mathematical framework. Hence, for example, we structure space via our own perceptions and memories. A representation of space in terms of idealized geometrical objects (e.g., polygons, lines, and points) is merely one rather simplified approach. Such standard models and representations are poorly suited to the representation of more complex spatial distributions such as ecological systems, characterized by continuous and dynamically changing interactions at many scales simultaneously. Furthermore, the uncertainties introduced into the data in order to mold them within existing representations are likely to be large, complex, and closely tied to conceptualizations of space and hence to cognition. In this chapter, we explore some of the links between spatial uncertainty and human perception and cognition. The different processes by which data are collected and transformed are briefly examined from this perspective, from field data collection and image acquisition and interpretation, through classification and generalization, right through to the decision-making stage. Different forms of spatial uncertainty that have a cognitive component and are likely to be pertinent to ecological data are highlighted at this stage, including boundaries, categories, aggregates, patterns, sampling, and decision making. Following this, currently emerging techniques for handling such uncertainties are surveyed along with both known limitations and areas where theory is still indeterminate. Finally, we examine new, longer-term research that seeks to formalize cognitive and perceptual representations and hence to provide new means of addressing spatial uncertainty and the role it plays in constraining data used for decision-support. These approaches include the development of a general theory for uniting perceptual and geometrical spaces; the use of image schemata, a concept developed in research in cognitive science over a decade ago; work on cognitive modeling; qualitative reasoning systems developed in artificial intelligence (AI), designed to operate in the presence

---

[1] The use of the concept of a nominal ground (see Vauglin 1997) provides a way around this difficulty—error becomes the difference between the description and the nominal ground, which is the image of the world produced through the filter of a set of specifications and around a given goal; however, the nominal ground does not remove the importance of cognition to the whole cycle of data processing—if anything, it adds to the importance of cognition.

of uncertainty; and category theory such as is present in work in cognitive science and AI.

Before proceeding further, it is worth noting that there are several possible views on the extent to which cognition plays a role in the study of uncertainty. One possible view is that cognition and perception are fundamental to the process of describing or measuring a thing, and therefore they are ubiquitous to the understanding of uncertainty (as outlined earlier). It is possible to argue, however, that cognition (what goes on in people's heads) can be eliminated by formally specifying the description and the measurement technique or techniques used to verify the nature of the thing being studied. This relegates cognition to the stages preceding the specification (i.e., to philosophy), and attempts to exclude it from the problem of comparing the description and the measurement. In fact, this is how most sciences excise cognition from their respective realms of study. From another perspective, however, even if one accepts the latter argument, cognition may retain a role in ensuring an adequate comparison between a specified description and measurement. This may occur for several reasons: The description involves elements that cannot be described in any other way than via their cognitive aspects, or the measurements involve significant contributions from cognitive processes, or the comparison cannot be achieved without invoking cognitive processes.

In the following chapter, we argue that this last view holds for spatial data in general, and spatially distributed ecological data in particular. The primary justification for this is that, for spatial data, comparison between measurements and descriptions cannot be performed without invoking cognitive processes. In addition, many measurements rely on procedures with a nonnegligible cognitive component (e.g., measuring the dimensions of a tree crown in an aerial photograph relies on visual perception of the limits of the tree crown). This is true for photointerpretation and certain forms of field sampling, as well as certain types of remote sensing analysis. Furthermore, a variety of transformations that are typically applied to spatial data also involve a significant cognitive component, including matching of phenomena across data sets and generalization procedures.

7.1 A Brief History of the Links Between GIS, Cognitive Science, and Ecology

Since the early 1980s ecologists have started to consider the spatial component part of ecological data. Before this, ecologists deliberately avoided heterogeneous study areas and focused only on homogeneous areas, largely because they lacked the statistical and theoretical tools to deal with spatial heterogeneity. The combination of the widespread use of aerial photographs, remote sensing imagery, spatial statistics, and computer developments such as GIS contributed to stimulate ecologists to incorporate the spatial component of data in their studies, analyses and theories. Such integration and analysis of the spatial context to understand better ecological processes spread so rapidly that there is now a new discipline of ecology that is concerned almost exclusively with spatial data: landscape ecology (Wiens 1989; Wiens 1992).

The links between GIS, geographic information science, and cognition have been made even more recently. The late 1980s saw early efforts by a number of researchers in geographic information science to link research on GIS and a variety of cognitive disciplines, including computer science and AI, psychology, linguistics, and philosophy (Frank and Mark 1988). This was motivated at the time by a number of factors, mostly intuitive—the means by which spatial relations are expressed in natural language seemed to have tantalizing properties and power compared with the standard geometrical relations used in GIS; it was recognized that popular access to GIS was limited partly by its complexity and the rigidity of its data models; it was believed that different cultural perceptions of space might enhance or limit the use of GIS by different groups of people; and so on. It was likewise felt that a complete theory of spatial information might rely more heavily on these questions than had been heretofore expected. These efforts evolved into a series of conferences designed to explore such links in more detail, under the title of the International Conference on Spatial Information Theory (COSIT) (Frank and Campari 1993; Frank and Kuhn 1995; Hirtle and Frank 1997).

This work is now beginning to pay off. At first, the link to cognition was considered to be a peripheral question of interest only to a small group of researchers. Over the years, awareness has grown that spatial data are tricky to handle and that serious conceptual issues with respect to these data are not understood. This has led to an emerging perception in the research community that perhaps cognition is not so peripheral after all, although many users of GIS still do not understand this (Raper 1996). Some of the early work on linking psychology and linguistics to GIS has begun to mature, resulting in quantitative methods for structuring or representing cognitive aspects of geographic data (Freksa 1992; Edwards and Moulin 1995; Couclelis and Gottsegen 1997; Edwards 1997; Raubal et al. 1997). The coming years are likely to see a lot more of this work, which is beginning to consolidate into a coherent subdiscipline.

7.2 A Survey of the Links Between Spatial Uncertainty and Human Cognitive and Perceptual Processes

7.2.1 The Definition and Nature of Spatial Data and Uncertainty

Our knowledge of space is obtained in two ways—by integrating local experiences of space into broader frames (called the *navigational* or *route perspective*) or by examining the whole of a space in an overview (called the *bird's eye* or *survey perspective*) (Taylor and Tversky 1992). Space is defined mathematically as sets of points exhibiting neighborhood relations (topological space) or as sets of points on which a distance metric may be defined (metrical space). All four of these methods of characterizing space are typically called upon when describing phenomena that are spatially referenced or spatially distributed. If only the latter two approaches were exploited (i.e., neighbor relations and distance metrics), then there would be no need to invoke cognitive operations in describing spatial data. The bird's eye perspective may likewise be accommodated by such a mathematical approach; however, for most types of spatial data of interest to geography, ecology, natural resource management, and many other applications, the integration of local perspectives into a broader framework also occurs, but is, however, poorly modeled.

The reason this method of acquiring knowledge about space occurs so frequently is that our understanding of space arises from such a procedure. We do not understand a scene via an overview. Even when such an overview is available, we break it down visually into regions with distinct characteristics and then integrate it back to obtain our "bird's eye view" understanding. Hence, an "adequate" description of spatially distributed data requires the incorporation of local perspectives into a larger view (where "adequacy" is determined by the ability to understand the resulting description). This is the primary reason why cognition is so important to the spatial description of data. Without an adequate understanding of how we understand spatial data, we will be bereft of adequate tools for characterizing it in understandable ways and for quantifying its uncertainty. This is the primary way we make sense out of spatial data, but spatial phenomena themselves are not necessarily structured in this way. Different spatial phenomena are structured in different ways, sometimes as simple gradient fields, other times as complex nested clusters of entities, and on still other occasions as dynamically changing and interacting processes.

7.2.2 The Production of Spatial Data

Aside from simple gradient fields, most spatial phenomena are too complex to measure in any simple way. Instead, we must average phenomena across intervals of space or time or both, and we must select and abstract only certain predetermined aspects of the phenomena of interest. For many spatial data, both the averaging and the selection operations are dependent on context-sensitive perceptions of phenomena. Hence, spatial data measurements implicitly involve perceptions of the phenomena being measured.

As a result, uncertainty (which is recognized when comparing descriptions of spatial data with measurements of the latter) is related to properties of human cognition both as a result of description and as a result of measurement. Although measurement can sometimes be formalized so as to preclude the cognitive element, this is true only for the description of simple spatial phenomena. For complex spatial phenomena, descriptions must be cognitively adequate, or they will not be understood. We shall come back to the problem of the cognitive adequacy of the description later. In the next few sections, we examine the cognitive component of measurements made on spatial data, and techniques for modeling and quantifying the uncertainties related to such measurements.

Data Sampling. Data for spatially distributed phenomena are obtained by appropriate sampling. Many different types of sampling are possible: random, systematic, stratified, nested, and so on. (Green 1979; Thompson 1992). Furthermore, data samples are transformed in a series of interpretation steps in order to obtain more complete descriptions of the phenomena of interest. The main sources of data are examined here, along with the transformation processes they may undergo.

Field sampling is the basic technique behind most other sampling procedures. Image-based techniques of acquiring data are all validated by field sampling. Furthermore, most methods of spatially referencing the data involve field measurements, whether via total stations used in surveying, GPS measurements of geodetic position, compass and chain methods in forestry, or the like. Field sampling requires adequate design (see Chap. 14) to ensure that all forms of bias may be properly accounted for. When a spatial pattern is to be characterized, the choice of sampling design is even more important (Fortin et al. 1989).

Sampling strategies are usually based on an assumed theoretical framework. One such framework that is widely used assumes that ecological processes draw energy off gradient fields that can be characterized independently of the processes themselves (e.g., local temperature or latent heat flux). Sampling under such a framework would involve attempting to locate samples so as to capture the possible variations and fluctuations in the gradient field. Another theoretical framework is that ecological processes are dominated by local interactions and are only weakly affected by interactions with processes occurring in neighboring regions. In this case, sampling would be designed to capture a large number of variables involved in local processes, and the spatial pattern of samples would be treated as less important. Both frameworks and their corresponding sampling strategies involve the selection of certain types of measurements in certain patterns, but both the measurements and patterns are different in the two examples given. The choice of measurements and patterns could lead to two different descriptions of a spatial phenomenon that might actually be the same. These descriptions would be complementary not conflictual. This again emphasizes the role played by perception in the description of spatial phenomena.

As indicated earlier, field data are often used to calibrate remotely sensed data. The major problem encountered in such a practice is that these two types of data are not generally measured at the same spatial resolution.

There are several ways to transfer data from one scale to another, including aggregation, filtering, and interpolation of field data. Furthermore, there are several interpolation techniques available, but not all deal with the inherent degree of spatial autocorrelation contained in the data. Kriging is the interpolation technique most frequently used by ecologists because it takes into consideration the spatial structure of the data (Cressie 1991). This kriging procedure is part of the geostatistics family of methods that are based on the assumption of stationarity.

A reliable kriged interpolation must be directly related to the sampling design used. Indeed, the rule-of-thumb in geostatistics is that a minimum of 50 samples is needed to provide a good estimate of the spatial function (variogram), which is then used to interpolate. Furthermore, both the number of samples is important as well as the way they are spatially distributed (Fortin et al. 1989). When the scale of the spatial pattern is unknown, Oliver and Webster (1986) as well as Fortin et al. (1989) suggested the use of a nested sampling design where the spatial step (lag) between samples varies. By having samples at different spatial spacing, it provides more chances to detect the spatial pattern of the data. It is interesting to note that human memory of space seems to be structured recursively, indicating that human perception may also rely on nested sampling of the environment (Hirtle and Heidorn 1993).

Image Interpretation. A second form of data sampling is to use images as a source of data. The images themselves are rarely used as such; rather, they are transformed into a more useful form. Images usually provide very little information over small regions, compared with field data that may involve many different kinds of measurements at each location; however, images contain a great deal of information across larger regions, and also when interpreted in context on the basis of past experience.

Field data are costly to obtain, and hence cannot be acquired at high density over large regions; on the other hand, field data usually contain abundant references to context. Image data are much cheaper and provide samples at high density over large regions, at the expense of adequate contextual information. Hence, image data are widely used to complement field data in measuring ecological phenomena at regional scales.

For most complex spatial phenomena, images cannot be easily converted to estimates of interest. Furthermore, even for the examples described in the following section, images are characterized by spatially varying context. It is difficult to determine these contexts correctly based on automated methods of image analysis. Hence, human interpretation is used to determine spatially variable context.

The only boundaries that can be accurately delineated as sharp lines are artificial: administrative and political boundaries, and those following precise land-use practices (clearcuts in forestry, crops in agriculture, etc.). Natural boundaries (ecotones) are more likely to be gradual and thus wider.

The degree of uncertainty in boundary delineation with spatial data is hence affected by the spatial resolution of the data, the extent of the study area, and the interpretation procedures used. Thus, if the study area is too small and contains not enough material to determine if the region is homogeneous, the delineation of its boundaries will be very difficult. Furthermore, as for pattern characterization, too small a spatial resolution will generate too many small discontinuous boundaries, whereas too large a spatial resolution will smooth out too much of the boundary signal. Even when resolution and the study area are optimal, however, interpretation procedures may yield wide variation in the characterization of ecotones (Edwards and Lowell 1996).

Interpretation usually involves separating regions judged to be characterized by different conditions with a boundary or line drawn onto the image (i.e., the aerial photograph). Different individuals tend to place the same boundary at slightly different locations, based on different evaluations of what constitutes the characteristic texture on each side, and on a variety of spatial phenomena in the neighborhood of the boundary (Edwards and Lowell 1996). As a general tendency, the more easily the two textures may be confused, the shallower the contrast and the wider the transition zone demarcated by different interpreters; however, a model developed to characterize the width of the transition zones based on properties of their defining textures showed that not more than half the variation in width can be attributed to the local characteristics of the textures. The remainder of the variation appears to be due to a variety of neighborhood contextual effects such as the presence or absence of regions of similar textural characteristics, the presence of small islands and peninsulas, and the heterogeneity of the textures. In addition, not all interpreters will perceive the same boundaries. The wider the transition zone, the higher the probability that an interpreter will fail to draw a boundary through it. Hence, a small number of interpretations will undersample wide transition zones, and most boundaries will be displaced with respect to that of a median interpretation. It is possible to obtain a fairly representative characterization of such a median interpretation by combining at least two, and preferably three or more, interpretations performed by different individuals on the same data set. For each boundary segment, such a combination of interpretations will provide a measure of the probability of identifying a given boundary (Aubert 1995).

Although human experts proceed by drawing boundaries, they nonetheless seek to identify regions, and it is the regions that are often imbued with significance. Interpreters usually assign a category to the fenced off regions once the polygon is complete. Interpreters may also move back to the boundaries, adding or suppressing certain boundaries based on refined region characterizations, later on in the interpretation process (most experienced photointerpreters do not do so; however, this is a tactic used more by beginner interpreters). Most efforts to evaluate the uncertainty in interpretations attempt to characterize the ambiguity in the categories associated with the regions. Field sampling is sometimes used for this. Our analysis of interpretation results suggests that this is not the correct approach. The difference with respect to boundary-based field sampling may be small if the polygons are large compared with the size of transition zones, but, in general, region-based field sampling will underestimate the uncertainty in the data, especially if the location of field samples is moved away from boundaries, as is frequently the case. Tests aimed at understanding and improving sampling procedures near boundaries are presently being explored (Lowell 1997).

Image Classification. For certain simple (e.g., gradient) phenomena, image data may be converted directly to useful information. Hence, for example, so-called thermal infrared image data may be converted directly to temperature and for many regions may be used as a direct measurement of surface temperature. A fair number of remote sensing satellites are used in this way.

In addition, for some complex phenomena, models have been developed that may be used to transform image data into useful estimates. Hence, radiative transfer models are sometimes used to transform radar data into dielectric estimates. Likewise, empirical models may be used to convert radar data of forested regions into biomass estimations. These transformations are almost always context sensitive, however—that is, they are valid only if certain local conditions occur. Hence, the thermal infrared data may be invalid across water bodies, and the radar data may produce reliable biomass estimates only for forested regions of certain moderate densities.

Remotely sensed satellite imagery is therefore used frequently, as with aerial photography, for determining the spatial contexts over which different processes might occur. Because satellite imagery is not available at the fine resolutions obtained with airborne photography, and because most remote sensing satellites use several spectral bands to characterize ground covers (if not hundreds in some cases), remote sensing scientists have developed or adapted statistical classification techniques that partition the image into the zones of interest. This is called *image classification.* They could have used image interpretation in ways analogous to those described in the previous section, but have developed alternative procedures for a number of reasons. Photointerpretation is based primarily on the interpretation of textures and the spatial information content, and uses spectral information to a much more limited extent. Satellite sensors acquire data with high spectral content, but for which the spatial information has been degraded. Hence, traditional photointerpreters operate at a disadvantage with such images (Leckie 1990). Furthermore, the high spectral content provides more information about local conditions and hence allows automated algorithms to function much more efficiently than is the case with scanned aerial photography.

Hence, although interpretation is sometimes used, satellite imagery is usually converted to "land use maps" or "land cover maps" via either automated analysis or human-assisted computer analysis. In the latter case, the human user selects and tests different statistical parameters and models over a limited region for which other information exists, in order to achieve the "best classifier," and this is then applied to the entire image to obtain a transformed image. For fully automated classifiers, little human involvement occurs (although results are usually significantly less accurate than when humans are included in the processing). Hence, the cognitive component of uncertainty is likely to be small. For the semi-automated classification procedures, however, humans are involved in selecting the appropriate classification model. This is known to reduce the overall error, but clearly increases the cognitive component of the uncertainty. In fact, the use of humans in the classification procedure introduces a variability in the classification results from one occasion to the next.

Classification in remote sensing is accompanied by a variant on the error matrix called a *confusion matrix* (see Chap. 14). The total accuracy of a classification will fluctuate based on the human component, but the result will be reported in the confusion matrix. Not reported, however, is the gain obtained by involving the human in the classification process and the loss involved in not obtaining the best model or parameters. Classification accuracies will therefore fluctuate across the middle ground of such a range, although no known study examines this question. Hence, human cognition affects classification results in remote sensing, but at an unknown level; however, it is likely to be no greater than 5–10%.

Generalization. Most spatial data undergo some form of generalization before they are used. Base maps are produced from surveying data or photointerpreted data at relatively fine scale, but are then generalized to produce maps for a variety of uses at coarser map scale. Photointerpretation involves some generalization directly during interpretation, usually as a consequence of operating within the constraint of a minimum mapping unit. Classifications obtained from remotely sensed data are usually noisy, and may be "cleaned" by passing a filter across them or manually redigitizing the polygon boundaries. The generalizations involved in changing map scale are traditionally carried out by expert cartographers drawing on long years of experience, working within scaling specifications and using aesthetic as well as cartographic criteria. Efforts to automate this process have met with difficulties not unlike those encountered in automating photointerpretation. The transformation of information (losses and gains) through such scaling is also not well understood.

Field data may be generalized via aggregation, filtering, interpolation, or subsampling. In general, field data are not acquired at the locations at which comparisons with other data are required; hence, some means must be found to transfer and modify these data in appropriate ways in order to perform such comparisons.

There are several ways to aggregate data from one scale to another; however, none of them is completely satisfying. The first problem with aggregating data is that there is inevitably a loss of information. The second problem is that all aggregation methods reflect one or two, at most, specific characteristics of the data, but not all of them. The most intuitive way to aggregate data is to compute the mean value of adjacent samples (pixels, cells, quadrats). This method is valuable when the data are normally distributed, but with very skewed data this method will produce misleading, if not meaningless information. With skewed data, the majority rule, to take the mode value, is more appropriate. These two methods take care of data values, but not how they are spatially distributed. Another way to aggregate would be to ensure the fractal dimension of the data across scales stays constant.

Up to now, aggregation methods have been used on data sets without taking into account that the inherent pattern at one scale may be anisotropic (i.e., may vary with orientation) and that the aggregation may transform it into an isotropic pattern. Furthermore, no verification is usually performed to test whether the whole study area is stationary or not. When the study area is not stationary, one should not aggregate adjacent samples that do not have the same mean, variance, and isotropy properties. This would have the effect of mixing two different signals and masking their specific information.

Generalization may be viewed as the process of abstracting the essential features of a spatial representation. This is precisely what humans do—it is cognition at work. The process may introduce biases with respect to the phenomenon studied, but it also represents an attempt to identify what is significant in the phenomenon studied; hence, in at least one sense generalization serves to *reduce* uncertainty rather than to increase it! This is one of the difficulties encountered in comparing classifications obtained from remotely sensed imagery to so-called ground truth. After generalizing, the remote sensing classifications may constitute a more faithful representation of the "ground truth" than before generalization.

For another example, generalization used to change map scales may serve to enhance the utility of the resulting map, compared to more detailed, supposedly more "accurate" fine-scale maps. The role of generalization in relation to uncertainty needs to be examined in more detail.

7.2.3 The Limitations of Existing Data Structures

Standard means for representing spatial data in GIS are largely determined by two families of data structures: raster and vector structures. *Rasters* are regular tilings of space, usually represented as a two-dimensional matrix, whereas *vectors* are collections of linear features represented by the coordinates of their end points. Polygons are represented as interior regions of connected line segments in the latter case, or as contiguous sets of tiles in the former case. Variants on standard tiled representations include quadtrees (Samet 1984), which partition space into different levels of aggregation, whereas variants on vector representations include TINs and other network structures.

These data structures allow for the representation of gradients (raster structures), surveyed features (vector structures), and categorized or classified regions (either raster or vector structures). They do not easily accommodate nested clusters of entities, dynamically changing and interacting processes, the recursive integration of local perspectives into broader views, the primary means adopted by human users to understand spatial data. Furthermore, although these data structures may be adapted to represent certain types of uncertainty, such as fuzzy membership in region-based categories, many of the sources of uncertainty discussed earlier cannot easily be modeled using these data structures. As a result, it is very difficult to characterize the uncertainty in spatial data within the framework of existing tools for geographic information. Under moderate modification, it may be possible to represent sets of interpreted boundaries and median boundaries in support of photointerpretation. Metadata may also be used to characterize the human component in remote sensing classifications, or to characterize the sensitivity of resampled field data to the aggregation kernel used (see Chap. 11); however, the decrease or increase in uncertainty during generalization and the uncertainty associated with complex, dynamic, or recursive data representations is unlikely to be handled correctly using existing tools for geographic information.

7.2.4 Uncertainty in Pattern Delineation

Another source of uncertainty that is poorly handled using existing methods is the uncertainty in pattern delineation. So-called pattern metrics have been found to correlate with a variety of animal habitat features, or to enhance the health or survival of different animal species. Pattern metrics are measures that characterize different types of patterns. Hundreds of pattern metrics are known (McGarigal and Marks 1995), of which only a few have been found to play a significant role in ecology (Riitters et al. 1995).

The degree of uncertainty in pattern characterization is related to the sampling design used, the spatial resolution of the data, and the pattern metric sensitivity. There are different types of pattern metrics, those that quantify the within patch heterogeneity (spatial statistics, geostatistics) and those that quantify patch geometry and relationships with adjacent patches (landscape metrics) (Li and Reynolds 1995). There is often more than one pattern in spatial data and our ability to detect them depends on the scale at which the data are analyzed: At local scales there could be too much noise to identify the pattern and at regional scales there may be too much environmental variability. Thus, one of the most common sources of pattern uncertainty is that an inappropriate scale is used to analyze the data. This may either smooth the pattern, render it noisier than it is in reality, or change the measured directionality of the pattern.

Landscape metrics have been developed to characterize the geometric complexity of patches (area, perimeter, boundary shape, etc.) and to quantify the degree of relationships among adjacent patches [connectivity, diversity, etc. (Li and Reynolds 1995)]. These metrics are mainly used to describe patterns. The accuracy of the pattern metrics depends directly on the degree of accuracy of the patch delineation and classification; hence, if the patch delineation is not precise, all the resulting pattern metrics will be biased and fail to characterize the patterns accurately. Furthermore, the landscape metrics are often derived from each other and do not provide independent information on pattern. Moreover, pattern uncertainty is also linked to the choice of landscape metric as some cannot differentiate among certain patterns. One therefore needs to be careful and critical about the interpretation of these metrics. It is not because they can be computed that they necessarily furnish new and relevant information or are pertinent to any ecological understanding of spatially distributed phenomena.

The determination of pattern is also one of the processes at which humans excel. Humans appear to use mechanisms related to the integration of local perspectives in a broader framework to characterize pattern; hence, the study of pattern metrics constitutes an example of a spatial description that is very close to that used by humans to understand spatial data. As a result, it should be possible to decrease the uncertainty in an adequate description to very low levels; however, several questions arise. Do existing pattern metrics describe patterns adequately for human understanding, and are they used in representations which help humans understand the data? What characteristics of pattern are not captured by these pattern metrics?

Furthermore, within the field of artificial intelligence, there is a great deal of interest currently in the development of qualitative spatial reasoning tools that may handle incomplete or uncertain information about spatial—environments. These developments are informed by results emerging from studies in cognitive linguistics and psychology; hence, even if landscape metrics contain some level of uncertainty, they may be incorporated into such a reasoning environment and used effectively to draw conclusions about the structure of the landscape. This is another area that bears closer inspection.

7.2.5 Decision Making and Spatial Uncertainty

Spatially distributed data are ultimately used to make decisions. Studies in ecology are increasingly being used to drive policy concerning environmental management (see Chaps. 2 and 3). Hence, it is pertinent to ask what the relationship is between spatial uncertainty and the decision-making process.

Spatial uncertainty is frequently viewed as an obstacle to be removed or reduced. From this perspective, spatial uncertainty inhibits the decision-making process; hence, any attempt to reduce or remove the uncertainty is viewed in a favorable light. Identifying sources of uncertainty is viewed as one of the principal means of reducing uncertainty. Once a source has been identified, it becomes possible to expend more effort or develop better quality control methods to ensure the best data possible are made available.

A second view, however, suggests that some uncertainty is intrinsic to the decision-making process. In fact, it may be argued that in the absence of uncertainty, no decision making is possible. Studies in cognitive science indicate that decision making requires uncertainty to operate efficiently (Bell et al. 1988). In models of decision making, decisions are viewed as part of the process of managing uncertainty; hence, in the absence of decision making, it is impossible to test for uncertainty. Decision making therefore constitutes a probe into uncertainty rather than a means of reducing uncertainty. Following such a probe, it may be possible to act so as to reduce the uncertainty entailed in a subsequent decision, but, again, this may not always be the case. Furthermore, sometimes the decision may serve to increase the uncertainty measured in subsequent probes.

The goal of managing spatial uncertainty is therefore not so much the reduction of uncertainty in decision making as it is the understanding of the role uncertainty plays. Some types of decision making may necessarily be associated with high degrees of uncertainty, whereas others may require lower uncertainty levels. It is important to understand the links in order to perform effective policy making; hence, the relationship between different types of uncertainty, different sources of data, and the decision-making process must also be explored. This requires an examination of both the cognitive aspects of decision making as well as the institutional, social, political and economic dimensions of decision making.

In an earlier time, when ecologists limited themselves to the study of homogeneous areas, population management decisions were simple; however, management practices based on these simple models failed to ensure sustainable populations. Ecologists realized the importance of considering the heterogeneous nature of ecosystems in their prediction of population size; hence, the notions of uncertainty (Ludwig et al. 1993) and risk assessment (Suter II 1995) have started to be taken seriously in decision making in ecology. These days, decision making also implies accommodating public concerns about the health of the environment and sustainable management (Commission 1991). These new constraints on the decision-making process imply that uncertain quantitative data must be balanced against qualitative public perception of risks while establishing policies. Here again, work in qualitative logics and spatial reasoning may play a useful role.

7.2.6 Metadata

Finally, as has been mentioned, metadata may provide a useful framework for tracking some sources of uncertainty. The reader is referred to Beard's chapter in the present volume for more details (Chap. 11).

7.3 Emerging Techniques for Handling Uncertainty due to Cognitive Factors

The analysis laid out earlier indicates that there is a role for existing representations, either as such or under moderate modifications, in handling certain aspects of spatial uncertainty due to cognitive factors. There are four broad issues that can be addressed here: the development of theory for understanding error, the use of computer-assisted methods to reduce uncertainty in image interpretation, the use of spatial data structures that may accommodate more sophisticated error characterizations, and the characterization and incorporation of context using knowledge representations and data mining techniques. None of these unfortunately has resulted in software for handling spatial uncertainty currently available on a widespread basis. These are, instead, areas where new approaches are emerging and which users such as ecologists should monitor as they develop over the next few years.

7.3.1 Perception-Based Error Models

The survey provided earlier on the links between spatial uncertainty and cognitive or perceptual factors is indicative of the extent to which these are interrelated. In one study, Edwards (1998) proposed a new theoretical framework for characterizing error for vector map coverages based on recognizing the importance of the cognitive component of spatial error. This new approach differs from standard methods used in the modeling of spatially referenced data that assume that uncertainty is linked to measurement errors of data produced by real-world processes. It is assumed instead that for data that have been produced via interpretation, transformed via generalization, or linked to other data sets via pattern-matching procedures, a perception-based error model should be used. Edwards (1998) proposed a three component perception-based error model. The three types of error—aggregation error, curve or boundary error and directional error—are only partially dependent on each other. This approach to error characterization differs in several ways from earlier attempts. Although it is widely recognized that boundary error is linked to polygon areal error, this is generally seen as a fairly strong dependence, so that one type of error could be determined from the other. Furthermore, although some authors have recognized that there are links between aggregation and boundaries, no explicit error model describing this link has been proposed. In Edwards (1998), polygon areal error is treated as a side effect of aggregation error or of boundary error. Aggregation error and boundary error are considered to be only loosely coupled; and the concept of directionality is introduced as a third, equally important component to the perception of areas and boundaries that is rarely mentioned in geographic information science.

All three error models may be simulated and represented using stochastic methods, in a manner analogous to Goodchild's categorical error model (Goodchild et al. 1992). Hence, aggregations can be modeled as chains of incremental fusions among primitive elements to form a given map of interest. By making hypotheses about what types of aggregation are likely to be favored by human perceptual processes (these hypotheses could, of course, be tested, using methods developed in cognitive psychology), it should be possible to simulate the range of partitions consistent with a given map under the assumption of some level of aggregational uncertainty. Second, directional error may likewise be modeled, using an approach based on understanding of the roles directionality and symmetry play in human visual processing. Finally, a curve or boundary model may be developed, again based on the understanding of human visual processing and using information derived from the aggregation and directional models, and boundaries may be stochastically perturbed to study their error properties.

Because these three models depend only partially on each other, their integration into a single, common error model is tricky, just as is the accommodation of spatial autocorrelation; however, this three-component perception-based error model appears to capture a great deal of the complexity of the perceptual process in a manner that is quantitatively tractable; however, the full functionality of how this model might be applied to spatial data still needs to be worked out. Error propagation in particular appears to be difficult within this framework (Edwards 1998).

The development of appropriate error models for perception-based data is still nascent. Although practical methods for characterizing boundary error based on multiple interpretations exist (Aubert 1995), methods for evaluating aggregation error and directional error, and for integrating these models, remain to be developed.

7.3.2 Computer-Assisted Image Interpretation

Although most photointerpretation is still performed by experts on paper photographs, there is growing interest in the use of computer systems to either facilitate the process, to render parts of the process more automated or more accessible to the casual user, or to combine image analysis algorithms with interactive image interpretation in new interpretation procedures; hence, current research efforts range from developing effective means of displaying image data to potential interpreters, to providing some low-level interaction tools, such as the kind available in commercial software like Photoshop^{TM}, to providing more sophisticated tools (Fortier et al. 1991), to fully automating parts of the interpretation procedure (Poirier and Edwards 1996; Barbazet and Jacot 1999). The determination of the spatial uncertainty related to human interpretation, automated interpretation, or hybrid procedures might also fall within the realm of computer-assisted image analysis methods.

Most remote sensing research since 1970 has focused on the analysis of low spatial resolution images such as that obtained by satellites. These have been characterized by the availability of several spectral bands, by the ability to combine images obtained at different times, and by their radiometric fidelity; however, human interpreters tend to rely primarily on high resolution spatial information, sometimes supplemented by color (although many photointerpreters prefer working with black and white images). Radiometry may be poor, and temporal coverage is not, in general, used in the interpretation process itself. Research aimed at studying human vision, and research aimed at developing machine vision, have converged on what appears to be a growing common understanding of how vision is achieved by biological organisms. This understanding is rooted in the recognition that multiscale filters are used, and the results grouped into primitive models from which distinctive objects are constructed (Cantoni et al. 1997).

Many groups are now returning to the study of high spatial resolution imagery within the framework of these new insights. In doing so, focus has shifted from attempts to automate full scene analysis, to attempts to isolate particular objects of interest within these images (object-based approach) on the one hand, and to identify particular systematic patterns on the other hand (pattern-based approach). The object-based approach allows developers to extract functional models of images relatively quickly, albeit at a fairly low level of sophistication; hence, individual tree crowns may be readily identified and extracted from high-resolution airborne imagery by a variety of techniques (Eldridge and Edwards 1993; Hill and Leckie 1999). Likewise, edges can be readily detected, and a human interpreter may then link selected edges into boundary zones or regions of interest (Reisfeld et al. 1995). On the other hand, an algorithm may be used to link edges in a similar way (Alquier and Montesinos 1996; Fitzback 1998).

Patterns, on the other hand, are collections of objects characterized by specific kinds of spatial relation. Patterns are found by comparing subsets of objects detected in the image with existing descriptions stored in memory. These may be determined from other images analyzed previously, or from auxiliary knowledge stored in an appropriate representation structure (see 7.3.4). Pattern detection is therefore a question of matching pattern descriptions stored in memory with elements from the scene under study. The problem of matching is a very complex problem, both for image analysis and for comparing maps.

Complementary to these feature detection and mapping algorithms, it is also possible to build and operate error estimation algorithms. Hence, for example, if tree crowns are merged into regions using a dilation operation (similar to the convolution of the target object's shape with a template shape such as a circle of fixed size), the size of the dilation may be used as an indication of uncertainty (Barbazet and Jacot 1999). Although this kind of computer-assisted image interpretation system is still emerging, it seems likely that the capability to compute simple error statistics to accompany interpretations should be available fairly quickly.

7.3.3 "Intelligent" Spatial Data Structures

In addition to image analysis, there are a growing number of spatial data structures that support more sophisticated operations on spatial data than the standard raster and vector structures in widespread use; hence, modified vector data structures that support multiple boundary representations are one example of what might be done. Such a system would allow a user to model variations in vector boundaries explicitly (Edwards 1994).

Nested data structures are being explored in a number of contexts (Dutton 1996; Yang and Gold 1996). These permit objects at different scales to be embedded in the same data structure. In some cases, multiple versions of objects at different scales may be supported. These kinds of data structures should permit better use of multiscale data and allow us better to track spatial uncertainty in such data.

Voronoï diagrams can be used as the basis for another kind of data structure which has many properties of interest, both for cognition and for modeling uncertainty. Voronoï diagrams partition space into regions determined by their proximity to an object (Okabe et al. 1992). All points nearest a given object are part of its Voronoï region. As such, Voronoï diagrams share characteristics with both object (vector) and field (raster) representations of space. The set of all Voronoï regions in a Voronoï diagram may be used to form a topology of neighbors, by linking adjacent Voronoï regions. For a set of point objects, this adjacency topology forms a Delaunay triangulation. Delaunay triangulations are characterized by some very powerful mathematical features and can be used to build a spatial data structure (Gold 1989).

Voronoï diagrams seem to offer the means to characterize qualitative spatial relations similar to those in natural language representations, and hence support the development of qualitative or fuzzy representations of space (Edwards et al. 1996). Qualitative representations of space may be viewed as having less uncertainty than quantitative representations, because often the latter must make assumptions about the distribution of spatial phenomena which are not made in qualitative representations. Hence, a Voronoï diagram representation of ecological field data (i.e., the field plot locations are used as generator points for the Voronoï diagram) does not assume the location of polygon boundaries, as would a polygonal representation; therefore, it does not require a boundary error model.

The Dempster-Shafer theory of evidence may also be used in combination with existing spatial data structures to provide a framework for developing cognitively sensitive applications. Hence maps may be represented in terms of beliefs concerning their relevance and the relevance of particular features contained in them (see Chap. 8).

7.3.4 *Knowledge Representations that Include Context*

Various kinds of knowledge representations have been developed in artificial intelligence. These include rule systems, frame representations, semantic networks, software agents, neural networks and other structures (Winston 1993). Knowledge representations may be used to store information concerning specialized contexts which affect uncertainty characterizations of spatial data. Hence, for example, rule systems may be employed to model changes in estimated uncertainty as a result of a variety of conditions. Frame representations, semantic nets, and neural nets are designed to store information concerning pattern. Knowledge representations are also increasingly being used to support data mining and knowledge extraction (Fayyad et al. 1996). Over the next few years, the relationships between data mining, knowledge extraction, and spatial uncertainty are likely to be more clearly established.

7.4 Future Approaches for Handling Cognitive Uncertainty

This final section presents research of a more fundamental, long-term nature.

In order to develop methods of representing spatial phenomena and the uncertainty associated with them that accommodate cognitive aspects of space, there is a need to develop theory that relates cognitive conceptualizations of space to geometry. This is because cognitive conceptualizations of space are known to be different than standard Euclidean geometric representations—they contain distortions, they are not fully metric, and they are characterized by incompleteness and varying kinds of uncertainty.

One approach to this problem is to exploit the topological or directional characteristics of spatial data only, not the metrical information. Indeed, topology provides a broader framework for handling spatial data than does metricity (Edwards et al. 1998); despite the fact that topology is treated as a derivative of metric information in existing GIS. Hence, for example, the connectedness of the World Wide Web is more a question of topology than metricity.

Several attempts have been made to develop a "theory" of space that is broad enough to accommodate perceptual space as well as geometrical aspects of space (Couclelis and Gale 1986; Couclelis 1998; Edwards et al. 1998). Although none of these are conclusive, they do constitute the beginnings of a theoretical approach to providing a framework for spatial concepts in a broader context. As a result, they may be used to structure representations which might combine the two perspectives.

A second area where cognitive science is impacting on geographic information science is in the area of the perception of spatial pattern. Johnson (1987) put forward the idea of image schemata as a formal model of spatial perception. These are prototypical spatial relations between phenomena (e.g., the relation "container/contained"). Lakoff and Johnson (1980) showed how these could be used to characterize spatial metaphors applied to abstract concepts (e.g., "fall in love" as if love were a container). Image schemata have fallen into disfavor among psychologists as an explanatory mechanism; however, as a formal tool for encapsulating spatial relations their use is on the increase (Couclelis and Gottsegen 1997; Raubal et al. 1997). It is still early, but there are indications that image schemata may provide a useful basis for specifying spatial pattern on a nonquantified basis, and hence may provide a means of controlling uncertainty in measures of landscape pattern such as are used in ecology.

A third development that holds promise for encapsulating some of the relationships between spatial uncertainty and cognition is the theoretical approach developed by Edwards (1996, 1997) called *Geocognostics*. This essentially constitutes a method of doing cognitive modeling using symbolic and/or neural network representations organized into an architecture. Geocognostics encapsulates the process of taking local sets of perceptions and nesting them into more and more sophisticated representations of an environment. These representations constitute a simulation of spatial memory. The method exploits the notion of trajectories, or sequences of perceptions, which are then grouped or chained in a variety of representations which are organized into a tree of abstractions. Using this type of cognitive model, it is possible to make decisions based on this nesting of perceptions into sets of abstractions, and to study the uncertainty structure around such decisions. Furthermore, the memory structure which results from this model appears to support both linguistic processing and computer vision. This appears to provide a framework for cognitively adequate descriptions of spatial phenomena, one of the requirements of an appropriate theory which was laid out earlier (Section 1.1.2). Although promising, the bottom-up nature of this modeling procedure means that concrete applications of the method are probably several years away.

An area of considerable interest is what has been called "categorization theory" (Lakoff 1987). The process of categorizing is fundamentally cognitive. There have been many studies that attempt to identify the categorization method that underlies, say, linguistic categories. Fuzzy set theory is an early example of a nonboolean categorization method; unfortunately, no one theory has yet emerged from category theory. Although a variety of methods (i.e., prototype theory and other nonlinear classification schemes) have been proposed, no clear computational model has been developed. One of the more interesting lines of inquiry views attractors in a nonlinear dynamical space as category "prototypes" (Lakoff 1987).

Finally, the area of qualitative spatial reasoning has been drawning a great deal of attention (Frank 1992; Freksa 1992; Ligozat 1993; Cohn et al. 1997; Kettani and Moulin 1998). Qualitative reasoning is based on the premise that humans avoid uncertainty by structuring knowledge qualitatively, and reasoning about knowledge similarly. In this approach, uncertainty is "absorbed" rather than reduced, eliminated, or modeled (similar to the use of the Voronoï diagram described earlier). It is implicit to a qualitative representation. Qualitative reasoning, over the long term, should permit the development of analysis or interpretation methods that allow conclusions to be drawn without making explicit assumptions about the nature of uncertainty. This is precisely what is required for many kinds of decision making.

7.5 Summary

In summary, cognition plays a very important role in the creation, interpretation, and transformation of spatial data, and hence in the production of spatial uncertainty. It is not unreasonable to assert that the primary source of uncertainty in most spatial data is introduced as a result of human cognitive and perceptual processes. Although a great deal can be done to characterize error without dealing with the cognitive aspects of spatial data, these methods largely pertain to certain forms of data (e.g., field samples, remotely sensed data, and digital elevation models). Interpreted data still dominate many fields, including ecology, and are likely to play an important role for many years to come.

Because the recognition of the links between spatial data and cognition and the development of appropriate tools are still relatively recent phenomena, there are not yet a great many practical tools that address this problem. Theory is beginning to emerge from a variety of areas of inquiry, however, and new tools will likely see the light of day over the coming few years. This research is the focus of scientists from a broad range of disciplines, including geographic information science, computer science, mathematics, and cognitive science. In this chapter, several existing trends in methods and tools were discussed, many of which may lead to concrete and practical solutions over the next few years. In addition, longer-term research is also surveyed, suggesting areas where new results may likely emerge that are of interest to ecologists studying spatially distributed data.

In addition to these more practical issues, there are important philosophical links to be made between ecology, space, and cognition. This was alluded to in the opening paragraphs of this chapter. At this deeper level, there appear to be many similarities between the kinds of processes that obtain and are studied in the field of ecology, and those that operate in the sphere of human cognition. This was highlighted several years ago in a groundbreaking treatise that has still to be fully understood (Bateson 1972). Both cognition and ecology are characterized by dynamic, organic phenomena organized into complex, nested systems. Both appear to use spatial segregation of processes to function effectively. Both appear to be characterized by nonlinear (chaotic) dynamics. As Bateson indicated, both appear to be characterized by an evolutionary process leading to increasing abstraction.

It should not be surprising, therefore, that cognition plays a very important role in the study of ecology, and that spatial phenomena should be linked to both cognitive and ecological factors. It is more surprising, perhaps, that it has taken so long to address these issues on a more quantitative basis.

References

Alquier, L., and P. Montesinos. 1996. Perceptual grouping and active contour functions for the extraction of road in satellite pictures. Proceedings of the SPIE 2955:153–63.

Aubert, E. 1995. Quantification de l'incertitude spatiale en photo-interprétation forestière à l'aide d'un SIG pour le suivi spatio-temporel des peuplements. Mémoire de maîtrise en sciences géomatiques, Faculté de Foresterie et de Géomatique, Université Laval, Québec.

Barbazet, V., and J. Jacot. 1999. The Clapa project: automated classification of forest with aerial photographs. In Proceedings of an international forum on automated interpretation of high spatial resolution digital imagery for forestry, Hill, D.A., and D.G. Leckie, eds. Natural Resources Canada, Canadian Forest Service, Victoria, B.C.

Bateson, G. 1972. Steps to an ecology of mind. Chandler Press, San Francisco.

Bell, D.E., H. Raiffa, and A. Tversky, eds. 1988. *Decision making: descriptive, normative, and prescriptive interactions.* Cambridge University Press, Cambridge.

Cantoni, V., S. Levialdi, and V. Roberto, eds. 1997. Artificial vision: image description, recognition and communication. Academic Press, New York.

Cohn, A.G., B. Bennett, J. Gooday, and N. Gotts. 1997. Qualitative spatial representation and reasoning with the region connection calculus. GeoInformatica 1(3): 1–42.

Commission sur la protection des forêts. 1991. Des forêts en santé: rapport d'enquête et d'audience publique sur la stratégie de protection des forêts, Gouvernement du Québec.

Couclelis, H., and N. Gale. 1986. Space and spaces. *Geografiska Annaler*, 68B:1–12.

Couclelis, H., and J. Gottsegen. 1997. What maps mean to people: denotation, connotation, and geographic visualization in land-use debates. In Proceedings of the international conference on spatial information theory (COSIT '97), S.C. Hirtle and A.U. Frank, eds. Lecture notes in computer science 1329:151–62.

Couclelis, H., 1998. Worlds of information: the geographic metaphor in the visualization of complex information. Cartography and Geographic Information Systems 25, no. 4:209–20.

Cressie, N.A.C. 1991. Statistics for spatial data. John Wiley & Sons, New York.

Dutton, G. 1996. Encoding and handling geospatial data with hierarchical triangular meshes. Pages 8B15–23 in Proceedings of the seventh international symposium on spatial data handling, Delft, The Netherlands.

Edwards, G. 1994. Modelling of fuzzy data: aggregation and disaggregation of fuzzy polygons for spatial-temporal modeling. Pages 141–54 in Proceedings of the advanced geographic data modeling workshop (AGDM'94): spatial data modelling and query languages for 2D and 3D applications. Delft, Netherlands, September 12–16.

Edwards, G., and B. Moulin. 1995. Towards the simulation of spatial mental images using the Voronoi model. Pages 63–73 in International joint conference on artificial intelligence (IJCAI-95), workshop on representation and processing of spatial expressions. Montreal, August 26–28.

Edwards, G. 1996. Geocognostics—A new paradigm for spatial information? Pages 6–14 in AAAI-96 spring symposium series: cognitive and computational models of spatial representation, Stanford, CA, March 25–27.

Edwards, G., G. Ligozat, A. Gryl, L. Fraczak, B. Moulin, and C.M. Gold. 1996. A Voronoï-based pivot representation of spatial concepts and its application to route descriptions expressed in natural language. Pages 7B1–16 in Proceedings of the seventh international symposium on spatial data handling. Delft, The Netherlands.

Edwards, G., and K.E. Lowell. 1996. Modelling uncertainty in photointerpreted boundaries. Photogrammetric Engineering and Remote Sensing 62(4):377–91.

Edwards, G. 1997. Geocognostics—a new framework for spatial information theory. Lecture notes in computer science 1329:455–72.

Edwards, G. 1998. Towards a theory of vector error characterisation and propagation. Pages 183–188 in Proceedings of the third international symposium on spatial data accuracy, Laval University, Québec, May.

Edwards, G., T. DeGroeve, A. Gryl, J. Kritter, M. Mostafavi, F. Anton, et al. 1998. Extending GIS—integrating multiple spaces into a single concept. Pages 123–37 in Proceedings of the eighth international symposium on spatial data handling. Vancouver, B.C.

Eldridge, N.R., and G. Edwards. 1993. Acquiring localized forest inventory information: extraction from high resolution airborne digital images. Pages 443–48 in sixteenth Canadian symposium of Remote Sensing—8e congrès de l'association Québécoise de Télédétection. Sherbrooke, June 7–10.

Fayyad, U.M., G. Piatetsky-Shapiro, P. Smyth, and R. Uthurusamy, eds. 1996. Advances in knowledge discovery and data mining. AAAI Press, Menlo Park, CA.

Fitzback, J. 1998. Détection des cours d'eau par images radar selon les stratégies de la vision humaine. M.Sc. Thesis, Université Laval.

Fortier, J.-J., A. Dupras, and B. Lachapelle. 1991. Super image: la classification des pixels et l'estimation des proportions de chaque classe avec applications à lamesure de la surface des couverts forestiers. Télédétection et gestion des ressources, volume VII, Paul Gagnon, P., ed.

Fortin, M.-J., P. Drapeau, and P. Legendre. 1989. Spatial autocorrelation and sampling design in plant ecology. Vegetatio, 83:209–22.

Frank, A.U., and D.M. Mark. 1988. Workshop on cognitive and linguistic aspects of geographical space. Technical report of the NCGIA, Buffalo, NY.

Frank, A.U. 1992. Qualitative spatial reasoning with cardinal directions. Journal of visual languages and computing 3:343–71.

Frank, A.U., and I. Campari, eds. 1993. Proceedings of the European conference on spatial information theory (COSIT '93). Elba Island, Italy. In Volume 716 of Lecture notes in computer science.

Frank, A.U., and W. Kuhn, eds. 1995. Proceedings of the international conference on spatial information theory (COSIT '95). Semmering, Austria. In Volume 988 of Lecture notes in computer science.

Freksa, C. 1992. Using orientation information for qualitative spatial reasoning. In Frank, A.U., I. Campari and U. Formentini, eds. Theories and methods of spatio-temporal reasoning in geographic space. International conference on GIS—from space to territory, Pisa. Pages 162–78 in volume 639 of Lecture notes in computer science.

Gold, C.M. 1989. Chapter 3—surface interpolation, spatial adjacency and GIS. Pages 21–35 in Three dimensional applications in geographic information systems. Raper, J., ed. Taylor & Francis, London.

Goodchild, M.F., G. Sun and S. Yang. 1992. Development and test of an error model for categorical data. International journal of geographic information systems 6(2):87–104.

Green, R.H. 1979. Sampling design and statistical methods for environmental biologists. John Wiley & Sons, New York.

Hill, D.A., and D.G. Leckie, eds. 1999. Proceedings of an international forum on automated interpretation of high spatial resolution digital imagery for forestry, Feb. 10–12, 1998. Natural Resources Canada, Canandian Forestry Service, Victoria, B.C.

Hirtle, S.C., and P.B. Heidorn. 1993. The structure of cognitive maps: representations and processes. Pages 170–92 in Gärling, T., and R.G. Golledge, eds. Behavior and environment: psychological and geographical approaches. North-Holland, Amsterdam.

Hirtle, S.C., and A.U. Frank, eds. 1997. Proceedings of the international conference on spatial information theory (COSIT '97). Lecture notes in computer science, volume 1329.

Johnson, M. 1987. The body in the mind: the bodily basis of meaning, imagination, and reasoning. The University of Chicago Press, Chicago.

Kettani, D., and B. Moulin. 1998. Modèle computationnel pour la simulation du raisonnement spatial humain. Informations in cognito 2:1–15.

Lakoff, G., and M. Johnson. 1980. Metaphors we live by. University of Chicago Press, Chicago.

Lakoff, G. 1987. Women, fire and dangerous things: what categories reveal about the mind. University of Chicago Press, Chicago.

Leckie, D. 1990. Advances in remote sensing technologies for forest surveys and management. Canadian journal of forest research 20:464–83.

Li, H., and J.F. Reynolds. 1995. On definition and quantification of heterogeneity. Oikos 73:280–84.

Ligozat, G. 1993. Models for qualitative spatial reasoning. Pages 35–45 in Proceedings of the workshop on spatial and temporal reasoning. Thirteenth international joint conference on artificial intelligence IJCAI '93, F. Anger, H.W. Güsgen, and Benthem V., eds. Chambéry, France.

Lowell, K.E., 1997. An empirical evaluation of spatially based forest inventory samples. Canadian journal of forest research 27:352–60.

Ludwig, D., Hilborn, R., and C. Walters. 1993. Uncertainty, resource exploitation, and conservation: lessons from history. Science 260(17):547–49.

McGarigal, K., and B.J. Marks. 1995. FRAGSTATS: spatial pattern analysis program for quantifying landscape structure. U.S. Forest Service General Technical Report PNW 351.

Okabe A., B. Boots, and K. Sugihara. 1992. Spatial tessellations—concepts and applications of Voronï diagrams. John Wiley & Sons, Chichester.

Oliver, M.A., and R. Webster. 1986. Combining nested and linear sampling for determining the scale and form of spatial variation of regionalized variables. Geografiska Annaler 18:227–42.

Poirier, S.P.E., and G. Edwards. 1996. Regroupement contextuel d'arbres individ-uels, basé sur la charactéristique des cimes dans les images aéroportées. Actes du 9e congrès de la'ssociation Québécoise de télédétection. Québec, 30 avril–3 mai, CD-ROM.

Raper, J. 1996. Unsolved problems of spatial representation. In Proceedings of the seventh international symposium on spatial data handling. August 12–16, Delft, The Netherlands, 2:14.1–11.

Raubal, M., M.J. Egenhofer, D. Pfoser and N. Tryfona. 1997. Structuring space with image schemata: wayfinding in airports as a case study. Proceedings of the international conference on spatial information theory (COSIT '97). Lecture notes in computer science, 1329:85–102.

Reisfeld, D., H. Wolfson, and Y. Yeshurun. 1995. Context free attentional operators: the generalized symmetry transform. International Journal of Computer Vision 14:119–30.

Riitters, K.H., R.V. O'Neill, C.T. Hunsaker, J.D. Wickham, D.H. Yankee, S.P. Timmins, et al. 1995. A factor analysis of landscape pattern and structure metrics. Landscape Ecology 10:23–29.

Samet, H. 1984. The quadtree and related hierarchical data structures. ACM Computer Surveys 16(2):187–260.

Suter II, G.W. 1995. Adapting ecological risk assessment for ecosystem evaluation. Ecological Economics 14:137–41.

Taylor, H.A., and B. Tversky. 1992. Spatial mental models derived from survey and route descriptions. Journal of Memory and Language 31:261–82.

Thompson, S.K. 1992. Sampling. Wiley Interscience, New York.

Vauglin, F. 1997. Modéles statistiques des imprécisions géométriques des objects géographiques linéaires, Ph.D. Thesis, Université de Marne-la-Vallée.

Wiens, J.A. 1989. Spatial scaling in ecology. Functional Ecology 3:385–97.

Wiens, J.A. 1992. What is landscape ecology, really? Landscape Ecology 7:149–50.

Winston, P.H. 1993. Artificial intelligence, third edition. Addison Wesley, New York.

Yang, W., and C.M. Gold. 1996. Managing spatial objects with the VMO-tree. Pages 11B15–30 in Proceedings of the seventh international symposium on spatial data handling. Delft, The Netherlands.

8
Delineation and Analysis of Vegetation Boundaries

MARIE-JOSÉE FORTIN and GEOFFREY EDWARDS

Vegetation boundaries such as riparian zones or forest stand edges are inherent features of landscapes that play more than one functional role in terrestrial ecosystems dynamics (Hansen and Di Castri 1992). The delineation of ecotones (i.e., boundary areas), either sharp or gradual, between adjacent vegetation communities is important in several aspects of land use management planning (e.g., the delineation of forest stands for estimating wood volume available for harvesting). The location and accuracy of the delimited vegetation boundaries, however, depend on the spatial and temporal resolution of the data available as well as the subsequent statistical methods used to detect them. In the context of forest stand delineation, various data types (e.g., field data, aerial photographs and remotely sensed images) are available. The type of data determines which edge detection methods (e.g., statistical methods, photointerpretation, or filter kernels) should be used. Accuracy of the delineated vegetation boundaries therefore depends both on the accuracy of the data and the sensitivity of the edge detection methods.

In fact, there are several steps where differences in data and methods can cause the location of a delineated boundary to vary: choice of species (Fortin 1997); choice of data type (e.g., field surveys, aerial photography, remotely sensed imagery); choice of sampling design (Fortin 1994); choice of edge detection methods (Fortin and Drapeau 1995). This chapter will highlight the methodological issues that need to be addressed in the detection of ecotones such that their management and monitoring can yield useful information about their dynamics and their functional role within ecosystems. We start by presenting how the inherent properties of spatial data (e.g., spatial dependency and autocorrelation, grain, or extent) affect data accuracy. We will then discuss the different accuracies associated with each data medium (e.g., field data, aerial photography, and remotely sensed images). Finally, we will illustrate the sensitivity of some boundary delineation methods to accurately delineate boundaries.

8.1 Inherent Properties of Spatial Data

Ecological phenomena are dynamic both in space and time; hence, ecological data are measured at a given location (x–y coordinates) and time (t). This dynamic property of ecological processes usually results in data that show more or less pronounced spatial patterns and temporal trends. This implies that nearby values of a variable are spatially more likely to be similar than would be expected by chance. Such data can be referred to as "spatial data," which implies that they exhibit topological or neighborhood relations. This spatial dependency (i.e., spatial autocorrelation) is therefore seen to be intrinsic to the definition of spatial data (Chaps. 9 and 10).

These spatial relations are one-dimensional, like temporal relations, as well as two-dimensional (areal) or even three-dimensional (volume filling). Furthermore, when a spatial pattern shows the same intensity regardless of the orientation, it is called *isotropic*; when the intensity varies according to the orientation, it is called *anisotropic*. By having x–y coordinates, spatial data can be mapped and rearranged (e.g., grouped, clustered, aggregated). Hence, spatial data measures at one scale can be aggregated for comparison with other data at a different scale.

Spatial ecological data are therefore characterized by the fact that more than one ecological process, and this at different spatial scales, may influence these data. Spatial information contained in the data, therefore, is a composite of several spatial scales: trends at macroscales; patches, gradients and patterns at meso- and local scales; random patterns at local and microscales. Spatial data are therefore a composite of this nested spatial structure. These different processes and patterns at different scales are not necessarily all linear and additive, which contributes to the intrinsic degree of uncertainty contained in spatial data.

8.1.1 Spatial Resolution: Grain and Extent

By studying nature using arbitrary data we are imposing arbitrary scales on it. This often results in distorting our representation of nature, what we are looking for, and what we are trying to understand. There are different ways in which we impose scale on nature. The two most obvious are the size of the sampling unit (known as the *grain*) used to gather the data and the delimitation of the size of the study area (known as the *extent*). The grain determines the minimum spatial resolution of the data, whereas the extent determines the scale of observation [i.e., the spatial region implicitly characterized by the entire collection of data (local, landscape, region, etc.)]. Information sampled at one scale can be aggregated to provide information at a coarser scale. The reverse, however, is not true: One cannot get local or detailed information from regional information. The combination of grain and extent will define the scale at which ecological interpretations of the data can be made. Researchers working on the same species often do not find the same spatial pattern simply because the data they use are not at the same spatial resolution.

The problem of selecting an appropriate grain size (quadrat or pixel) is not new in ecology (Greig-Smith 1964; Palmer and Dixon 1990), and is a source of measurement error and lack of accuracy. Grain size should be big enough to contain at least one specimen sampled (e.g., big enough to contain at least one tree crown). When the grain is too small, the field sampling effort to detect spatial pattern requires too much effort; however, when the grain is too big, it incorporates so much environmental heterogeneity that the spatial pattern can be smoothed out. This may lead to misclassification of pixels in remotely sensed images [e.g., (Dixon and Palmer 1990; Fortin 1992)]. The size of the study site (the extent) is also important. A study area that is too large may incorporate several spatial scales (e.g., small local patches and a regional trend), resulting in complex spatial structure that is harder to detect and characterize. In the context of ecotone detection, the accuracy (grain and extent) of the data will affect the location of the delimited boundary, as will be shown in the methods section.

8.1.2 Data Sampling

Reliable data analysis of spatially distributed data requires both the use of appropriate statistical tools, as well as a sound data sampling strategy. Generic data sampling involves making several decisions, such as the number of samples to take, their size, their shape, their layout in space, and their frequency in time. Shape and size refer to the notion of grain introduced earlier, whereas the layout in space refers indirectly to the extent. The layout in space characterizes the sampling design used (i.e., the spatial arrangement of the samples: random, systematic, stratified, nested, adaptive, etc.) (Thompson 1992). In general, the number of samples, and how frequently they should be sampled, are often limited by cost constraints.

Once all of these basic decisions have been considered, there are other decisions that need to be made when spatial data are studied. Indeed, because of the dependencies present in spatial data, how data are sampled is as critical as what is sampled. Data acquired in a systematic way that does not take into account the nature of spatial dependencies can lead to erroneous results in the analysis phase, regardless of the sophistication of the analytical tools. It is important, therefore, to define an appropriate sampling spatial lag (step) first. When a spatial pattern is to be quantified, the spacing between samples must be smaller than the size of the spatial pattern; however, when parametric inferences are to be performed, sample spacing should exceed the size of the spatial pattern. Data ideally should be collected at several scales using a spatially nested design (Fortin et al. 1989; Webster and Oliver 1990) to detect spatial patterns at more than one spatial scale. The data should cover the entire extent, and be obtained at different orientations allowing for the testing of anisotropy. Depending on the sampling design used, the spatial pattern identified can be more or less distorted (Fortin et al. 1989). The accuracy of the spatial pattern identified increases as the number of samples increases. This unfortunately also implies an increase in the cost of field work. In order to reduce the cost of field work, data sampling strategies usually involve several different data-collection methods, including a blend of field data, aerial photography, and satellite imagery.

8.1.3 Field Data

Field sampling is the most costly part of the sampling strategy, but it results in the richest data set and is also necessary to validate the measurements obtained from both the aerial photography and the remotely sensed imagery. The types of data gathered in the field are either quantitative (e.g., species abundance per quadrat, percent cover, continuous, discrete) or qualitative (e.g., presence–absence). Even though the latter is faster to record in the field, it provides less information about the spatial structure than the former, which unfortunately requires more sampling effort. Hence, the major problem with field data is that one is always making compromises between the data type, the grain size, the spacing between sampling units, the number of sampling units, and the size of the study site. All of these factors affect the degree of uncertainty in the data and the accuracy of any estimation of spatial structure in the data (Fortin et al. 1989; Fortin 1999).

8.1.4 Aerial Photographs

Aerial photography is widely used as a source of data in applications concerned with characterizing vegetation, topography, or urban regions. Many of the variables for which field samples are acquired may also be studied using aerial photography, although usually in considerably less detail. Species may therefore be mapped, but the specific mix of species might be obtainable only through field surveys. On the other hand, over many decades of practice, methods have been developed for characterizing complex mixtures of vegetation via contextual image interpretation. Percent cover, height, density, and, to some extent, biomass and diversity indicators may also be derived from aerial photographs via interpretation. Abundance, concentration, and moisture, however, as well as other specific variables, cannot be so readily obtained.

Aerial photographs are usually obtained in stereo pairs at map scales appropriate to the information sought. These photographs are then examined via an optical system (called a *stereo plotter*) designed to present the stereo image to the user and permit the latter to fix (for points) or trace (for lines) the locations of features of interest, in both the x–y plane and in altitude (z). These devices usually incorporate the ability to determine a stereo model and hence to obtain terrain coordinates from the image locations. These have traditionally been optical–mechanical devices, but since 1990 there has been an emergence of a wide variety of digital computer-based stereo plotters. Less sophisticated devices are sometimes used, which allows stereo viewing, but no control over the z-axis.

8.1.5 Remotely Sensed Data

Aerial photography, although still the dominant source of ecological information at scales beyond that of field sampling, relies on the work of human photointerpreters to extract this information; hence, it is subject to the problems of consistency and error that characterize human interpretation. Although research has increasingly focused on automated information extraction methods in high-resolution imagery (e.g., the automated identification of individual tree crowns) (Lowell et al. 1997), operational techniques are not yet available. As a result, there has been widespread interest in the use of satellite remote sensing for obtaining coverage over large regions. Furthermore, remotely sensed data may cover scales from hundreds of square kilometers up to an entire continent—in other words, the scales at which most ecological studies are traditionally focused.

Satellite-based remotely sensed data are usually transformed to map data via classification procedures. Although both supervised and unsupervised methods are used, experience indicates that supervised methods generally yield higher accuracy than unsupervised methods; hence, the former are generally preferred over the latter. Supervised classification requires the presence of a human expert both to choose data sets that are used to train the classifier, decision criteria for selecting among competing criteria, and appropriate class aggregation levels. The presence of a human expert, although yielding lower uncertainties, may affect the consistency of the classification results (see Chap. 7).

Hence, both aerial photographs and remote-sensing data sources may be used to develop information on landscape pattern; however, the spatial resolution of an imaging system will set fundamental limits on the observed spatial variability. The strength of relationships between species will vary in time and space, so it may be difficult to determine an optimal spatial precision for identifying units under study. If we consider a situation of a continuous transition from grassland through wooded savanna to forest, then one approach might be to use a low-resolution sensor that aggregates a large enough area that individual trees are integrated into a coarse, but reasonably accurate, representation of percent cover. This representation may then be thresholded, and the resulting map used as a basis for determining pattern metrics of the region under study. An alternate approach might use a high-resolution sensor that is sensitive to individual tree canopies, and then to rely on a neighborhood analysis to develop information on spatial pattern. The use of spatial statistics falls within this latter approach (Cressie 1991).

8.2 Detection of Ecotones

Ecologists have historically studied homogeneous regions to characterize and understand ecosystem processes and have avoided the heterogeneous areas between ecosystems. As a result, gradual ecotones were often reduced to lines on a map; however, these transitional zones, or ecotones, are dynamic and play several functional roles in ecosystem dynamics [e.g., controlling the flux of materials between ecosystems and influencing biodiversity (Naiman and Décamps 1990)]. In fact, because species may be at the limits of their tolerance when encountering these transitional zones, characteristics of ecotones may be especially sensitive to environmental change. As the number of ecotones and edges increases due to the human use of land, better understanding of the dynamics of these zones is needed (Holland et al. 1991; Hansen and di Castri 1992). There is also a greater need for objective tools to detect them (Johnston et al. 1992; Fortin 1994; Fortin and Drapeau 1995); therefore, to study and understand the functional roles of ecotones and their dynamics within ecosystems better, we need quantitative methods to identify their location and to characterize them (Gosz 1993).

Although the factors and processes involved in edge and ecotone formation may be quite different, their detection is related to the identification of the spatially homogeneous adjacent areas that they separate. The delineation of edges and ecotones can therefore be carried out either by forming spatially homogeneous clusters or by detecting boundaries. To create spatially homogeneous clusters, spatial contiguity constraints may be added to clustering algorithms (Legendre and Fortin 1989). This process is similar to the process called *segmentation* in image analysis (Edwards 1995); hence, boundaries are formed as a by-product between sampled sites belonging to different spatial clusters. Clustering with spatial constraints is appropriate when boundaries are sharp. When boundaries are more gradual and do not necessarily completely divide an area in two, however, edge detection methods are more appropriate. Indeed, with edge detection algorithms, boundaries are identified as spatially contiguous locations characterized by high rates of change. They allow us to gain information about both the location of boundaries as well as their width, shape, and intensity (Fortin 1994; Fortin and Drapeau 1995). In either case, the accuracy of the delineated boundaries is strongly dependent on the selected scale of analysis, the spatial resolution of the data, and the type of data source used (Fortin 1992; Fortin 1999). In general, the edge-detection methods are more flexible in that they allow the identification of transition zones, and not just boundary lines like the spatial clustering methods.

8.2.1 Edge Detection Methods for Field Data

Edge detection algorithms were first developed for remotely sensed imagery, and thus for raster data. Given that field data are usually not sampled in a continuous and regular pixel-like fashion as they are in remotely sensed images, but that they come instead in a discontinuous way, these algorithms cannot be used directly with field data. Another problem encountered with field data is that the data are either qualitative (e.g., presence–absence), semi-quantitative (e.g., age class, percent coverage), or quantitative (e.g., density, biomass). In consequence, edge detection methods need either to be modified or new ones developed to be able to deal with such irregular and nonadditive data.

With quantitative regularly spaced data, an edge detection algorithm called *lattice-wombling* can be used (Fortin 1994; Fortin and Drapeau 1995). This algorithm requires that the values of the variable be mapped on the nodes of a rectangular lattice, as is the case for remotely sensed data. The rate of change is computed for the first-order partial derivatives of each of four quadrats forming a square (see Fortin 1994 for the mathematical details). When the values in the four quadrats are similar, the magnitude of the rate of change will be close to zero; when the values at the four quadrats change abruptly, the magnitude of the rate of change is high. A boundary is identified from the spatially adjacent locations that are characterized by high values of the rate of change. When several variables are available for study, the mean rate of change is defined as the average of the rate of change for the given assemblage of variables.

When field data are quantitative but irregularly spaced, a triangulation-wombling edge detection algorithm is more appropriate (Fortin 1994). Indeed, as mentioned earlier, field data are usually sampled using either random, stratified, or systematic sampling designs that require less sampling effort than a complete survey of an area. To bypass the regularly spaced data requirement, Fortin (1994) modified the lattice-wombling algorithm in such way as to deal directly with irregularly spaced samples. The algorithm also finds first-order partial derivatives, but it uses the three nearest points that form a triangle rather than four nearby points that form a square. The Delaunay algorithm can be used to find the list of nearby samples that form triangles (Upton and Fingleton 1985).

Ecological data include quantitative variables as well as semi-quantitative and qualitative ones (e.g., presence–absence of species, type of soil, geomorphologic formation). With such qualitative data, boundaries can be established by computing a dissimilarity, or distance value, using the data of several variables at once, and looking for the highest dissimilarity between adjacent sampled points (Oden et al. 1993; Fortin and Drapeau 1995). With this approach, qualitative data (or semi-quantitative or quantitative data) are transformed into quantitative coefficients such as dissimilarity measures, or a simple mismatch measure (Oden et al. 1993). To detect boundaries, only the dissimilarities between pairs of spatially adjacent samples are considered. Adjacent samples are defined as those directly connected by the links of a Delaunay network (Upton and Fingleton 1985), although other network structures can be used. An overall high-dissimilarity value indicates the presence and location of a boundary.

We illustrate the sensitivity of lattice-wombling to detect ecotones as a function of grain resolution and data type for woody species sampled in a northern hardwood forest at the David Weld Sanctuary (Nissequogue, New York). The study site (60 m × 140 m) was surveyed along the slope of a hill where hardwood (e.g., red maple, alder, spice bush and viburnum) dominated the bottom and the slope of the hill, whereas red cedar dominated the ridge and the top of the hill. All woody species with diameter at breast height more than 5 cm were recorded using quadrats of 5 × 5 m. To study the effect of grain resolution, the data from the 5 × 5 m quadrats were combined into the following larger size quadrats: 10 × 10 and 20× 20 m. The highest 10% rate of change are arbitrarily considered to be boundary elements (BEs) (Fortin 1994); hence, the number of potential boundary elements decreases as the quadrat sizes increase: 5 × 5 m ⇒ 30, 10 × 10 m ⇒ 6, and 20 × 20 m ⇒ 2. Boundaries were detected using abundance and presence–absence data (Fig. 8.1).

Boundary locations based on the abundance data were quite consistent across quadrat sizes: All grain resolutions highlight the boundary at the foot of the hill, and some boundaries were found at the ridge of the hill, depending on the quadrat size (Fig. 8.1). The exact location of the boundaries delineated, however, varies slightly according to quadrat size.

The boundary locations based on the presence–absence data are less consistent across quadrat sizes (Fig. 8.1). The three quadrat sizes find the bottom of the hill boundary (5 × 5 m, 10 × 10 m), but only the ridge boundary was found with the 20 × 20 m quadrat size. The differences among the locations of boundaries between the abundance and the presence–absence data are more the result of the nature of the data than of the edge detection algorithm itself. Indeed, abundance can change without any species composition change (Fortin 1997); therefore, it is interesting to compare boundaries based on these different measurements in order to understand ecotone dynamics better. Such comparisons can be performed using overlap statistics that compute the minimum distance between boundaries based on different data (Fortin et al. 1996).

8.2.2 Methods for Aerial Photographs

Experienced photointerpreters will examine the photos under stereo magnification and draw in a network of lines considered to partition the image into meaningful units. In practice, most production photointerpretation is carried out under additional constraints (e.g., the existence of a minimum mapping unit, the use of interpretation keys, and criteria for ensuring errors

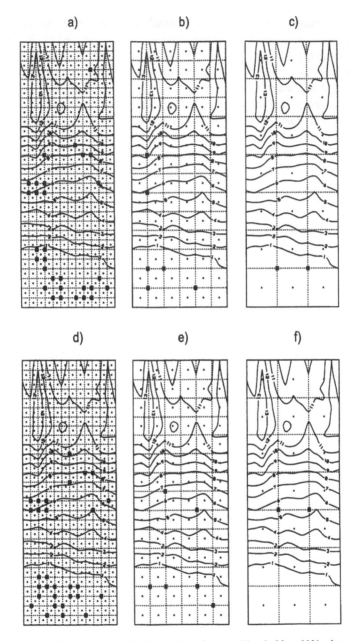

FIGURE 8.1. The lattice-wombled boundary elements (threshold at 10%; shown as filled squares) of tree and shrub abundance data: (a) 5 × 5 m, (b) 10 × 10 m, and (c) 20 × 20 m; and the presence/absence data: (d) 5 × 5 m, (e) 10 × 10 m, and (f) 20 × 20 m. The contour lines represent the relative elevation in meters.

stay within certain bounds). A minimum mapping unit specifies that no polygon should have an area smaller than a given threshold value. A key specifies which textures correspond to which ground conditions—experienced interpreters rely less on keys, but they are a useful training tool. A coarse error tolerance allows the interpreter to smooth a boundary or otherwise modify the boundary to decrease the time taken to perform the interpretation. All these considerations will affect data quality.

Most photointerpreters will claim high accuracy for the resulting maps; however, there are several kinds of accuracy that could be measured, and the interpreter refers to only a subset of these measures. For example, accuracy may refer to the ability of the interpreter repeatedly to assign a polygon to a given class. In general, photointerpretation accuracy of this type is indeed relatively high and errors will be confined to confusion between texturally similar classes. A second type of accuracy concerns the number of occurrences of extreme boundary deviations (i.e., deviations beyond a fixed threshold as illustrated in Fig. 8.2, see detailed explanation later). This type of accuracy measure is similar to the criteria used to specify the desired accuracy of an interpretation, and hence interpretation results are generally characterized by good accuracy in this regard. With the arrival of GIS, however, different maps must be compared with each other via precise overlays (Burrough 1986; Anselin and Getis 1992). As a result, the location of the boundary lines themselves has become important. For many maps, the fraction of the map that is on or near a boundary line may be a significant part of the entire map. For forest maps, fractions of the order of 50% typically may be obtained even for relatively modest boundary transition zones. Studies on the reliability of forest cover types determined by photointerpretation at random sample locations indicate that accuracy is in general no better than 60% and maybe as low as 40% (Biging et al. 1992, Edwards and Lowell 1996). This appears to be at least partially a result of the high surface area occupied by boundary regions.

It is useful to attempt to identify what variation is present in the location of interpreted boundaries, as well as to study the variation of other parameters of interest near these boundaries. One method for doing this consists of superimposing several photointerpretations of the same region, preferably using the same photographs, but different interpreters. In a study of artificially generated forestlike textures, a method for characterizing the median interpretation and the spatial variation of a set of interpretations around the median was explored (Aubert 1995). The method consists of overlaying the set of interpretations using an adjustable corridor width, determining the median interpretation location from the medial axis transform of the corridors, and determining the uncertainty width from sampling the distribution of interpretations orthogonal to the median interpretation (see Fig. 8.2). A proportion of 39% of the map was found within two standard deviations of the boundaries, for the artificial images studied. Preliminary studies with the interpretation of real aerial photographs indicate these results are likely

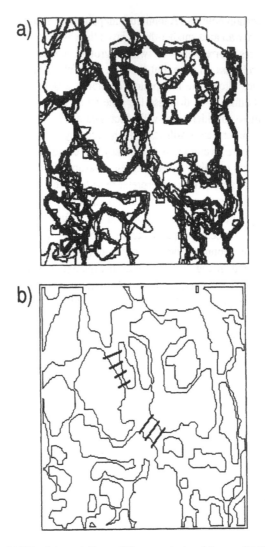

FIGURE 8.2. (a) Nine interpretations of the same textured image. (b) Corridors determined for the nine interpretations and some examples of locations of orthogonal samples of boundary width.

to underestimate the uncertainty for real vegetation maps, where texture contrasts are a lot more subtle than for this contrived example, and vegetation zones are more heterogeneous. Thus, the spatial variation of ecotone boundaries determined by photointerpretation is extremely high.

It is tempting to equate the width of the uncertain zone arising from the study of several interpretations with the width of the ecotone transition zone; unfortunately, the interpretations are affected by a number of factors that do not relate to the local properties of the ecotone. In the study on artificial images cited earlier, an interpretation model based on local properties was developed. This model was able to explain less than 50% of the variation in width of the boundary regions in the set of interpretations. The remainder of the variation appears to be due to complex contextual factors (e.g., the presence or absence of similar features or textures nearby, the spatial configuration or pattern of local textures, and previous experience of the interpreters). Because ecotones are themselves a perceived boundary, some of these contextual factors may be pertinent for ecotone definition as well; hence, considerably more work must be done on untangling the different effects involved in human photointerpretation if photointerpretation is to be used to characterize ecotones and their transition zones. See Chapter 7 for more discussion of cognitive aspects of spatial uncertainty.

Overlaying two interpreted maps usually results in the creation of a large number of what are generally called *sliver polygons*. These are an artifact of the uncertainty of boundary placement. Some GIS permit the automated reassignment of these polygons to one or the other of the neighboring polygons in order to eliminate these polygons; unfortunately, they do so at the expense of increasing the boundary error. A more common method is to use so-called epsilon bands (Perkal 1956; Chrisman 1982), which are fixed size buffer zones or corridors defined around each boundary to perform the overlay operation; hence, boundaries whose epsilon bands overlap are assigned a single boundary (e.g., a median boundary) at output. These usually use a single global width for the entire region.

The use of multiple interpretations allowing the characterization of uncertainty widths (i.e., one standard deviation of variation) can also be exploited during overlay to determine when boundaries from different maps are consistent with each other within interpreter uncertainty and when they are different (Aubert 1995). This is similar to the epsilon-band method described earlier, but specifies a different width for each boundary and associates this width with a statistical model. The normality of the distribution of interpretations around the median value has been tested, as has the relationship between the mean and median values. Photointerpretations of forestry data appear to be consistent with normal distributions, although there are indications that they are not (Harvey and Vauglin 1996) in other application domains (e.g., when producing an actual map). Aggregation of regions may also serve to reduce the impact of the uncertainty along the boundaries by reducing the relative surface area occupied by near-boundary regions.

8.2.3 Methods for Remotely Sensed Images

As indicated earlier, both spatially constrained clustering techniques and edge detection methods are used with remotely sensed images to extract boundaries and transition zones, just as has been done in ecology. Edge detection has been appreciated as a computational problem, and numerous techniques have been proposed for its solution (Pitas 1993). Although many edge algorithms have been developed most focus on edge enhancement (i.e., visually emphasizing the boundaries in a picture). This can be done by analyzing the texture of an image. In this context, the term *texture* refers to the brightness variations of an image. Texture can be described using a structural or statistical approach. In the structural approach, an image is assumed to be composed of primitive elements (i.e., groups of pixels) that can be characterized by their shape and size as well as their pattern of repetition; however, a statistical approach is often preferred because image processing encounters problems similar to those met in field ecology (i.e., misclassification or unclear repetitive patterns). Such an approach consists of analyzing the intensity of the gradient among neighborhood pixels with various techniques such as autocorrelation functions, autoregressive models, spatial intensity co-occurrence probabilities, textural edges, and structural element filtering (Pitas 1993).

The major problem with the detection of edges in image processing is the presence of noise due to poor resolution. This is similar to the situation in field data where the finest level at which an edge can be detected depends on the resolution of the samples or pixels, the quadrat size, and, at the limit, the crown canopy size. Furthermore, the texture itself adds to the noise, as may any small discontinuities present in the image. If the discontinuities do not contrast sufficiently with the background (texture), they are considered as noise.

The most common edge detection methods take into consideration locally neighboring pixels, and are called *parallel methods* because in theory all the pixels are processed at the same time. The simplest methods are those that are based on linear differences [e.g., the Sobel (Fig. 8.3; see color insert) and Kirsch operators or local first-order derivatives between adjacent pixels], and these are called *edge detector kernels* or simply *edge filters* (Pitas 1993). These kernel operators can be either windows of 2×2 or 3×3 pixels. Gradients computed from 3×3 windows are smoother than those computed from 2×2 windows; hence, they reduce more of the noise. The size of the window is critical, as is the block size in field data, for overcoming noise and for the ability of detecting small edges.

These algorithms, based on linear differences and first-order derivatives, detect edges by identifying adjacent pixels characterized by the highest rates of change. When an edge is wide, it is important to detect its starting and ending locations. This can be accomplished by using second-order derivatives where the derivative values equal zero except at the locations where the boundary begins and ends. The Laplacian operator (Fig. 8.3; Pitas 1993) is such a second-order derivative operator. The major problem with the Laplacian operator is that it is sensitive to noise, such that it is necessary to smooth the data first. There are several other parallel edge detection algorithms: nonlinear filters that use kernels based on polynomials; edge-preserving smoothing techniques that use nonlinear filters; global thresholding that segments the pixels based on their spectral brightness; adaptive filters that correct for random noise as well as additive or multiplicative noise in data related to the image scene (Pitas 1993).

Spatial clustering techniques in image analysis are grouped under the heading of image segmentation. They have been used for several decades to carry out automated image analysis in a variety of fields, including remote sensing, medical imaging, and industrial imaging. Three broad categories of segmentation techniques have emerged. These are region-growing techniques, edge-detection techniques, and hybrid techniques. Region-growing techniques rely on "seed" regions that may have been determined ahead of time, either by another algorithm or a human operator, which are then grown outward using a homogeneity criterion based on spatial contiguity until the regions meet at boundary zones. On the other hand, boundary techniques use edge detection methods to identify boundary elements, and then attempt to connect discontinuous boundaries together to form a spatial partition. Hybrid methods use both region-growing and edge detection strategies.

A large variety of segmentation strategies have been developed over the past several decades. At least several hundred different types of segmentation exist. Aggregation methods, hierarchical strategies, Fourier techniques, context-based methods, methods that statistically iterate toward a better partition, fuzzy set theory approaches, texture-based methods, and so on have all been developed. In principle, methods have been developed to extract transition zones that have largely been applied to applications in medical and industrial imaging, and only infrequently to remote sensing data. This is an area where more work might be usefully attempted. Despite the large number of segmentation techniques that have been developed, relatively little attention has been paid to the need to develop error measures for the resulting partitions (Edwards 1995). Beauchemin et al. (1997) surveyed different methods that have been used to characterize the error or uncertainty associated with segmentation techniques. None of these are fully satisfactory, but they do provide some insight into the errors involved.

Over the past several years, the use of image segmentation techniques has evolved toward methods that mimic human vision more closely (Cantoni et al. 1997), but little effort to study error has accompanied these efforts. Nonetheless, these methods of identifying boundaries and regions in images represent the beginning of a convergence toward the kinds of processes used by humans in photointerpretation tasks, and may well provide new insights into the sources of error and uncertainty found in the latter (Story and Congalton 1986; Edwards and Lowell 1996).

8.3 Summary

As we have presented, ecological data are inherently spatial, thereby exhibiting a certain level of spatial autocorrelation. According to the sampling procedure used to measure ecological data (i.e., quadrat, aerial photographs, and remote sensing), various degrees of spatial uncertainty will be contained in the data. Uncertainty may also vary according to neighborhood (i.e., by not having the same stationarity). There may be uncertainty in the estimation of angular effects or differences in angular dependencies (anisotropy), and there may be a need to estimate uncertainty after aggregation (error propagation).

Methods for estimating boundary uncertainty and the uncertainty associated with the delineation of transition zones have been developed for photointerpretation results and, to a lesser extent, for image segmentation methods. Furthermore, a parallel has been drawn between the edge detection and image segmentation methods found in remote sensing and those applied to ecological data. It seems reasonable, therefore, to exploit these cross-disciplinary links to develop and characterize the delineation of transition zones further across a wide variety of sources of data and methods of analysis. This multidisciplinary effort should lead to a greater understanding of spatial uncertainty present in ecotone delineation, systematizing the results from smaller studies to a broader context.

In the context of boundary detection all the sources of spatial uncertainty may result in inaccuracy in the location of the boundary identified. Boundaries derived from field data using algorithms like lattice or triangular wombling, from aerial photography by photointerpretation, and from remotely sensed images by edge-detection or image segmentation are all characterized by uncertainty of different forms. In sampling strategies that combine these data sources, such uncertainty will accumulate, affecting strongly the ecological relations derived from the data. To palliate these cumulative effects of inaccuracy, statistical methods such as randomization tests can be performed as suggested by Fortin and Drapeau (1995). With aerial photographs and remotely sensed images, boundary errors based on a polygon-specific error (Edwards 1995) or a measure of boundary similarity (Beauchemin et al. 1997) might be used.

Finally, more research is needed to develop:

1. An objective definition of boundary regardless of the sources of the data and the algorithms/methods used to determine the boundaries.
2. The means to determine the cumulative effects of combining errors and uncertainty of different types on derived ecological relations.
3. Statistical methods to assess the degree of uncertainty in the location of a boundary.

References

Anselin, L., and A. Getis. 1992. Statistical analysis and geographic information systems. The Annals of Regional Science 26:19–33.
Aubert, E. 1995. Quantification de l'incertitude spatiale en photo-interprétation forestière à l'aide d'un SIG pour le suivi spatio-temporel des peuplements.

Mémoire de maîtrise en sciences géomatiques, Faculté de Foresterie et de Géomatique, Université Laval, Québec.

Beauchemin, M., K.P.B. Thomson, and G. Edwards. 1997. On the hausdorff concept of distance used for the evaluation of segmentation results in remote sensing. Proceedings of the international symposium: geomatics in the Era of Radarsat (GER'97), Ottawa, 25–30 mai, CD-ROM.

Biging, G.S., R.G. Congalton, and E.C. Murphy. 1992. A comparison of photo-interpretation and ground measurements of forest structure. Proceedings of the ASPRS annual meeting, Baltimore 3:6–15.

Burrough, P.A. 1986. Principles of geographical information systems for land resources assessment. Monographs on soil and resources survey no. 12. Clarendon Press, Oxford.

Cantoni, V., S. Levialdi, and V. Roberto, eds. 1997. Artificial vision: image description, recognition and communication. Academic Press, New York.

Chrisman, N.R. 1982. A theory of cartographic error and its measurement in digital databases. Pages 159–68 in Proceedings of auto-carto 5, Crystal City, VA.

Cressie, N.A.C. 1991. Statistics for spatial data. John Wiley & Sons, New York.

Edwards, G. 1995. Methods for assessing local map accuracy in thematic classifications derived from remotely sensed imagery. Pages 1521–30 in Proceedings of the seventeenth international cartographic conference, Barcelona.

Edwards, G., and K.E. Lowell. 1996. Modelling uncertainty in photo-interpreted boundaries. Photogrammetric Engineering and Remote Sensing 62:377–91.

Fortin, M.-J., P. Drapeau, and P. Legendre. 1989. Spatial autocorrelation and sampling design in plant ecology. Vegetatio 83:209–22.

Fortin, M.-J. 1992. Detection of ecotones: definition and scaling factors. Ph.D. Thesis. Department of Ecology and Evolution, State University of New York at Stony Brook.

Fortin, M.-J. 1994. Edge detection algorithms for two-dimensional ecological data. Ecology 75:956–65.

Fortin, M.-J., and P. Drapeau. 1995. Delineation of ecological boundaries: comparison of approaches and significance tests. Oikos 72:323–32.

Fortin, M.-J. 1997. Effects of data types on vegetation boundary delineation. Can J For Res 27:1851–58.

Fortin, M.-J. 1999. The effects of quadrat size and data measurement on the detection of boundaries. Journal of Vegetation Science 10:43–50.

Gosz, J.R. 1993. Ecotone hierarchies. Ecological Applications 3:369–76.

Greig-Smith, P. 1964. Quantitative plant ecology, second ed. Butterworth, London.

Hansen, A., and F. di Castri, eds. 1992. Landscape boundaries: consequences for biotic diversity and ecological flows. Springer-Verlag, New York.

Harvey, F., and F. Vauglin. 1996. Geometric match processing: applying multiple tolerances, Proceedings of the seventh international symposium on spatial data handling. Delft, The Netherlands, Vol. 1: 4A13–29.

Holland, M.M., P.G. Risser, and R.J. Naiman, eds. 1991. Ecotones. Chapman and Hall, New York.

Johnston, C.A., J. Pastor, and G. Pinay. 1992. Quantitative methods for studying landscape boundaries. Pages 107–28 in Hansen, A., and F. di Castri, eds. Landscape boundaries: consequences for biotic diversity and ecological flows. Springer-Verlag, New York.

Legendre, P., and M.-J. Fortin. 1989. Spatial pattern and ecological analysis. Vegetatio 80:107–38.

Lowell, K.E., G. Edwards, and P. Bolduc. 1997. Nouvelle méthode pour estimer le volume local à partir de photographies à grande échelle et de données de terrain. Project #3054, Technical Report for the Testing, Experimenting and Technology Transfer Program, Canadian Forestry Service.

Naiman, R.J., and H. Décamps, eds. 1990. The ecology and management of aquatic-terrestrial ecotones. The Parthenon Publishing Group, Paris.

Oden, N.L., R.R. Sokal, M.-J. Fortin, and H. Goebl. 1993. Categorical wombling: determining regions of abrupt change in categorical variables. Geographical Analysis 25:315–36.

Palmer, M.W., and P.M. Dixon. 1990. Small-scale environmental heterogeneity and the analysis of species distributions along gradients. Journal of Vegetation Science 1:57–65.

Perkal, J. 1956. On epsilon length. Bulletin of the Polish Academy of Sciences 4:399–403.

Pitas, I. 1993. Digital image processing algorithms. Prentice Hall, New York.

Story, M., and R. Congalton. 1986. Accuracy assessment: a user's perspective. Photogrammetric Engineering and Remote Sensing 52:397–99.

Thompson, S.K. 1992. Sampling. Wiley Interscience, New York.

Upton, G.J.G., and B. Fingleton. 1985. Spatial data analysis by example. Volume 1: point pattern and quantitative data. John Wiley & Sons, New York.

Webster, R., and M.A. Oliver. 1990. Statistical methods in soil and land resource survey. Oxford University Press, Oxford.

9
Geostatistical Models of Uncertainty for Spatial Data

Phaedon C. Kyriakidis

Spatial databases are increasingly used to "drive" environmental models by providing input parameters for spatially explicit process simulators. Global biogeochemical models, for example, are based on parameter maps of multiple spatially distributed variables, [e.g., temperature, precipitation, and land cover (Potter et al. 1993)]; these are typically provided in the form of information layers (coverages) in geographical information systems (GIS). Such parameter maps, however, are inherently uncertain because their elements represent derived, not directly measured, quantities.

Parameter maps are typically synthesized from a small set of direct measurements, possibly taking into account secondary indirect information, such as that provided by remote sensing. Such a synthesis involves prediction of parameter values at unsampled locations (i.e., is a data expansion–integration procedure), and calls for the use of interpolation–extrapolation algorithms to obtain parameter estimates on a usually regular grid. Interpolation, however, is inevitably accompanied with uncertainty regarding the predicted parameter values. It is therefore essential to provide a model of uncertainty associated with each reported estimate, as well as to evaluate the impact of such uncertainty on the output response of the process simulator (i.e., on the ecological model predictions).

Notwithstanding its importance in spatial uncertainty characterization, a set of uncertainty models, each associated with a single parameter estimate, suffers from an important limitation: Uncertainty models specific to single map elements or pixels do not characterize the impact of the uncertainty regarding the input spatial data (the parameter map as a whole) on the overall uncertainty regarding the output from an environmental process simulator. In the case of spatially explicit ecological models, assessment of model response uncertainty calls for processing through the ecological model alternative input parameter maps, the construction of which is accomplished via stochastic simulation.

Stochastic simulation amounts to generating alternative, equiprobable joint realizations (simulated outcomes) of the unknown parameter values over the domain of interest [e.g., see Journel (1989), Englund (1993), Rossi et al.

175

(1993), and Deutsch and Journel (1998)]. All generated realizations reproduce the available data at their measurement locations (in which case the simulation is termed *conditional*), and, on average, they reproduce the data histogram and a model of spatial correlation between the observations. Complex relations with relevant secondary information can also be reproduced.

The set of alternative parameter realizations (multiple maps), after being processed by an environmental process simulator (ecological model), allows transferring uncertainty regarding the input spatial data into uncertainty regarding process response variables (i.e., uncertainty in ecological model predictions) (see Fig. 9.1). For example, alternative land cover maps can be input into a crop yield model, resulting in a distribution of net primary production values (see Kyriakidis and Dungan 2001). Availability of a histogram of potential outcomes of the environmental model provides a much richer input than a single answer for risk analysis studies and decision making (Massmann and Freeze 1987).

From Figure 9.1, it can be seen that such an integrated modeling approach consists of three steps: (1) assessment of the relevance of any piece of information (e.g., both ground-based data and remotely sensed imagery) to the environmental variable of interest: This task is performed via a calibration procedure based on the available data, which accounts for the varying degree of reliability of the different sources of information, (2) generation of alternative simulated realizations of possible spatial numerical models (parameter maps), which are consistent with all available information, and (3) input of this set of simulated parameter maps into process simulators (ecological models) for evaluating the impact of spatial data uncertainty on the output process response (ecological model predictions).

The geostatistical paradigm for spatial uncertainty assessment within the framework described in Figure 9.1 is presented in this chapter. Emphasis is placed on the first two steps; modeling decisions are highlighted and their tradeoffs are discussed in an effort to render uncertainty models for spatial data transparent to the users of geospatial information. Some preliminary concepts regarding spatial uncertainty modeling are introduced in Section 9.1, along with related aspects of spatial correlation. Section 9.2 presents two procedures for updating prior regional (location-independent) uncertainty models, as established from the histogram of the available data, into posterior local (location-dependent) uncertainty models, which account for the data values in their vicinity. In Section 9.3, various integration avenues are introduced to account for secondary (indirect) information (e.g., parameter estimates derived from satellite imagery) in order to constrain local uncertainty models further. In Section 9.4, the sequential simulation paradigm for assessing *joint* spatial uncertainty (i.e., uncertainty regarding unknown parameter values at a set of locations *simultaneously*), and for propagating such uncertainty to ecological model predictions is described. Last, some conceptual and methodological issues are discussed in Section 9.5.

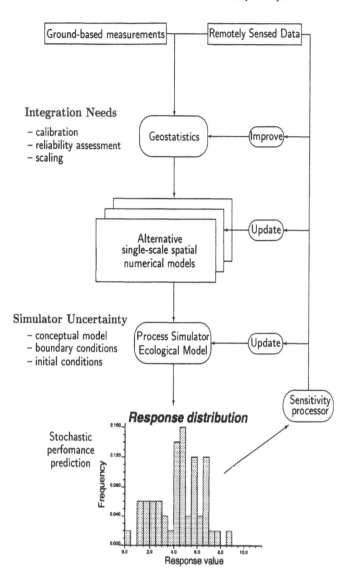

FIGURE 9.1. Geostatistics for data integration and stochastic simulator-performance assessment. Ground-based measurements and remotely sensed information are integrated via geostatistical algorithms to generate maximally constrained simulated realizations, which are input in an environmental process simulator (ecological model).

9.1 Local Uncertainty Models

Local (location-specific) uncertainty models characterize uncertainty regarding the unknown levels of environmental attributes (e.g., nutrient content or soil type) at unsampled locations. Such local uncertainty models depend only on the level of information available (i.e., the amount and quality of the relevant data found in the vicinity), not on the particular estimated value reported at that location. Hence, local models of uncertainty should be established prior to deriving any estimate for the unknown attribute values (Journel 1989). In addition, the goal at hand might require evaluation of the risk associated with reporting a specific estimate versus another, itself requiring the prior determination of a model of local uncertainty. For example, in a regional scale pollution problem, the risk of misclassifying a location as polluted while it is actually clean might depend on whether that location belongs to a residential or agricultural area. Even if the unknown pollution level in both areas is thought to be the same, reported pollution estimates in residential areas might be higher than those in agricultural areas. This difference in the reported estimates could aim at mitigating the corresponding higher impact of misclassification (i.e., the higher cost associated with potential health problems incurred due to underestimation of the actual pollution level). Such a risk-conscious decision requires the prior determination of a model of local uncertainty regarding the unknown contaminant concentration level before actually computing the concentration estimate. Modern geostatistics [e.g., Goovaerts (1997)] is primarily concerned with building such local uncertainty models for the unknown attribute values at unsampled locations.

Consider the case where a continuous environmental attribute (e.g., the content of a nutrient or the concentration of a pollutant) is directly measured at n locations $\{\mathbf{u}_\alpha, \ \alpha = 1,\ldots,n\}$, with $\mathbf{u}_\alpha$ denoting the vector of coordinates $(x_\alpha, y_\alpha, z_\alpha)$ in the three-dimensional space. Within the geostatistical framework [e.g., see Isaaks and Srivastava (1989)] the unknown attribute value $z(\mathbf{u})$ at a location $\mathbf{u}$ within the domain of interest D is modeled as an outcome of a random variable (RV) $Z(\mathbf{u})$ defined at that location.

A model of uncertainty regarding the unknown value $z(\mathbf{u})$ at location $\mathbf{u}$ is the local cumulative distribution function (cdf) $F_Z(\mathbf{u}; z)$ of the RV $Z(\mathbf{u})$ (i.e., the probability that the RV $Z(\mathbf{u})$ takes an outcome no greater than any given threshold z):

$$F_Z(\mathbf{u}; z) = Prob\{Z(\mathbf{u}) \leq z\}, \quad \forall z \tag{9.1}$$

where z is expressed in data units.

In the case where $\mathbf{u}$ designates any location within the domain of interest D, a continuous random function (RF) model $\{Z(\mathbf{u}), \mathbf{u} \in D\}$ is defined as the set of spatially dependent RVs $Z(\mathbf{u})$, one at each location $\mathbf{u} \in D$. The joint uncertainty regarding any number N of actual values $\{z(\mathbf{u}_1),\ldots,z(\mathbf{u}_N)\}$ is modeled by the multivariate or N-point cdf:

$$F_Z(\mathbf{u}_1,\ldots,\mathbf{u}_N; z_1,\ldots,z_N) = Prob\{Z(\mathbf{u}_1) \leq z_1, \ldots, Z(\mathbf{u}_N) \leq z_N\} \tag{9.2}$$

where the term N-point cdf is used to indicate that all N RVs relate to the same attribute Z at N different locations. In the case of N different attributes measured at the same location, the distribution (Eq. 9.2) is an N-variate cdf.

Within the geostatistical framework, the n data values $\{z(\mathbf{u}_\alpha), \alpha = 1, \ldots, n\}$ are viewed as a joint outcome of n respective RVs $\{Z(\mathbf{u}_\alpha), \alpha = 1, \ldots, n\}$. The objective is then to model the uncertainty regarding the unknown attribute value $z(\mathbf{u})$ at any unsampled location $\mathbf{u}$ conditional to the joint outcome of the n RVs $\{Z(\mathbf{u}_\alpha), \alpha = 1, \ldots, n\}$ (i.e., conditional to the n data values).

9.1.1 Stationarity: An Inference Necessity

Direct determination of a model of uncertainty $F_Z(\mathbf{u}; z)$ at location $\mathbf{u}$, would require L multiple outcomes $\{z^{(l)}(\mathbf{u}), \ l = 1, \ldots, L\}$ of the RV $Z(\mathbf{u})$. At best, if the location $\mathbf{u}$ coincides with a sample location, i.e., if $\mathbf{u} \equiv \mathbf{u}_\alpha$, only one outcome of the RV $Z(\mathbf{u})$ is available, that is the datum value $z(\mathbf{u}_\alpha)$. The idea is to replace the unavailable repetitive outcomes $z^{(l)}(\mathbf{u})$ of the RV $Z(\mathbf{u})$ for other outcomes $z(\mathbf{u}')$ found elsewhere within the domain D. These outcomes could be the data values $z(\mathbf{u}_\alpha)$ measured at some or all of the n sampling locations. Such a pooling decision, which is termed *stationarity*, is a statistical inference necessity and is graphically depicted in Figure 9.2.

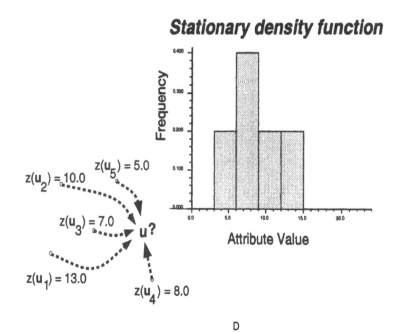

FIGURE 9.2. Stationarity decision for inference. Data values $z(\mathbf{u}_\alpha)$ found elsewhere in D are substituted for the unavailable outcomes $z^{(l)}(\mathbf{u})$ of the RV $Z(\mathbf{u})$.

The decision (not a hypothesis) of stationarity allows computation of the sample distribution, or cumulative histogram, since all data $\{z(\mathbf{u}_\alpha),\ \alpha = 1, \ldots, n\}$ are now considered outcomes of RVs with the same distribution, i.e.:

$$F_Z(\mathbf{u}; z) = Prob\{Z(\mathbf{u}) \leq z\} = F_Z(z), \quad \forall \mathbf{u} \in D, \ \forall z \qquad (9.3)$$

where the cdf $F_Z(z)$ is location-independent within D. This cdf is a model of regional uncertainty: It cannot discriminate uncertainty regarding the unknown value $z(\mathbf{u})$ at a location $\mathbf{u}$ surrounded entirely by high values, from uncertainty regarding $z(\mathbf{u}')$ at another location $\mathbf{u}'$ surrounded by both high and low values.

The cdf $F_Z(z)$ can be estimated by the proportion of data values no greater than any given threshold z, with such proportion being expressed as an average of the indicator data $i(\mathbf{u}_\alpha; z)$:

$$F_Z(z) \simeq \frac{1}{n} \sum_\alpha^n i(\mathbf{u}_\alpha; z), \quad \forall z$$

where $i(\mathbf{u}_\alpha; z)$ is the indicator transform of the datum $z(\mathbf{u}_\alpha)$ defined as:

$$i(\mathbf{u}_\alpha; z) = \begin{cases} 1 & \text{if } z(\mathbf{u}_\alpha) \leq z, \quad \forall z \\ 0 & \text{otherwise} \end{cases} \qquad (9.4)$$

Likewise, modeling the bivariate distribution of any two RVs, $Z(\mathbf{u})$ and $Z(\mathbf{u}')$ separated by a vector $\mathbf{h} = \mathbf{u}' - \mathbf{u}$, would require L repetitive joint outcomes of the data pairs $\{z^{(l)}(\mathbf{u}), z^{(l)}(\mathbf{u}')\}, l = 1, \ldots, L$. At best, if $\mathbf{u}$ and $\mathbf{u}'$ coincide with two sampling locations $\mathbf{u}_\alpha$ and $\mathbf{u}_\beta$, only one joint outcome of the RVs $Z(\mathbf{u})$ and $Z(\mathbf{u}')$ is available, that is the data pair $\{z(\mathbf{u}_\alpha), z(\mathbf{u}_\beta)\}$. The unavailable repetitive joint outcomes of the RVs $Z(\mathbf{u})$ and $Z(\mathbf{u}')$ are replaced by other pairs of data $\{z(\mathbf{u}_\alpha), z(\mathbf{u}_\alpha + \mathbf{h})\}$ separated by the same vector $\mathbf{h}$ and found elsewhere within the domain D. Such a pooling decision is the two-point version of stationarity and leads to the construction of $\mathbf{h}$-scattergrams, see Figure 9.3.

By convention, the value $z(\mathbf{u}_\alpha)$ at the origin of the vector $\mathbf{h}$ is called the *tail value*, whereas the value $z(\mathbf{u}_\alpha + \mathbf{h})$ at the end is called the *head value*. The $\mathbf{h}$-scattergrams can be used to depict the following three types of information:

1. Samples $z(\mathbf{u}_\alpha)$ and $z(\mathbf{u}_\alpha + \mathbf{h})$ of the same environmental attribute Z measured at different locations $\mathbf{u}_\alpha$ and $\mathbf{u}_\alpha + \mathbf{h}$ separated by vector $\mathbf{h}$ (this is the case of modeling spatial dependence of a single attribute for the specific lag $\mathbf{h}$).
2. Samples $z(\mathbf{u}_\alpha)$ and $y(\mathbf{u}_\alpha)$ of two different attributes Z and Y measured at the same locations $\mathbf{u}_\alpha$ (this is the traditional bivariate scatterplot for $|\mathbf{h}| = 0$).
3. Samples $z(\mathbf{u}_\alpha)$ and $y(\mathbf{u}_\alpha + \mathbf{h})$ of two different attributes Z and Y measured at different locations $\mathbf{u}_\alpha$ and $\mathbf{u}_\alpha + \mathbf{h}$ separated by vector $\mathbf{h}$ (this is the case

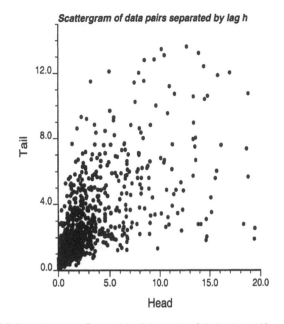

FIGURE 9.3. **h**-scattergram. Cross-plot of data pairs $\{z(\mathbf{u}_\alpha), z(\mathbf{u}_\alpha + \mathbf{h})\}$ separated by the same vector **h**. Head and tail values for the pairs can correspond to data of the same or different attribute sampled at the same or different locations.

of a cross-**h**-scattergram used to model dependence between two spatially distributed attributes separated by lag **h**).

Stationarity implies that there is no strong spatial trend in the data, such as a pattern of "average" spatial variation specific to a particular direction. Stationarity also implies that the data pooled for inference within the domain D originate from the same population, and no subpopulations can be distinguished. If the existence of such subpopulations is postulated, then one should consider subdividing the study area in various subregions (or strata) and recompute all above statistics from the data specific to each subregion (this procedure is also known as *stratification*). Beware, though, that a very fine subdivision of the study area could result in unreliable statistics, if the latter are computed from very few data within each stratum.

In all cases, stationarity can be seen as the license to export statistical inference made from the data to the underlying population, whether this inference relates to the whole study area or is specific to strata within it. Stationarity is a property of the model characterizing the data, not of the data themselves; hence, it cannot be proven wrong or right; stationarity can only be characterized appropriate or inappropriate. When an inappropriate model is used, the statistical properties of the actual estimates may be significantly different from their model counterparts.

9.1.2 Two-Point Dependence and Linear Correlation

The dependence between two RVs $Z(\mathbf{u})$ and $Z(\mathbf{u}+\mathbf{h})$ separated by vector $\mathbf{h}$ is fully characterized by the stationary two-point cdf $F_Z(\mathbf{h};z,z')$ defined as:

$$F_Z(\mathbf{h};z,z') = Prob\{Z(\mathbf{u}) \le z, Z(\mathbf{u}+\mathbf{h}) \le z'\}, \quad \forall z,z' \qquad (9.5)$$

That is, the probability that any two pairs of RVs $Z(\mathbf{u})$ and $Z(\mathbf{u}+\mathbf{h})$ take outcomes that are jointly no greater than any two thresholds z and z'.

For the case of a single threshold $z = z'$, centering the two-point cdf $F_Z(\mathbf{h};z)$ leads to the indicator covariance $C_I(\mathbf{h};z)$ for the given lag $\mathbf{h}$:

$$
\begin{aligned}
F_Z(\mathbf{h};z) - F_Z^2(z) &= E\{I(\mathbf{u};z) \cdot I(\mathbf{u}+\mathbf{h};z)\} - [E\{I(\mathbf{u};z)\}]^2 \\
&= Cov\{I(\mathbf{u};z), I(\mathbf{u}+\mathbf{h};z)\} = C_I(\mathbf{h};z) \\
&\simeq \frac{1}{n(\mathbf{h})} \sum_{\alpha=1}^{n(\mathbf{h})} i(\mathbf{u}_\alpha;z) \cdot i(\mathbf{u}_\alpha+\mathbf{h};z) - F_Z^2(z) \qquad (9.6)
\end{aligned}
$$

where $n(\mathbf{h})$ is the number of indicator data pairs $\{i(\mathbf{u}_\alpha;z), i(\mathbf{u}_\alpha+\mathbf{h};z)\}$ defined at threshold z and separated by vector $\mathbf{h}$, and $F_Z(z)$ is the regional stationary cdf.

The two-point cdf $F_Z(\mathbf{h};z)$; consequently, its centered version, the indicator covariance $C_I(\mathbf{h};z)$, measures how often two values of the same attribute separated by a vector $\mathbf{h}$ are jointly less than the threshold value z (see Fig. 9.4).

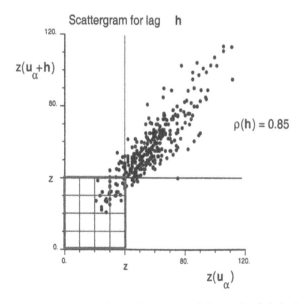

FIGURE 9.4. Two-point dependence. Proportion of data pairs $\{z(\mathbf{u}_\alpha), z(\mathbf{u}_\alpha+\mathbf{h})\}$ separated by a vector $\mathbf{h}$ that are jointly no greater than any threshold z (i.e., that lie in the corresponding hatched area).

For example, in the case of a large threshold value z, the two-point cdf $F_Z(\mathbf{h}; z)$ measures the spatial connectivity of large attribute values: The larger the value of $F_Z(\mathbf{h}; z)$, the better the spatial connectivity of such large attribute values for the specific lag $\mathbf{h}$.

The linear correlation between data pairs $z(\mathbf{u}_\alpha)$ and $z(\mathbf{u}_\alpha + \mathbf{h})$ separated by vector $\mathbf{h}$, is often summarized via the Z-covariance $C_Z(\mathbf{h})$:

$$C_Z(\mathbf{h}) \simeq \frac{1}{n(\mathbf{h})} \sum_{\alpha=1}^{n(\mathbf{h})} z(\mathbf{u}_\alpha) \cdot z(\mathbf{u}_\alpha + \mathbf{h}) - m_Z^2 \qquad (9.7)$$

where $n(\mathbf{h})$ is the number of data pairs separated by the vector $\mathbf{h}$, and m_Z is the stationary mean of the RF model, inferred from the sample histogram.

It is important to realize that the dependence between any two RVs $Z(\mathbf{u})$ and $Z(\mathbf{u} + \mathbf{h})$, as provided by the set of all indicator covariances $C_I(\mathbf{h}; z, z')$ for all possible thresholds z and z', is not equivalent to the linear correlation, as measured by the Z-covariance $C_Z(\mathbf{h})$. In fact, the covariance $C_Z(\mathbf{h})$ is the average of all indicator covariances for all thresholds z and z' (e.g., Journel 1989):

$$C_Z(\mathbf{h}) = \int_{z_{min}}^{z_{max}} dz \int_{z_{min}}^{z_{max}} C_I(\mathbf{h}; z, z') dz'$$

where z_{min} and z_{max} are the minimum and maximum z-values, respectively.

Absence of linear spatial correlation for lag $\mathbf{h}$ [i.e., $C_Z(\mathbf{h}) = 0$] does not entail absence of spatial dependence for that lag [i.e., $C_Z(\mathbf{h}) = 0 \nRightarrow C_I(\mathbf{h}; z) = 0$, $\forall z$]. Indeed, a nonlinear dependence could entail $C_Z(\mathbf{h}) = 0$, whereas $C_I(\mathbf{h}; z) \neq 0$ for some threshold z. The absence of spatial dependence [i.e., $C_I(\mathbf{h}; z) = 0$, $\forall z$], however, does entail absence of linear spatial correlation for that lag [i.e., $C_I(\mathbf{h}; z) = 0$, $\forall z \Rightarrow C_Z(\mathbf{h}) = 0$].

The linear correlation between data pairs separated by a vector $\mathbf{h}$ is also quantified by the unit-free, standardized equivalent of the covariance $C_Z(\mathbf{h})$ [i.e., the correlogram $\rho_Z(\mathbf{h})$]:

$$\rho_Z(\mathbf{h}) = \frac{C_Z(\mathbf{h})}{\sqrt{Var\{Z(\mathbf{u})\}}}, \in [-1, +1] \qquad (9.8)$$

where $Var\{Z(\mathbf{u})\}$ is the stationary variance of the RF model, inferred from the sample histogram.

The average dissimilarity between data pairs separated by the vector $\mathbf{h}$ is quantified by the moment of inertia of the $\mathbf{h}$-scattergram, called the *semi-variogram*:

$$\gamma_Z(\mathbf{h}) = \frac{1}{2} Var\{Z(\mathbf{u}) - Z(\mathbf{u} + \mathbf{h})\} \simeq \frac{1}{2n(\mathbf{h})} \sum_{\alpha=1}^{n(\mathbf{h})} [z(\mathbf{u}_\alpha) - z(\mathbf{u}_\alpha + \mathbf{h})]^2 \qquad (9.9)$$

The covariance, semivariogram, and correlogram of a stationary RF $\{Z(\mathbf{u}), \mathbf{u} \in D\}$ are linked by the following relations:

$$\gamma_Z(\mathbf{h}) = C_Z(\mathbf{0}) - C_Z(\mathbf{h})$$

$$\rho_Z(\mathbf{h}) = 1 - \frac{\gamma_Z(\mathbf{h})}{C_Z(\mathbf{0})}$$

where $C_Z(\mathbf{0}) = Var\{Z(\mathbf{u})\}$ is the stationary variance of the RF model.

By varying the magnitude, and possibly the direction of the vector $\mathbf{h}$ (anisotropic case), a series of $\mathbf{h}$-scattergrams can be constructed. The set of summary spatial continuity or variability measures calculated from many such $\mathbf{h}$-scattergrams constitute the experimental covariance, correlogram, or semivariogram measures, as shown for the case of the correlogram in Figure 9.5.

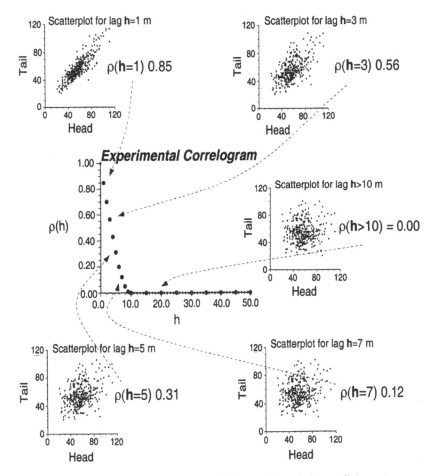

FIGURE 9.5. Experimental correlogram $\rho_Z(\mathbf{h})$. A set of correlation coefficient values, calculated from a series of $\mathbf{h}$-scattergrams, and plotted as a function of the magnitude of vector $\mathbf{h}$.

Covariance values, for example, are then obtained for all possible lag distances $\mathbf{h}$ (and possibly directions) by fitting a covariance function to the experimentally available covariance values. The distance at which model covariance values tend to zero constitutes the range of correlation between any two data within D no matter their actual locations. In other words, a value $z(\mathbf{u}_\alpha)$ measured at a location $\mathbf{u}_\alpha$ further away than the range of correlation from a location $\mathbf{u}$, does not influence the uncertainty model at location $\mathbf{u}$. For an in-depth discussion of the various measures of spatial (cross)correlation and their application to ecological modeling the reader can refer to Rossi et al. (1992). Note that the fitted semivariogram or covariance models must be permissible (i.e., result in posterior models of local uncertainty that do not exhibit negative variance) (e.g., see Cressie 1991).

The possibly nonstationary RF model $Z(\mathbf{u})$ is often decomposed into a deterministic trend component $m_Z(\mathbf{u})$ and a stochastic, zero mean, stationary residual RF $R(\mathbf{u})$:

$$Z(\mathbf{u}) = m_Z(\mathbf{u}) + R(\mathbf{u}), \quad \forall \mathbf{u} \in D \qquad (9.10)$$

where the trend component $m_Z(\mathbf{u})$ models some "average" spatial variation of Z, and the stationary residual RF component $R(\mathbf{u})$ models random, yet possibly correlated in space, fluctuations around that trend.

Note that the residual component is zero in expected value (on average) [i.e., $E\{R(\mathbf{u})\} = 0$, $\forall \mathbf{u} \in D$]. Hence, the mean of the RV $Z(\mathbf{u})$ at location $\mathbf{u}$ is identified to the value of the trend component at that location:

$$E\{Z(\mathbf{u})\} = m_Z(\mathbf{u}), \quad \forall \mathbf{u} \in D$$

Under stationarity, the mean of the prior cdf $F_Z(\mathbf{u}; z)$ of any RV $Z(\mathbf{u})$ is the same over D [i.e., $E\{Z(\mathbf{u})\} = m_Z$, $\forall \mathbf{u} \in D$], in which case the covariance $C_R(\mathbf{h})$ of the residual RF $R(\mathbf{u})$ is equal to the Z-covariance $C_Z(\mathbf{h})$.

It should be stressed that no actual data are available on either $m_Z(\mathbf{u})$ or $R(\mathbf{u})$, hence any modeling decomposition of type Equation 9.10 relies entirely on the decision of identifying some smooth component of the attribute spatial variation to the trend component. A different decision about what constitutes a smooth or "average" spatial variation would result in a different trend component $m_Z(\mathbf{u})$; hence, a different residual component $R(\mathbf{u})$. Any spatial variation not attributed to the trend component is absorbed (modeled) by the residual component. If a constant (stationary) mean component is adopted, all spatial variation is modeled by the residual component. Likewise, if a linear trend component is adopted [e.g., $m_Z(\mathbf{u}) = m_Z(x, y) = a_0 + a_1 x + a_2 y$] only the remaining spatial variation is modeled by the resulting residual component.

Practice has shown (e.g., see Goovaerts 1997) that in interpolation mode (i.e., when location $\mathbf{u}$ is surrounded by data locations $\mathbf{u}_\alpha$) a complex trend model does not yield better estimation results than a simple restriction of the stationarity decision regarding the mean $m_Z(\mathbf{u})$ within a local neighborhood $W(\mathbf{u})$ centered at location $\mathbf{u}$ [i.e., $m_Z(\mathbf{u}) = c$ (constant) $\forall \mathbf{u} \in W(\mathbf{u})$]; this latter decision is termed *local stationarity*. On the other hand, a spatially varying trend model $m_Z(\mathbf{u})$ has paramount influence on estimation results in extrapolation mode (i.e., when location $\mathbf{u}$ is not surrounded by data locations $\mathbf{u}_\alpha$). In such an extrapolation mode, misspecification of the trend model $m_Z(\mathbf{u})$ can lead to erroneous results. Favoring local stationarity over a complex spatially varying trend model is therefore recommended. A spatially varying trend model $m_Z(\mathbf{u})$ ideally should have a physical justification (i.e., the particular trend model adopted should correspond to some physically meaningful process).

9.2 Posterior Local Uncertainty Models

Stationary uncertainty models of the type $F_Z(\mathbf{u}; z) = F_Z(z)$, $\forall \mathbf{u} \in D$, are also termed *prior cdfs* because they are determined per stationarity before taking into account any nearby information (e.g., the fact that data around location $\mathbf{u}$ might all be high). Such a prior local cdf is updated into a local conditional cumulative distribution function (ccdf) (i.e., into a local posterior cdf given the nearby information) by weighting the contribution of each datum according to its correlation with the unknown. A posterior, nonstationary, local ccdf represents a model of uncertainty regarding the unknown attribute value $z(\mathbf{u})$ at location $\mathbf{u}$ given the joint outcomes (data) $\{z(\mathbf{u}_\alpha), \alpha = 1, \ldots, n\}$ of the n nearby RVs $\{Z(\mathbf{u}_\alpha), \alpha = 1, \ldots, n\}$. It is defined as:

$$F_Z(\mathbf{u}; z|(n)) = Prob\{Z(\mathbf{u}) \le z|(n)\}, \quad \forall z \qquad (9.11)$$

where (n) denotes the conditioning information, typically limited to the number of data $n(\mathbf{u})$ found in neighborhood $W(\mathbf{u})$ centered at $\mathbf{u}$.

The construction of uncertainty models of type Equation 9.11, and the subsequent generation of simulated attribute realizations will be illustrated using the set of data shown in Figure 9.6. These data are in the public domain, and the reader is referred to Deutsch and Journel (1998) for a detailed description regarding their origin. No explicit attribute or distance units are attached to Figure 9.6 in order to emphasize the generality of the concepts being presented.

Note that the (cross)covariance models used (but not shown here) were inferred from a larger data set because spatial correlation measures (covariances or variograms) computed from a sample of only 29 data would

---

FIGURE 9.6. Illustration of data set used in chapter examples. (Top) A sample data set of 29 measurements regarding an attribute Z over an area D of size 50×50. (Middle) A sample data set of 29 labels of a categorical attribute S with three possible states/classes. (Bottom) A set of 2500 (soft) data of a secondary attribute Y covering exhaustively the area D, used for illustrating data integration procedures.

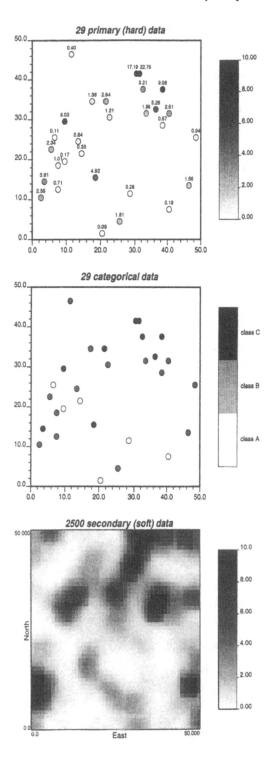

be too erratic to be reliable. The decision to use spatial correlation measures not based on the sample set was made in order to focus attention solely on the procedures employed for determining local uncertainty models, independently of issues regarding statistical inference of semivariogram or covariance functions. Note also, that using a small data set allows a better appreciation of the impact of integrating more abundant secondary (indirect or soft) information.

9.2.1 The Nonparametric Framework

Consider K threshold values $\{z_k, k = 1, \ldots, K\}$, discretizing the range of variability of the attribute Z, such as the deciles of the sample distribution $F_Z(z_k)$. Within the nonparametric framework (Journel 1983), the local ccdf $F_Z(\mathbf{u}; z_k)|(n))$ is modeled explicitly one threshold z_k at a time, without any parametric assumption regarding the distributional type of $F_Z(\mathbf{u}; z_k)|(n))$. This local ccdf is interpreted as the conditional expectation of an indicator RV $I(\mathbf{u}; z_k)$ defined at location $\mathbf{u}$, given the conditioning information (n):

$$F_Z(\mathbf{u}; z_k|(n)) = E\{I(\mathbf{u}; z_k)|(n)\}, \quad k = 1, \ldots, K$$

with $I(\mathbf{u}; z_k) = 1$ if $Z(\mathbf{u}) \leq z_k$, 0 otherwise. Such an interpretation allows estimating the local ccdf by a weighted proportion of nearby indicator data $i(\mathbf{u}_\alpha; z_k)$ (see hereafter).

The range of variability of the 29 sample data shown in Figure 9.7 is discretized using $K = 9$ thresholds corresponding to the nine deciles of the sample distribution. The resulting indicator data for the second and eighth deciles, $z_2 = 0.35$ and $z_8 = 4.53$, are shown in Figure 9.7.

The conditional cdf $F_Z(\mathbf{u}; z_k|(n))$ at location $\mathbf{u}$ for threshold z_k [i.e., the conditional expectation of $I(\mathbf{u}; z_k)$] is estimated by a linear combination of the n indicator RVs $I(\mathbf{u}_\alpha; z_k)$ defined at the n data locations for cutoff z_k and the regional prior cdf $F_Z(z_k)$ as:

$$F_Z^*(\mathbf{u}; z_k|(n)) - F_Z(z_k) = \sum_{\alpha=1}^{n} \lambda_\alpha(\mathbf{u}; z_k)[I(\mathbf{u}_\alpha; z_k) - F_Z(z_k)] \quad (9.12)$$

where $\lambda_\alpha(\mathbf{u}; z_k)$ are weights assigned to the n data RVs $I(\mathbf{u}_\alpha; z_k)$ retained within a neighborhood $W(\mathbf{u})$ centered at $\mathbf{u}$, and obtained per solution of a system of normal equations termed *simple indicator kriging* (SIK) system:

$$[C_I(\mathbf{u}_\alpha - \mathbf{u}_\beta; z_k)][\lambda_\beta(\mathbf{u}; z_k)] = [C_I(\mathbf{u}_\alpha - \mathbf{u}; z_k)] \quad (9.13)$$

where $[C_I(\mathbf{u}_\alpha - \mathbf{u}_\beta; z_k)]$ is $(n \times n)$ matrix of covariance values between any two indicator RVs $I(\mathbf{u}_\alpha; z_k)$ and $I(\mathbf{u}_\beta; z_k)$, $[\lambda_\beta(\mathbf{u}; z_k)]$ is the $(n \times 1)$ vector of SIK weights, and $[C_I(\mathbf{u}_\alpha - \mathbf{u}; z_k)]$ is the $(n \times 1)$ vector of covariance values between the indicator RVs $I(\mathbf{u}_\alpha; z_k)$ and the RV $I(\mathbf{u}; z_k)$.

For the example at hand, the domain D is discretized using a set of $N = 2500$ nodes on a 50×50 regular grid with a unit grid cell. Local

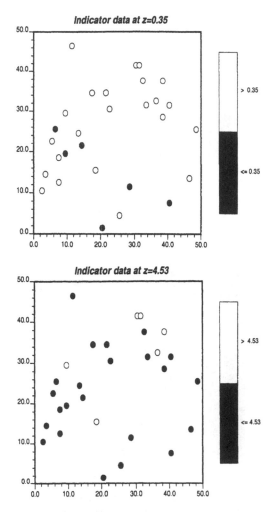

FIGURE 9.7. Indicator data defined at thresholds $z_2 = 0.35$ (top), and $z_8 = 4.53$ (bottom), corresponding to the twentieth and eightieth percentiles of the 29 data distribution.

uncertainty models of type Equation 9.12 are computed only at these nodes (i.e., location **u** denotes anyone of theses N nodes). Using the two sets of 29 indicator data defined at thresholds $z_2 = 0.35$ and $z_8 = 4.53$, local ccdf values for these two thresholds are derived via SIK and shown in Figure 9.8.

Note that SIK (as all variants of kriging) reproduces the indicator data at their locations [i.e., $F_Z^*(\mathbf{u}_\alpha; z_k|(n)) = i(\mathbf{u}_\alpha; z_k)$, $\forall \mathbf{u}_\alpha \in D$] a characteristic termed exactitude property of kriging. For example, it can be seen from the top graph of Figure 9.8 that dark colored areas (i.e., areas where the local conditional probability that the unknown attribute value $z(\mathbf{u})$ be no greater

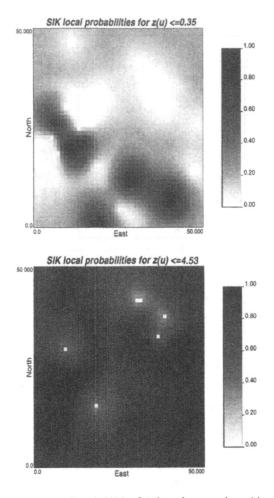

FIGURE 9.8. Posterior local probabilities for the unknown values $z(\mathbf{u})$ not to exceed 0.35 (top) and 4.53 (bottom), given the corresponding $n = 29$ indicator data, derived via simple indicator kriging (SIK).

than $z_2 = 0.35$ is one) correspond to data locations where the z-value is indeed known to be no greater than z_2 [i.e., $Prob\{Z(\mathbf{u}_\alpha) \leq 0.35|(n)\} = i(\mathbf{u}_\alpha; 0.35) = 1$]. Note also that away from the data locations, at distances $\mathbf{h}$ greater than the correlation range for threshold z_2 (which is about 10 map units), the posterior local conditional probability reverts to the prior regional proportion $F_Z(0.35) = 0.2$.

The SIK system thus provides a general procedure for updating prior uncertainty models into posterior models of local uncertainty, conditioned to nearby data. This procedure amounts to modifying the prior local cdf $F_Z(\mathbf{u}; z_k) = F_Z(z_k)$ into a posterior local ccdf model $F_Z(\mathbf{u}; z_k|(n))$ by weighting the nearby indicator data $i(\mathbf{u}_\alpha; z_k)$ according to their degree of correlation with the unknown indicator value $i(\mathbf{u}; z_k)$ at location $\mathbf{u}$. This degree of correlation is identified to the two-point indicator covariance $C_I(\mathbf{u}_\alpha - \mathbf{u}; z_k)$ inferred through the decision of stationarity. Data redundancy (i.e., excessive influence of data clusters) is accounted for by the left-hand side, data-to-data indicator covariance matrix $[C_I(\mathbf{u}_\alpha - \mathbf{u}_\beta; z_k)]$ in Equation 9.13.

Note that the spatial continuity of various classes of z-data values (e.g., values around the median as opposed to extremely high values) need not be the same. The indicator approach allows explicit modeling of the spatial correlation of different classes of data values via different covariance models $C_I(\mathbf{h}; z_k)$ for different threshold values z_k. For example, in a regional pollution setting, the spatial continuity of values above a regulatory threshold z_c modeled by $C_I(\mathbf{h}; z_c)$ may not be the same as the average spatial continuity modeled by the Z-covariance $C_Z(\mathbf{h})$, especially if z_c corresponds to an extreme threshold separating different spatial patterns of pollution transport and deposition.

For the case of K threshold values $\{z_k, k = 1, \ldots, K\}$ discretizing the range of variability of Z, a set of K SIK systems of type Equation 9.13 needs to be solved, one for each threshold value z_k. This also calls for consistent inference of the corresponding K indicator covariance models $C_I(\mathbf{h}; z_k)$. Because local ccdf values are estimated only for each threshold z_k, the resolution of the local ccdf model $F_Z(\mathbf{u}; z_k)$ depends on the number K of cutoffs considered. In practice, no more than $K = 9$ cutoffs are retained, and the local ccdf $F_Z(\mathbf{u}; z_k|(n))$ is approximated by first computing the K ccdf values and then interpolating within each class $(z_k, z_{k+1}]$, and extrapolating beyond the two extreme cutoffs z_1 and z_K up to the minimum z_{min} and maximum z_{max} values, as shown in Figure 9.9.

Local departures from a stationary regional cdf $F_Z(z_k)$ can be accounted for by limiting the domain of stationarity to each neighborhood $W(\mathbf{u})$ centered at the location $\mathbf{u}$ where local uncertainty is modeled. In this case, the cdf $F_Z(z_k)$ used in Equation 9.12 varies from one neighborhood $W(\mathbf{u})$ to another and is unknown. The solution is to filter that unknown cdf value from Equation 9.12 by constraining the kriging weights to sum to 1, leading to a variant of SIK called *ordinary indicator kriging* (OIK). For this and other variants of indicator kriging, the reader can consult Goovaerts (1997).

Categorical-type information (e.g., maps of soil or land cover types derived from indirect satellite measurements via classification algorithms) is frequently provided in the form of GIS coverages. The nonparametric (indicator) framework can readily be adopted for modeling local uncertainty regarding such categorical-type information.

Let $S(\mathbf{u})$ denote a categorical random variable that can be K mutually exclusive and exhaustive outcomes/states $\{s_k, k = 1, \ldots, K\}$. Regional

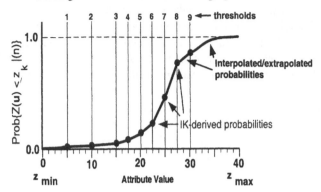

FIGURE 9.9. Local conditional cdf. Local conditional probabilities (gray dots) are estimated via SIK for K thresholds $\{z_k, k = 1, \ldots, K\}$. Interpolation and extrapolation (black lines) of these probability values increase the resolution of the local uncertainty model.

uncertainty regarding the unknown state $s(\mathbf{u})$ at any location $\mathbf{u}$ is modeled by the stationary probability distribution function (pdf) of the categorical RV $S(\mathbf{u})$:

$$p_S(\mathbf{u}; s_k) = Prob\{S(\mathbf{u}) = s_k\} = p_k, \quad k = 1, \ldots, K, \;\; \forall \mathbf{u} \in D$$

Given a set of n observations $\{s(\mathbf{u}_\alpha) = s_k, \; \alpha = 1, \ldots, n\}$ of the categorical attribute S measured at n locations $\{\mathbf{u}_\alpha, \; \alpha = 1, \ldots, n\}$, the objective is to update the prior stationary pdf model p_k into a posterior (conditional) local cpdf model $p_S^*(\mathbf{u}; s_k|(n))$:

$$p_S^*(\mathbf{u}; s_k|(n)) = Prob\{S(\mathbf{u}) = s_k|(n)\}, \quad k = 1, \ldots, K$$

That is, the probability that category s_k prevails at location $\mathbf{u}$, given the n neighboring categorical data $\{s(\mathbf{u}_\alpha), \; \alpha = 1, \ldots, n\}$.

Such a local cpdf model can be established via SIK, as in Equation 9.12, using the indicator transforms $i(\mathbf{u}_\alpha; s_k)$ of the data $s(\mathbf{u}_\alpha)$ for category s_k, defined as:

$$i(\mathbf{u}_\alpha; s_k) = \begin{cases} 1 & \text{if } s(\mathbf{u}_\alpha) = S_k \\ 0 & \text{if not} \end{cases} \tag{9.14}$$

and the corresponding model of spatial correlation, i.e., the indicator covariance $C_I(\mathbf{h}; s_k)$ for category s_k.

The indicator framework can be adopted for both continuous and categorical data, with only slight modifications. Indeed, all pieces of information can be coded as indicator values, irrespective of their origin, and processed in this common format with well-established generalized regression techniques, see for example Goovaerts (1997).

9.2.2 The Parametric Framework

The indicator approach provides a nonparametric framework for modeling local uncertainty regarding the unknown value $z(\mathbf{u})$ at location $\mathbf{u}$, in that the local ccdf $F_Z(\mathbf{u}; z|(n))$ need not be of any parametric (analytical) form. In the case of two-parameter distributions (e.g., the Gaussian distribution) the local ccdf $F_Z(\mathbf{u}; z|(n))$ is fully characterized by its first two moments: the conditional expectation and conditional variance.

Within the parametric framework, determination of the entire local ccdf $F_Z(\mathbf{u}; z|(n))$ typically amounts to estimating only its conditional mean and conditional variance at location $\mathbf{u}$. The conditional mean $E\{Z(\mathbf{u})|(n)\}$ of the local ccdf $F_Z(\mathbf{u}; z|(n))$ is estimated by a weighted linear combination of the n data RVs $Z(\mathbf{u}_\alpha)$ and the prior regional mean value m_Z:

$$Z_{SK}^*(\mathbf{u}) = E_{SK}^*\{Z(\mathbf{u})|(n)\} = m_Z + \sum_{\alpha=1}^{n} v_\alpha(\mathbf{u})[Z(\mathbf{u}_\alpha) - m_Z]$$

where $v_\alpha(\mathbf{u})$ are weights assigned to the n data RVs $Z(\mathbf{u}_\alpha)$ considered within a neighborhood $W(\mathbf{u})$ centered at $\mathbf{u}$, and obtained per solution of the *simple kriging* (SK) system:

$$[C_Z(\mathbf{u}_\alpha - \mathbf{u}_\beta)][v_\beta(\mathbf{u})] = [C_Z(\mathbf{u}_\alpha - \mathbf{u})] \qquad (9.15)$$

where $[C_Z(\mathbf{u}_\alpha - \mathbf{u}_\beta)]$ is $(n \times n)$ data-to-data covariance matrix, $[v_\beta(\mathbf{u})]$ is the $(n \times 1)$ vector of simple kriging weights, and $[C_Z(\mathbf{u}_\alpha - \mathbf{u})]$ is the $(n \times 1)$ vector of data-to-unknown covariance values.

The SIK system (Eq. 9.13) is a simple kriging system using the covariance $C_I(\mathbf{h}; z_k)$ of the indicator data $i(\mathbf{u}_\alpha; z_k)$ defined at cutoff z_k, instead of the Z-covariance $C_Z(\mathbf{h})$ modeled from the original data values $z(\mathbf{u}_\alpha)$ and used in Equation 9.15. The resulting weights are therefore different; hence, the notation $\lambda_\alpha(\mathbf{u})$ and $v_\alpha(\mathbf{u})$ for SIK and SK, respectively. Similar to the OIK case, local departures from a stationary regional mean m_Z can be accounted for by *ordinary kriging* (OK).

The posterior simple kriging variance $\sigma_{SK}^2(\mathbf{u})$ of the local ccdf $F_Z(\mathbf{u}; z|(n))$ is a result of the SK system (Eq. 9.15) and is written as:

$$\sigma_{SK}^2(\mathbf{u}) = Var_{SK}^*\{Z(\mathbf{u})|(n)\} = C_Z(\mathbf{0}) - \sum_{\alpha=1}^{n} v_\alpha(\mathbf{u})C_Z(\mathbf{u}_\alpha - \mathbf{u}) \qquad (9.16)$$

where $C_Z(\mathbf{0}) = Var\{Z(\mathbf{u})\}$ is the prior stationary regional variance of Z.

Posterior mean $z_{SK}^*(\mathbf{u})$ and variance $\sigma_{SK}^2(\mathbf{u})$ values of the local conditional distributions $F_Z(\mathbf{u}; z|(n))$ defined at all 2500 nodes discretizing the domain D were computed via SK, and are shown in Figure 9.10. Because the sample distribution appears to be lognormal, the local ccdfs could be modeled as lognormal, with mean and variance equal to the local SK mean and variance.

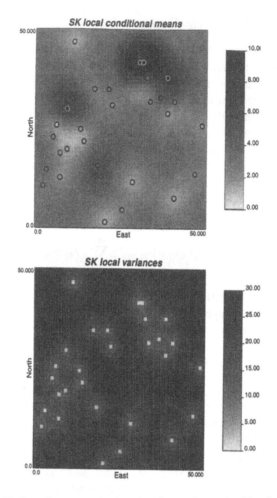

FIGURE 9.10. Posterior means (top) and variances (bottom) of local uncertainty models, given the $n = 29$ primary z-data (shown in circles), derived via simple kriging (SK).

At the limiting case of absence of spatial correlation [i.e., $C_Z(\mathbf{h}) = 0$, $\forall \mathbf{h} \neq 0$] the ccdf of the RV $Z(\mathbf{u})$ given any number n of nearby data has mean equal to the prior regional mean m_Z and variance equal to the stationary regional variance $C_Z(0)$. For example, it can be seen from Figure 9.10 that away from the data locations (i.e., at a distance $\mathbf{h}$ greater than the correlation range, which is about 10 map units), the SK mean is equal to the regional mean of the 29 data $m_Z = 3.38$ and the SK variance is maximum and equal to the regional variance $C_Z(0) = 25.81$. Again, accounting for spatial correlation appears as the means for updating the prior regional uncertainty into posterior models of local uncertainty.

Note that the data values are not directly involved in Equation 9.16. Only their geometry is accounted for through the covariance model $C_Z(\mathbf{h})$. Both the kriging weights and the kriging variance depend only on the data configuration, not on the actual data values. For example, the kriging variance for two locations surrounded by identical data configurations will be the same no matter the respective data values. Thus, the kriging variance is a measure of average distance between the location $\mathbf{u}$ where estimation is performed and the n data locations $\{\mathbf{u}_\alpha, \alpha = 1, \ldots, n\}$ considered. This average distance is not the traditional Euclidean distance $|\mathbf{h}|$, instead it is calculated from the covariance distance $C_Z(\mathbf{h})$ modeling the specific patterns of spatial continuity of the phenomenon under study.

When the RF $\{Z(\mathbf{u}), \mathbf{u} \in D\}$ is assumed to be multi-Gaussian (i.e., the joint distribution of the component RVs is N-point Gaussian) any local ccdf also is Gaussian (Anderson 1958); hence it is fully characterized by the SK-derived conditional mean $E^*_{SK}\{Z(\mathbf{u})|(n)\}$ and variance $Var^*_{SK}\{Z(\mathbf{u})|(n)\}$. Even if the regional sample distribution $F_Z(z_k)$ is not Gaussian, it can be transformed to a standard normal distribution via a graphical, rank preserving transform called *the normal scores transform* (e.g., see Deutsch and Journel 1998). SK is then performed on the normal score values for determining the conditional mean and variance of the local Gaussian ccdfs; however, note that transformation of the original one-point sample cdf $F_Z(z_k)$ into a Gaussian univariate distribution does not entail that the distribution of any set of RVs is jointly multivariate Gaussian. The assumption that any set of two RVs is joint normally distributed can be checked by comparing experimental indicator covariance values to those derived from theoretical expressions of the bivariate normal distribution (e.g., see Deutsch and Journel 1998).

Compared with the nonparametric approach, whereby inference of K indicator covariances $C_I(\mathbf{h}; z_k)$ and solution of K SIK systems is required, the parametric framework calls for inference of a single covariance model $C_Z(\mathbf{h})$ and solution of an SK system. This lesser modeling effort, however, comes at the expense of less flexibility in accommodating different patterns of spatial continuity for low, median, or high values, and amounts to adopting local uncertainty models whose variance (SK variance) does not depend on the data values. The congenial theoretical properties of the multi-Gaussian RF model, which lead to a lesser modeling effort, are associated with specific and possibly restrictive spatial characteristics, such as maximum spatial entropy for a given covariance model (Journel and Deutsch 1993) (i.e., purely random patterns of extremely high or low values similar to those associated with diffusive processes). This very specific characteristic need not be shared by the actual spatial distribution of the phenomenon under study. When the impact of spatially connected extreme values on the process response is known to be significant (e.g., paths of connected high hydraulic conductivity values conditioning first arrival time of pollutants at a compliance well) then the nonparametric approach should be favored for establishing local uncertainty models.

9.3 Constraining Local Uncertainty Models to Secondary Information

Secondary (indirect or soft) information, which is considered to be important because of its correlation with the attribute of interest, can be assimilated with direct (hard) data to establish maximally constrained (data-charged) models of local uncertainty. For example, indirect information obtained via remote sensing, either in its unprocessed format (e.g., reflectivity values recorded at different spectral bands) or in its processed format [e.g., the normalized vegetation index (NDVI)] can be integrated with ground-based direct observations for mapping [e.g., see Atkinson et al. (1992); Gohin and Langlois (1993); or Dungan et al. (1993)]. On the other hand, soft information could consist of abundant attribute measurements obtained by low accuracy sampling devices. In this case, data assimilation entails integration of such soft information with sparse more accurate (hard) data (e.g., Kyriakidis et al. 1999 for the conflation of elevation measurements with different accuracies). Integration algorithms are therefore essential for incorporating diverse sources of information into the assessment of local uncertainty; some frequently used procedures are presented.

In this section, it is assumed (without loss of generality) that only a single secondary continuous variable Y is available, which exhaustively covers the domain D, as is the case of information provided by remote sensing or geophysical imaging. The term *exhaustive coverage* implies that the secondary attribute Y is sampled at all nodes $\{\mathbf{u}_i, i = 1, \ldots, N\}$ where local uncertainty regarding the primary variable Z is modeled.

9.3.1 Secondary Data as Prior Local Means

Recalling the RF decomposition (Eq. 9.10) in a trend and residual component, one possible avenue for integration is to regard the secondary data as related to the prior mean $m_Z(\mathbf{u})$ of the local cdf $F_Z(\mathbf{u}; z)$ (e.g., see Goovaerts 1997). This relation could be modeled through a regression procedure between the co-located primary and secondary data pairs $\{z(\mathbf{u}_\alpha), y(\mathbf{u}_\alpha)\}$, resulting in estimates $m_Z^*(\mathbf{u}_\alpha)$ of the prior local mean values at all n primary data locations:

$$m_Z^*(\mathbf{u}_\alpha) = f(y(\mathbf{u}_\alpha)), \quad \alpha = 1, \ldots, n \ll N$$

where $f(\cdot)$ denotes a function, linear or not, of the secondary value $y(\mathbf{u}_\alpha)$ co-located at $\mathbf{u}_\alpha$. In this case, the preceding regression procedure does not depend on the actual data locations $\mathbf{u}_\alpha$ [i.e., function $f(\cdot)$ is the same from one location to another]; location-dependent regression procedures could also be considered.

Under the previous decision of stationarity for the regression function $f(\cdot)$ over the domain D, the prior local mean $m_Z(\mathbf{u})$ at a location $\mathbf{u} \neq \mathbf{u}_\alpha$, can be estimated from the co-located secondary datum $y(\mathbf{u})$ using the same regression function $f(\cdot)$ as:

$$m_Z^*(\mathbf{u}) = f(y(\mathbf{u})), \quad \forall \mathbf{u} \in D$$

Note that this regression model $f(\cdot)$ can be extended to include more than one secondary variable (e.g., longitude and latitude, elevation, distance from the coast, or reflectance values recorded at various spectral bands).

The SK estimator for the conditional mean $E\{Z(\mathbf{u})|(n)\}$ of the local ccdf $F_Z(\mathbf{u}; z|(n))$ is now written in its nonstationary version, termed *simple kriging with varying local means* (SKLM), as:

$$Z_{SKLM}^*(\mathbf{u}) = E_{SKLM}^*\{Z(\mathbf{u})|(n)\} = m_Z^*(\mathbf{u}) + \sum_{\alpha=1}^{n} \xi_\alpha(\mathbf{u})[Z(\mathbf{u}_\alpha) - m_Z^*(\mathbf{u}_\alpha)]$$

(9.17)

where the local mean values $\{m_Z^*(\mathbf{u}_\alpha), \alpha = 1, \ldots, n\}$ of the n data RVs $\{Z(\mathbf{u}_\alpha), \alpha = 1, \ldots, n\}$ and the local mean value $m_Z^*(\mathbf{u})$ of the RV $Z(\mathbf{u})$ need not be the same, and the weights $\xi_\alpha(\mathbf{u})$ are determined per solution of a SK- type system:

$$[C_R(\mathbf{u}_\alpha - \mathbf{u}_\beta)][\xi_\beta(\mathbf{u})] = [C_R(\mathbf{u}_\alpha - \mathbf{u})] \qquad (9.18)$$

where $C_R(\mathbf{h})$ is the covariance of the residual data RVs $\{Z(\mathbf{u}_\alpha) - m_Z^*(\mathbf{u}_\alpha), \alpha = 1, \ldots, n\}$, instead of the primary Z-covariance $C_Z(\mathbf{h})$ used in the SK system (Eq. 9.15). This is also the reason for the different notation $\xi_\alpha(\mathbf{u})$ used to denote the corresponding weights, instead of $\lambda_\alpha(\mathbf{u})$.

Equation 9.17 amounts to estimate the unknown residual $r(\mathbf{u})$ from the residual data $\{r(\mathbf{u}_\alpha), \alpha = 1, \ldots, n\}$ using SK with zero mean and covariance model $C_R(\mathbf{h})$, and to add the resulting estimate $r_{SK}^*(\mathbf{u})$ to the prior local mean $m_Z^*(\mathbf{u})$.

9.3.2 Secondary Data as Correlated Variables

As an alternative, the information brought by the secondary variable Y can be weighted by a model of its cross-correlation with the primary variable Z, which is the cross-covariance $C_{ZY}(\mathbf{h})$ estimated by:

$$C_{ZY}(\mathbf{h}) \simeq \frac{1}{n(\mathbf{h})} \sum_{\alpha=1}^{n(\mathbf{h})} z(\mathbf{u}_\alpha) \cdot y(\mathbf{u}_\alpha + \mathbf{h}) - m_Z \cdot m_Y$$

where $n(\mathbf{h})$ is the number of data pairs $\{z(\mathbf{u}_\alpha), y(\mathbf{u}_\alpha + \mathbf{h})\}$ separated by vector $\mathbf{h}$, and m_Z and m_Y are the stationary means of the primary and secondary variables.

The kriging formalism can be extended to include the secondary information through a procedure termed *cokriging* (e.g., see Goovaerts 1997). The simple cokriging (SCK) estimator $Z_{SCK}^*(\mathbf{u})$ for the conditional expectation $E\{Z(\mathbf{u})|(n+n')\}$ is expressed as a linear combination of the n primary RVs $\{Z(\mathbf{u}_\alpha),\ \alpha = 1,\ldots,n\}$, the n' secondary RVs $\{Y(\mathbf{u}_i),\ i = 1,\ldots,n'\}$, and the regional means m_Z and m_Y:

$$Z^*_{SCK}(\mathbf{u}) = E^*_{SCK}\{Z(\mathbf{u})|(n+n')\} = m_Z + \sum_{\alpha=1}^{n} \eta_\alpha(\mathbf{u})[Z(\mathbf{u}_\alpha) - m_Z]$$

$$+ \sum_{i=1}^{n'} \eta_i(\mathbf{u})[Y(\mathbf{u}_i) - m_Y] \qquad (9.19)$$

where the conditioning information $(n+n')$ typically refers to the $n(\mathbf{u})+ n'(\mathbf{u})$ data within a local neighborhood $W(\mathbf{u})$ centered on $\mathbf{u}$, and the weights $\eta_\alpha(\mathbf{u})$ for the n primary data, and $\eta_i(\mathbf{u})$ for the n' secondary data, are obtained per solution of the SCK system:

$$\begin{bmatrix} [C_Z(\mathbf{u}_\alpha - \mathbf{u}_\beta)] & [C_{ZY}(\mathbf{u}_\alpha - \mathbf{u}_{i'})] \\ [C_{YZ}(\mathbf{u}_i - \mathbf{u}_\beta)] & [C_Y(\mathbf{u}_i - \mathbf{u}_{i'})] \end{bmatrix} \begin{bmatrix} [\eta_\beta(\mathbf{u})] \\ [\eta_{i'}(\mathbf{u})] \end{bmatrix} = \begin{bmatrix} [C_Z(\mathbf{u}_\alpha - \mathbf{u})] \\ [C_{YZ}(\mathbf{u}_i - \mathbf{u})] \end{bmatrix} \qquad (9.20)$$

where the $(n+n') \times (n+n')$ left-hand side matrix contains auto and cross-covariance values between primary and secondary data, the $(n+n') \times 1$ vector of cokriging weights is comprised of the n weights $[\eta_\beta(\mathbf{u})]$ assigned to the primary data and the n' weights $[\eta_{i'}(\mathbf{u})]$ assigned to the secondary data, and the $(n+n') \times 1$ right-hand side covariance vector contains n covariance values between the primary data and the unknown and n' cross-covariance values between the secondary data and the unknown. Note again that all auto and cross-covariance models must be jointly modeled to ensure that Equation 9.20 has a unique solution.

The corresponding SCK posterior variance $\sigma^2_{SCK}(\mathbf{u})$ is written as:

$$\sigma^2_{SCK}(\mathbf{u}) = Var^*_{SCK}\{Z(\mathbf{u})|(n+n')\} = C_Z(\mathbf{0}) - \sum_{\alpha=1}^{n} \eta_\alpha(\mathbf{u})C_Z(\mathbf{u}_\alpha - \mathbf{u})$$

$$- \sum_{i=1}^{n'} \eta_i(\mathbf{u})C_{YZ}(\mathbf{u}_i - \mathbf{u}) \qquad (9.21)$$

where the term $\sum_{\alpha=1}^{n} \eta_\alpha(\mathbf{u})C_Z(\mathbf{u}_\alpha - \mathbf{u})$ represents the reduction in the prior regional variance $C_Z(\mathbf{0})$ brought by the n primary data, and the term $\sum_{i=1}^{n'} \eta_i(\mathbf{u})C_{YZ}(\mathbf{u}_i - \mathbf{u})$ represents the reduction brought by the secondary information. Note again that neither the primary nor the secondary data values are directly involved in Equation 9.21, only their geometry is accounted for through the (cross)covariance models. The posterior SCK variance is therefore independent of the data values, as is its univariate version, the SK variance (Eq. 9.16).

Posterior mean $z^*_{SCK}(\mathbf{u})$ and variance $\sigma^2_{SCK}(\mathbf{u})$ values for local ccdfs $F_Z(\mathbf{u}; z|(n+n'))$ at all 2500 nodes were computed via SCK and are shown in Figure 9.11. The SCK-derived variances are plotted using the same gray

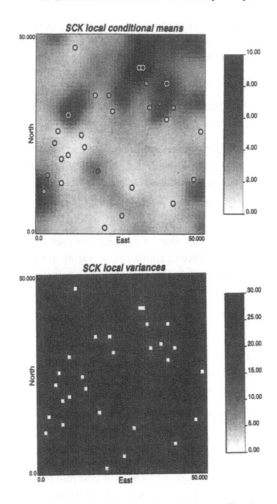

FIGURE 9.11. Posterior means (top) and variances (bottom) of local uncertainty models, given the $n = 29$ primary z-data (shown in circles) and the $N = 2500$ secondary y-data, derived via simple cokriging (SCK).

scale as the SK-derived variances of Figure 9.10, in order to appreciate the variance reduction brought by the secondary variable. Note that the pattern of spatial variation of the secondary data (see Fig. 9.6) has been imprinted on the local posterior mean estimates (top graph of Fig. 9.11), yet at the primary data locations these local means are equal to the primary data values.

9.3.3 Secondary Data as Prior Probabilities

Secondary information can be also accommodated within the indicator framework (Journel 1986). The discussion is hereafter focussed to the categorical case, whereas extension to the continuous case is straightforward. Within the indicator framework, each piece of information, whether related to the primary categorical attribute S or to a secondary attribute Y, is coded into a set of K local prior probabilities, each associated with the k-th state/outcome s_k at location $\mathbf{u}$:

$$Prob\{S(\mathbf{u}) = s_k | \text{local information at } \mathbf{u}\}, \quad k = 1, \ldots, K$$

Various types of information can be coded into indicator data, resulting in different local prior probabilities (e.g., see Goovaerts 1997):

1. Local hard prior probabilities resulting directly from precise measurements of the state s_k at location $\mathbf{u}_\alpha$ (e.g., ground-based samples) expressed as a vector of K hard indicator data valued 0 or 1:

$$i(\mathbf{u}_\alpha; s_k) = \begin{cases} 1 \text{ if category } s_k \text{ prevails at location } \mathbf{u}_\alpha, \quad k = 1, \ldots, K \\ 0 \text{ if not} \end{cases}$$

2. Local information in the form of zero probabilities of occurrence for certain states $s_{k'}$ (e.g., incompatibility of a vegetation type with a given elevation or climatic condition) expressed as an incomplete vector of hard indicator data:

$$i(\mathbf{u}_\alpha; s_k) = \begin{cases} 0 & \text{if } k = k', \ k = 1, \ldots, K \\ \text{(missing)} & \text{otherwise} \end{cases}$$

3. Fuzzy (soft) continuous or categorical information provided, for example, by calibration of remotely sensed imagery at n' locations $\{\mathbf{u}'_{\alpha'}, \alpha' = 1, \ldots, n'\}$, expressed as soft indicator data:

$$v(\mathbf{u}'_{\alpha'}; s_k) = p(s_k | y \in (y_{l-1}, y_l]) \in [0, 1], \quad k = 1, \ldots, K$$

where $p(s_k | y \in (y_{l-1}, y_l])$ is a calibrated frequency of observing state s_k, given that the co-located secondary y-value lies in the range $(y_{l-1}, y_l]$.

Updating these different types of local prior pdfs into posterior models of local uncertainty [i.e., local cpdfs $p_S^*(\mathbf{u}; s_k | (n))$] is accomplished through an SIK or OIK procedure or their extension to the multivariate case, indicator cokriging [e.g., see Zhu and Journel (1993) or Goovaerts (1997)]. All variants of indicator (co)kriging reproduce exactly the hard indicator data at their locations [i.e., $p_S^*(\mathbf{u}_\alpha; s_k | (n)) = i(\mathbf{u}_\alpha; s_k) = 0$ or 1], because kriging is an exact interpolator. Hence, no updating takes place at a primary datum location, which is where a complete vector of K hard indicator data is defined (case 1 earlier). Missing entries of incomplete vector of hard indicator data (case 2), and soft prior probabilities resulting from calibration (case 3) are updated into posterior local cpdfs.

Data integration should always be proceeded by a calibration step before the process of updating the local prior uncertainty models into posterior models of local uncertainty. In the context of integrating ground-based data with remotely sensed imagery, such calibration is critical for evaluating the contribution of the fuzzy (soft) information in reducing uncertainty regarding the unknown primary value. The two approaches described in this section for integrating continuous secondary information (i.e., SKLM and SCK) differ in the way the secondary data are treated. In the case of SKLM, the influence of the secondary information is limited to the mean $m_Z(\mathbf{u})$ of the prior local cdf $F_Z(\mathbf{u}; z)$ at location $\mathbf{u}$. This local mean is determined via a regression model, and the remaining spatial variation is modeled by a residual RF. In the case of SCK, the secondary information influences the local posterior ccdf $F_Z(\mathbf{u}; z|(n))$ directly, according to a specific model of spatial cross-correlation.

The decision to favor one data integration approach versus another depends on the nature of the secondary variables considered, their relation to the ecological attribute of interest, and the complexity of the model being constructed. If the secondary variable is considered to relate to an "average" spatial variation of the primary ecological attribute, then kriging with locally varying means (SKLM) should be preferred; however, if the secondary variable is thought to control the spatial variability of the primary attribute closely, beyond a mere "average" spatial pattern, then cokriging (SCK) should be favored instead. This latter situation is typically encountered when dealing with measurements of the same attribute obtained by different sampling devices (hard vs. soft data); however, beware that as more secondary variables are considered in the integration exercise, the cokriging framework becomes more cumbersome because one has to model an increasing number of auto- and cross-covariances jointly with permissible functions. Multiple (>3) variables, therefore, can be more easily integrated with the SKLM approach rather than with the SCK approach (although several approximations exist for alleviating the modeling effort in the case of SCK) (see Goovaerts 1997).

Differences in the volume and spatial scales of the various types of data can be accommodated by block kriging and factorial (co)kriging. Block kriging (e.g., see Goovaerts 1997) accounts explicitly for different data supports (e.g., the few cubic centimeters of ground-based soil samples vs. the few square meters of pixels of remotely sensed images). Factorial (co)kriging (e.g., see Wackernagel 1995) is used to establish local uncertainty models at specific spatial scales. The fundamental paradigm here consists of: (1) decomposing the primary attribute spatial variability into different components at distinct spatial scales, and (2) filtering (removing) specific spatial components in the estimation procedure. Factorial (co)kriging is the spatial equivalent of frequency domain filtering techniques, with the advantage that the data need not be gridded as in the case of frequency domain approaches.

9.4 Joint Spatial Uncertainty and Its Propagation to Ecological Model Predictions

Local uncertainty models, established either within an indicator or a multi-Gaussian framework, relate only to a single location $\mathbf{u}$ in the domain of

interest D; hence, the term *local*. A series of local uncertainty models of type
(Eq. 9.11), however, does not allow assessment of multipoint or joint spatial
uncertainty [e.g., the probability that the attribute values at $J \leq N$ locations
$\{\mathbf{u}_j, \dots, \mathbf{u}_J\}$ are *jointly* no greater than a given threshold value z_c]. In a land
cover classification scenario, for example, the probability that a particular
land cover type prevails *simultaneously* over a set of locations is required for
assessing classification accuracy about spatial features read from a thematic
map. In a pollution control scenario, uncertainty regarding paths along
which contaminant concentration values are *jointly* greater than a regulatory threshold is critical for designing effective remedial measures.

In addition, local uncertainty models of type (Eq. 9.11) cannot be used for
assessing the impact of the uncertainty regarding the input spatial data
on the uncertainty regarding the outcome of the environmental process
simulator (ecological model). In the case of spatially explicit ecological
models, which consider parameter values at many locations simultaneously,
assessment of model response uncertainty calls for processing alternative
input parameter maps and establishing an associated set (distribution) of
alternative model predictions. For example, alternative input land cover
maps can be input into a crop yield model, resulting in a distribution of net
primary production values (see Fig. 9.1). Such a probabilistic prediction
allows evaluating the impact of the imprecise input spatial information on
the ecological model response, and can be used in ecological risk assessment
studies. The generation of multiple alternative parameter maps, all of which
are consistent with the information (data) available, is accomplished via
stochastic simulation.

Joint spatial uncertainty is modeled by first generating a set of L-simulated
realizations of possible attribute values $\{z^{(l)}(\mathbf{u}_j), j = 1, \dots, J\}, l = 1, \dots, L$
at the J locations of interest, and then computing the joint probability that
all J attribute values are simultaneously no greater than the threshold z_c as:

$$Prob\{Z(\mathbf{u}_j) \leq z_c, j = 1, \dots, J|(n)\} \simeq \frac{1}{L} \sum_{l=1}^{L} \prod_{j=1}^{J} i^{(l)}(\mathbf{u}_j; z_c)$$

where the symbol $\prod$ denotes a product, and $i^{(l)}(\mathbf{u}_j; z_c)$ is the indicator value
defined as $i^{(l)}(\mathbf{u}_j; z_c) = 1$ if $z^{(l)}(\mathbf{u}_j) \leq z_c$, zero otherwise. Note that it suffices
that only one simulated value $z^{(l)}(\mathbf{u}_j)$ at location $\mathbf{u}_j \in (J)$ be greater than the
critical threshold z_c for the set of J values not to be jointly greater than z_c.

The task of generating a set of L-alternative, joint-simulated realizations
(images) $\{z^{(l)}(\mathbf{u}_i), i = 1, \dots, N\}, l = 1, \dots, L$, can be accomplished by sampling L times the N-point ccdf modeling the joint uncertainty regarding the
unknown values of the attribute Z at the N locations $\{\mathbf{u}_i, i = 1, \dots, N\}$,
conditional to the n sample data $\{z(\mathbf{u}_\alpha), \alpha = 1, \dots, n\}$:

$$F_Z(\mathbf{u}_1, \dots, \mathbf{u}_N; z_1, \dots, z_N|(n)) = Prob\{Z(\mathbf{u}_1) \leq z_1, \dots, Z(\mathbf{u}_N) \leq z_N|(n)\}$$

$$(9.22)$$

The N-point ccdf (Eq. 9.22) can be inferred only if the entire N-point cdf is completely known, as is the case for a multi-Gaussian RF fully determined by its covariance function. More general procedures exist for sampling an N-point ccdf, no matter the multivariate distribution type. Such procedures rely on the decomposition of the multipoint ccdf (Eq. 9.22) into a series of one-point ccdfs $F_Z(\mathbf{u}_i; z|(n))$, and the subsequent generation of simulated z-values at each location from these local uncertainty models (Journel 1989).

9.4.1 Monte Carlo Simulation

Once the local conditional model of uncertainty $F_Z(\mathbf{u}; z_k|(n))$ is established at a location $\mathbf{u}$, either within an indicator or multi-Gaussian framework, a large number L of alternative simulated realizations $\{z^{(l)}(\mathbf{u}), \; l = 1, \ldots, L\}$ of the unknown attribute value $z(\mathbf{u})$ can be generated via a quantile reading procedure (see hereafter and Goovaerts 1997). This Monte Carlo simulation procedure for generating L alternative realizations $\{z^{(l)}(\mathbf{u}), \; l = 1, \ldots, L\}$ of a single unknown value $z(\mathbf{u})$ at location $\mathbf{u}$ is graphically depicted in Figure 9.12, and consists of the following steps:

1. A set of L random numbers $\{r^{(l)}, \; l = 1, \ldots, L\}$ uniformly distributed in $[0,1]$ is generated.

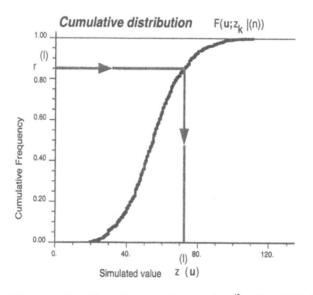

FIGURE 9.12. Monte Carlo Simulation. A random number $r^{(l)}$ uniformly distributed in $[0,1]$ representing a probability value is drawn, and the corresponding $r^{(l)}$th quantile of the local ccdf $F_Z(\mathbf{u}; z_k|(n))$ is read as the simulated value at location $\mathbf{u}$ [i.e., $z^{(l)}(\mathbf{u}) = F_Z^{-1}(\mathbf{u}; r^{(l)}|(n))$].

2. The lth simulated value $z^{(l)}(\mathbf{u})$ is identified to the $r^{(l)}$th quantile of the ccdf $F_Z(\mathbf{u}; z_k|(n))$ as:

$$z^{(l)}(\mathbf{u}) = F_Z^{-1}(\mathbf{u}; r^{(l)}|(n)), \quad l = 1, \ldots, L \qquad (9.23)$$

Each simulated value $z^{(l)}(\mathbf{u})$ represents a possible realization (outcome) of the RV $Z(\mathbf{u})$ modeling local uncertainty at $\mathbf{u}$, and the L-simulated values are distributed according to $F_Z(\mathbf{u}; z_k|(n))$, as:

$$Prob\{Z^{(l)}(\mathbf{u}) \leq z_k|(n)\} = Prob\{F_Z^{-1}(\mathbf{u}; r^{(l)}|(n)) \leq z_k\}$$
$$= Prob\{r^{(l)} \leq F_Z(\mathbf{u}; z_k|(n))\} = F_Z(\mathbf{u}; z_k|(n))$$

because the ccdf $F_Z(\mathbf{u}; z_k|(n))$ is a nondecreasing function of z_k, and the L random numbers $\{r^{(l)}, l = 1, \ldots, L\}$ are uniformly distributed in $[0,1]$.

For the categorical variable case, the Monte Carlo simulation procedure requires building a local ccdf from the estimated local conditional pdfs $p_S^*(\mathbf{u}; s_k|(n))$ for each category s_k. For any ordering of the K possible categories $\{s_k, k = 1, \ldots, K\}$, adding the estimated conditional probabilities of occurrence $p_S^*(\mathbf{u}; s_k|(n))$ results in the local ccdf:

$$F_S^*(\mathbf{u}; s_k|(n)) = \sum_{k'=1}^{k} p_S^*(\mathbf{u}; s_{k'}|(n)), \quad k = 1, \ldots, K$$

which gives the probability that any one of the categories $s_{k'}$, ordered lower or equal to s_k, prevails at location $\mathbf{u}$.

The simulated category $s^{(l)}(\mathbf{u})$ at location $\mathbf{u}$ is then the category that corresponds to the probability interval that includes the random number $r^{(l)}$:

$$s^{(l)}(\mathbf{u}) = s_k, \quad \text{such that } F_S^*(\mathbf{u}; s_{k-1}|(n)) < r^{(l)} \leq F_S^*(\mathbf{u}; s_k|(n))$$

The Monte Carlo simulation procedure provides a set of L alternative outcomes of a continuous or categorical variable at each location $\mathbf{u}$, irrespective of simulated values at any other location $\mathbf{u}' \in D$. The objective, however, is to generate L *joint* realizations of the N component RVs $\{Z(\mathbf{u}_i), i = 1, \ldots, N\}$ of the RF model $\{Z(\mathbf{u}), \mathbf{u} \in D\}$, over the N nodes $\{\mathbf{u}_i, i = 1, \ldots, N\}$ discretizing the domain D. The generation of L such alternative, equiprobable sets of simulated values $\{z^{(l)}(\mathbf{u}_i), i = 1, \ldots, N\}, l = 1, \ldots, L$ calls for conditioning the simulated outcomes at node $\mathbf{u}$ to previously simulated values at neighboring nodes $\mathbf{u}' \neq \mathbf{u}$. This requirement is accomplished by using previously simulated values, as well as the original data, when establishing a local uncertainty model $F_Z(\mathbf{u}; z_k|(n))$ at any node $\mathbf{u} \in D$.

9.4.2 The Sequential Simulation Paradigm

In the sequential simulation algorithm (e.g., see Deutsch and Journel 1998 or Goovaerts 1997) modeling of the N-point ccdf (Eq. 9.22) is approximated by a sequence of N univariate (one-point) ccdfs at each node along a random path. Bayes axiom allows decomposing any N-variate, or N-point, ccdf into a product of N one-point ccdfs as:

$$F_Z(\mathbf{u}_1,\ldots,\mathbf{u}_N|(n)\} = F_Z(\mathbf{u}_N;z_N|(n+N-1))$$

$$\cdot F_Z(\mathbf{u}_{N-1};z_{N-1}|(n+N-2))$$

$$\vdots$$

$$\cdot F_Z(\mathbf{u}_2;z_2|(n+1)) \cdot F_Z(\mathbf{u}_1;z_1|(n)) \qquad (9.24)$$

where, for example, $F_Z(\mathbf{u}_N;z_N|(n+N-1))$ is the conditional cdf of $Z(\mathbf{u}_N)$ given the n sample data and the $N-1$ realizations $\{z^{(l)}(\mathbf{u}_i),\ i=1,\ldots, N-1\}$. The target covariance model $C_Z(\mathbf{h})$ is reproduced by conditioning each one-point ccdf to the n sample data as well as to simulated values generated at all previously visited nodes.

The sequential simulation algorithm proceeds with the following steps:

1. Establish a random path visiting once and only once, all nodes, $\{\mathbf{u}_i,\ i = 1,\ldots,N\}$ discretizing the domain of interest D (these N nodes need not lie on a regular grid). A random visiting sequence ensures that no artifact of spatial continuity is introduced into the simulation by a specific path visiting the N nodes.
2. At the first visited node $\mathbf{u}_1$:
   a. Model, using either a parametric or nonparametric approach, the local ccdf of $Z(\mathbf{u}_1)$ conditional to the n original data $\{z(\mathbf{u}_\alpha),\ \alpha = 1,\ldots,n\}$:

   $$F_Z(\mathbf{u}_1;z_1|(n)) = Prob\{Z(\mathbf{u}_1) \leq z_1|(n)\}$$

   b. Generate, via the Monte Carlo drawing relation (Eq. 9.23), a simulated value $z^{(l)}(\mathbf{u}_1)$ from this ccdf $F_Z(\mathbf{u}_1;z_1|(n))$, and add it to the conditioning data set, now of dimension $n+1$, to be used for all subsequent local ccdf determinations.
3. At the ith node $\mathbf{u}_i$ along the random path:
   a. Model the local ccdf of $Z(\mathbf{u}_i)$ conditional to the n original data and the $i-1$ nearby previously simulated values $\{z^{(l)}(\mathbf{u}_j),\ j=1,\ldots,i-1\}$:

   $$F_Z(\mathbf{u}_i;z_i|(n+i-1)) = Prob\{Z(\mathbf{u}_i) \leq z_i|(n+i-1)\}$$

   b. Generate a simulated value $z^{(l)}(\mathbf{u}_i)$ from this ccdf and add it to the conditioning data set, now of dimension $n+i$.
4. Repeat step 3 until all N nodes along the random path are visited.

The result is a set of simulated values $\{z^{(l)}(\mathbf{u}_i),\ i=1,\ldots,N\}$, which represents the lth discrete realization of the RF $\{Z(\mathbf{u}),\ \mathbf{u} \in D\}$ over the N nodes $\{\mathbf{u}_i,\ i = 1,\ldots,N\}$. A set of L realizations $\{z^{(l)}(\mathbf{u}_i),\ i=1,\ldots,N\}, \ l= 1,\ldots,L$ can be generated by repeating the entire procedure, with a different random visiting path for each realization. Note that local uncertainty models might relate to either continuous (ccdf) or categorical (cpdf) attributes.

A set of $L = 50$ alternative realizations (images) of the attribute spatial distribution, conditioned to both the $n = 29$ primary z-data and the $N = 2500$ secondary y-data (shown in Fig. 9.6), was generated via sequential Gaussian simulation (SGSIM) (e.g., see Goovaerts 1997). The conditional distributions at each location $\mathbf{u}$ (local ccdfs) were established via SCK in a multi-Gaussian framework, and two out of the $L = 50$ simulated realizations are shown in Figure 9.13.

Note that both realizations in Figure 9.13 inherit spatial characteristics of the secondary y-data, and reproduce the $n = 29$ primary data values at their

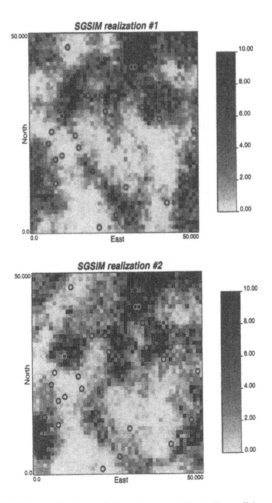

FIGURE 9.13. Two realizations of the primary attribute Z conditioned to the 29 primary z-data (circles) and the 2500 secondary y-data, generated via sequential Gaussian simulation (SGSIM).

sampling locations. All realizations reproduce, on average, the histogram of the primary data and the covariance model adopted for the simulations (not shown). Because the conditioning data set could eventually become very large if N is large, only the $n(\mathbf{u}_i)$ original data and the $N(\mathbf{u}_i)$ previously simulated values within a local neighborhood $W(\mathbf{u}_i)$ centered on $\mathbf{u}_i$ are considered for simulation at node $\mathbf{u}_i$. Long-range spatial correlation structures can be reproduced by a multiple-grid simulation procedure, whereby simulated values on a coarse grid are first generated, and nodes on a finer grid are subsequently visited (Deutsch and Journel 1998).

The one-point conditional probability that the unknown primary value $z(\mathbf{u})$ at any location $\mathbf{u}$ within D is greater than any threshold z can be calculated from the L realizations as:

$$Prob\{Z(\mathbf{u}) > z|(n)\} \simeq \frac{1}{L}\sum_{l=1}^{L} i^{(l)}(\mathbf{u};z), \quad \forall z \qquad (9.25)$$

where $i^{(l)}(\mathbf{u};z)$ is the indicator transform of the lth simulated value $z^{(l)}(\mathbf{u})$ at location $\mathbf{u}$, defined as $i^{(l)}(\mathbf{u};z) = 1$, if $z^{(l)}(\mathbf{u}) > z$, zero if not. Such local conditional probabilities for the unknown attribute value $z(\mathbf{u})$ at any location $\mathbf{u} \in D$ to exceed the threshold $z = 4.53$ were calculated from the $L = 50$ SGSIM-generated realizations, and are shown in Figure 9.14.

A set of $L = 50$ alternative images of the spatial distribution of a categorical attribute, conditioned to the $n = 29$ categorical data (see Fig. 9.6), was generated via sequential indicator simulation (SISIM) (e.g., see Journel

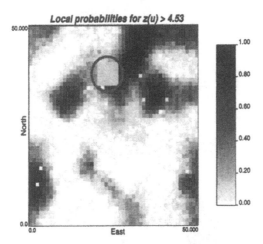

FIGURE 9.14. Local conditional probabilities for $z(\mathbf{u}) > 4.53$ calculated from 50 simulated realizations generated via SGSIM. The gray scale of the rectangle area in the region highlighted by the circle corresponds to the joint probability for the attribute value at all locations in this area to be simultaneously greater than threshold $z = 4.53$.

208   Phaedon C. Kyriakidis

1989; Bierkens and Burrough 1993; Goovaerts 1997). Local conditional probabilities for each category to prevail at location **u** (local cpdfs) were derived through simple indicator kriging, and two out of the $L = 50$ SISIM-generated realizations are shown in Figure 9.15. In this particular example, possible cross-correlations between data of different categories were not taken into account. In other words, only the indicator covariance $C_I(\mathbf{h}; s_2)$ between any two indicator data $i(\mathbf{u}_\alpha; s_2)$ and $i(\mathbf{u}_\alpha + \mathbf{h}; s_2)$ defined for category s_2 was considered, not the indicator cross-covariances $C_I(\mathbf{h}; s_2, s_1)$ and $C_I(\mathbf{h}; s_2, s_3)$ quantifying transition probabilities from category s_1 to s_2 and from category s_3 to s_2, respectively.

The one-point conditional probability that the unknown state $s(\mathbf{u})$ at any location **u** within D belongs to category s_2 [i.e., $Prob\{S(\mathbf{u}) = s_2|(n)\}$ given

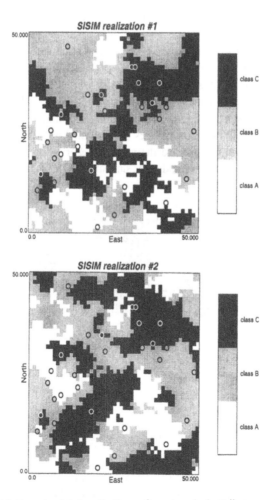

FIGURE 9.15. Two simulated realizations of a categorical attribute, conditioned to 29 class labels, generated via sequential indicator simulation (SISIM).

the $n = 29$ class labels (shown in Fig. 9.6)] is calculated from the $L = 50$ SISIM realizations, and shown in the top graph of Figure 9.16.

All realizations reproduce, on average, the regional proportions of the three categories and the indicator covariance models adopted for the simulations. For example, the histogram of simulated areal proportions occupied by category s_2 from the $L = 50$ realizations, is shown in the bottom graph of Figure 9.16. The sample proportion $p_S(s_2) = 0.48$ for category s_2 is

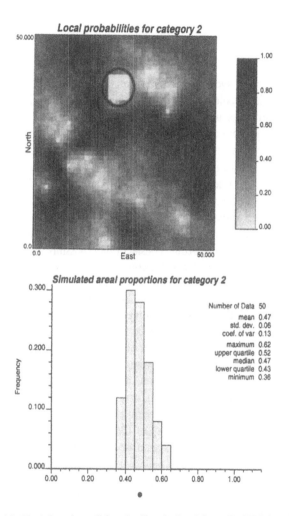

FIGURE 9.16. (Top) Local conditional pdfs calculated from 50 SISIM realizations. The gray scale of the rectangular area in the region highlighted by the circle corresponds to the joint probability for the attribute value at all locations in this area to belong simultaneously to category s_2. (Bottom) Histogram of simulated areal proportions occupied by category s_2 obtained from the 50 SISIM realizations. The black dot corresponds to the sample regional proportion $p_S(s_2) = 14/29 = 0.48$.

well reproduced by the average simulated proportion $\bar{p}_S(s_2) = 0.47$ calculated over the 50 realizations.

These simulated realizations of the categorical attribute $S(\mathbf{u})$ could be seen as alternative maps of land cover type, and could serve as input parameter maps in a spatially explicit biogeochemical model, for example. The outcome of running the biogeochemical model on this set of simulated land cover maps would be a histogram of alternative net primary production values, for example, which can be used in ecological risk assessment (see Kyriahidis and Dungan 2001).

9.5 Summary

Spatial databases, such as GIS coverages, providing input parameters for spatially explicit ecological models, are inherently uncertain; they should not be regarded as unquestionable pieces of information. Estimation of classification procedures yielding parameter maps in the form of GIS data products involve a set of prior decisions. Within the stochastic modeling framework presented in this chapter, reporting a single estimated value $z^*(\mathbf{u})$ for the unknown value $z(\mathbf{u})$ at location $\mathbf{u}$ should be seen as the consequence of the decision to retain a specific outcome or quantile of the uncertainty model at $\mathbf{u}$ [i.e., a quantile of the local ccdf $F_Z(\mathbf{u}; z|(n))$]. The widely used least squares criterion dictates that the estimate $z^*(\mathbf{u})$ be the mean of that local ccdf $F_Z(\mathbf{u}; z|(n))$. One could consider other, possibly more appropriate, optimality criteria, in which case the optimal estimates would differ from the mean of the local ccdf $F_Z(\mathbf{u}; z|(n))$. Local uncertainty models are prerequisites for any risk-qualified prediction, and should be established prior and independently of the particular estimate reported. These sorts of local uncertainty models should be provided to the users of geospatial information instead of forcing a particular set of estimated values corresponding to a possibly arbitrary optimality criterion.

A set of locally optimal estimates, when input into an environmental process simulator (ecological model), may not result in an optimal estimate for the unknown value of the simulator response (ecological model prediction). For example, the map of the conditional means of the local ccdfs, when input into a process simulator, does not yield the mean of all possible response values due to nonlinearities in the process simulator. In addition, the map of conditional means does not reflect the same pattern of spatial correlation displayed by the sample data. Maps of estimated values typically exhibit an oversmooth spatial variation. Covariance values $Cov\{Z^*(\mathbf{u}), Z^*(\mathbf{u} + \mathbf{h})\}$ between any two estimates $Z^*(\mathbf{u})$ and $Z^*(\mathbf{u} + \mathbf{h})$ are greater than those derived from the covariance model $C_Z(\mathbf{h})$ for the same lag distance $\mathbf{h}$. Such an artificially smooth spatial character can severely bias the process simulator responses; in certain cases, the resulting estimated responses might even be physically implausible.

Stochastic simulation has a goal different from deriving a set of locally accurate estimates. Stochastic simulation aims at modeling joint spatial uncertainty, and transferring that uncertainty through the ecological model to characterize uncertainty regarding model predictions. Conditional stochastic simulation "freezes" (reproduces exactly) reliable pieces of information (i.e., the hard data) and allows any histogram and spatial (cross)-correlation model to be reproduced within expected fluctuations. The remaining "space" of spatial uncertainty beyond these data constraints is explored via a random number generator for simulating attribute outcomes and a procedure for arranging these simulated values in space.

Sequential simulation algorithms rely on local uncertainty models established prior to Monte Carlo selection; hence, it is important to understand the various assumptions and tradeoffs involved in building these models of local uncertainty. For example, one should bear in mind the extremely specific characteristics of multi-Gaussian RF models, such as maximum spatial entropy. The congenial inference of a multi-Gaussian RF model should be weighted against the reduced ability to accommodate significant patterns of spatial continuity for extreme attribute values. In many applications, this spatial distribution of extreme values is of critical importance for the process simulator being considered.

Uncertainty models, regarding either spatial data or ecological model predictions should be continuously updated as more relevant information becomes available. For example, if more ground-based data are acquired, then the remotely sensed information should be calibrated anew in order to adjust the local uncertainty models, the simulated realizations, and the uncertainty model for the process simulator response. Uncertainty models, therefore, are not static; rather, they are instead dynamical entities that evolve continuously as a function of the information available. Each uncertainty assessment step should involve a prior evaluation of the reliability of each piece of information, and a subsequent updating of all prior uncertainty models to account for this additional information. The stochastic framework shown in Figure 9.1 and analyzed in this chapter proposes an approach for such reliability assessment and integration.

Spatial uncertainty assessment should result from an interdisciplinary and dynamic modeling philosophy, and account for all relevant expert knowledge and scientific evidence. All uncertainty models reflect the series of expert decisions involved at various steps of the modeling effort; hence, they can never be considered objective. It is precisely the sequence of all well-developed and properly documented decisions that constitutes a model. Spatial uncertainty models should be therefore transparent to the users of geospatial information (i.e., the latter should have access to all decisions and tradeoffs involved in the modeling exercise). Only then would they be able to check the relevance and impact of all such decisions on their modeling objective.

Acknowledgments. This work was completed while the author was at the Department of Geological and Environmental Sciences of Stanford University. Funding was partially provided by the Stanford Center for Reservoir Forecasting. The author would like to thank Professor Andre Journel and Professor Pamela Matson for their constructive reviews of this manuscript.

References

Anderson, T. 1958. An introduction to multivariate statistical analysis. John Wiley & Sons, New York.

Atkinson, P.M., R. Webster, R., and P.J. Curran. 1992. Cokriging with ground-based radiometry. Remote Sensing of the Environment 41:45–60.

Bierkens, M., and P. Burrough. 1993. The indicator approach to categorical soil data. II. Application to mapping and land use suitability analysis. Journal of Soil Science 44:369–81.

Cressie, N. 1991. Statistics for spatial data. John Wiley & Sons, New York.

Deutsch, C.V., and A.G. Journel. 1998. GSLIB: geostatistical software library and user's guide, second edition. Oxford University Press, New York.

Dungan, J.L., D.L. Peterson, and P.J. Curran. 1993. Alternative approaches for mapping vegetation quantities using ground and image data. Pages 237–61 in W.K. Michener, J.W. Brunt, and S.G. Stafford, eds. Environmental information management and analysis: ecosystem to global scales. Taylor and Francis, London.

Englund, E. 1993. Spatial simulations: environmental applications. Pages 432–37 in M. Goodchild, B. Parks, and L. Steyaert, eds. Environmental modeling with GIS. Oxford University Press, New York.

Gohin, F., and G. Langlois. 1993. Using geostatistics to merge in situ measurements and remotely-sensed observations of sea surface temperature. International Journal of Remote Sensing 14(1):9–19.

Goovaerts, P. 1997. Geostatistical for natural resources evaluation. Oxford University Press, New York.

Graham, R.L., C.T. Hunsaker, R.V. O'Neill, and B.L. Jackson. 1991. Ecological risk assessment at the regional scale. Ecological Applications 1(2):196–206.

Isaaks, E., and R. Srivastava. 1989. An introduction to applied geostatistics. Oxford University Press, New York.

Journel, A.G. 1983. Non-parametric estimation of spatial distributions. Mathematical Geology 15(3):445–68.

Journel, A.G. 1986. Constrained interpolation and quanlitative information. Mathematical Geology 18(3):269–86.

Journel, A.G. 1989. Fundamentals of geostatistics in five lessons. Volume 8: short course in geology. American Geophysical Union, Washington, DC.

Journel, A.G., and C.V. Deutsch. 1993. Entropy and spatial disorder. Mathematical Geology 25(3):329–55.

Kyriakidis, P.C., A.M. Shortridge, and M.F. Goodchild. 1999. Geostatistics for conflation and accuracy assessment of digital elevation models. International Journal of Geographical Information Science 13(7):677–707.

Kyriakidis, P.C., and J.L. Dungan. 2001. A geostatistical approach for mapping thematic classification accuracy and evaluating the impact of inaccurate spatial data on ecological model predictions. Environmental and Ecological Statistics. In press.

Massmann, J., and R. Freeze. 1987. Groundwater contamination from waste management sites: the interaction between risk-based engineering design and regulatory policy: 1. Methodology. Water Resources Research 23(2):351–67.

Potter, C.S., J.T. Randerson, C.B. Field, P.A. Matson, P.M. Vitousek, H.A. Mooney, et al. 1993. Terrestrial ecosystem production: a process model based on global satellite and surface data, Global Biogeochemical Cycles 7(4):811,841.

Rossi, R.E., D.J. Mulla, A.G. Journel, and E.H. Franz. 1992. Geostatistical tools for modeling and interpreting ecological spatial dependence. Ecological Monographs 62(2):277–314.

Rossi, R.E., P.W. Broth, and J.J. Tollefson. 1993. Stochastic simulation for characterizing ecological spatial patterns and appraising risk. Ecological Applications 3(4):719–35.

Wackernagel, H. 1995. Multivariate geostatistics. Springer-Verlag, Berlin.

Zhu, H., and A.G. Journel. 1993. Formatting and integrating soft data: stochastic imaging via the Markov-Bayes algorithm. Pages 1–12 (Volume 1) in A. Soares, ed. Geostatistics troia 1992. Kluwer Academic Publishers, Dordrecht, The Netherlands.

10
Uncertainty and Spatial Linear Models for Ecological Data

Jay M. Ver Hoef, Noel Cressie, Robert N. Fisher, and Ted J. Case

All models are wrong ... we make tentative assumptions about the real world which we know are false but which we believe may be useful ... the statistician knows, for example, that in nature there never was a normal distribution, there never was a straight line, yet with normal and linear assumptions, known to be false, he can often derive results which match, to a useful approximation, those found in the real world.
(Box 1976)

Models are not perfect; they do not fit the data exactly and they do not allow exact prediction. Given that models are imperfect, we need to assess the uncertainties in the fits of the models and their ability to predict new outcomes. The goals of building models for scientific problems include (1) understanding and developing appropriate relationships between variables, and (2) predicting variables in the future or at locations where data have not been collected. Ecological models range in complexity from those that are relatively simple (e.g., linear regression) to those that are very complex (e.g., ecosystem models, forest-growth models, and nitrogen-cycling models). In a mathematical model, parameters control the relationships between variables in the model. In this framework of parametric modeling, *inference* is the process whereby we take output (data) and estimate model parameters, whereas *deduction* is the process whereby we take a parameterized model and obtain output (data) or deduce properties. We often add random components in both inference and deduction to reflect a model's lack-of-fit and our uncertainty about predicting outcomes. Complex models in ecology have largely been of the deductive type, where the scientist takes some values of parameters (usually obtained from an independent data source or chosen from a reasonable range of values) and then simulates results based on model relationships. These models may be quite realistic, but the manner in which their parameters are obtained for the simulations is questionable. On the other hand, statistical models like linear regression posit simple relationships among the variables that may be unrealistic, but have the virtue that the uncertainty of the estimated parameters in the model can typically be quantified.

214

This chapter will demonstrate the usefulness of the spatial linear model for making inferences from ecological data. The spatial linear model is at the heart of many spatial methods, from optimal prediction to designed experiments (e.g., Ver Hoef 1993; Ver Hoef and Cressie 1993). Classical linear regression both assumes a linear relation on the explanatory variables and, conditional on these variables, that the responses are independent. For many ecological applications, this last assumption is simply not appropriate. In this chapter, we relax the assumption that the responses are independent.

We finish this introduction with a general philosophical framework for spatial models in ecology. Next, we describe how the spatial linear model uses simple linear relationships and then absorbs the uncertainty caused by unobserved explanatory variables, even residual nonlinear relationships, through spatial autocorrelation in the responses. In Section 10.2, we define spatial autocorrelation for the linear model and give properties, models, and methods of estimation. We also consider the effect of autocorrelation on prediction and estimation. We give examples in Section 10.3, including a simulation example to clarify trend versus autocorrelation, a simulation to show estimation methods and properties, and finally an example on real data. In Section 10.4, we move beyond the spatial linear model, and outline how hierarchical models can be used for more complex ecological situations.

10.1.1 Spatial Models in Ecology

Consider the following very general model:

$$y_i(\mathbf{s}, t) = f_i(\mathbf{x}_i(\mathbf{s}, t), \boldsymbol{\theta}), \tag{10.1}$$

where $y_i(\mathbf{s}, t)$ is considered a response for the ith variable located in space, with the spatial coordinates contained in the vector $\mathbf{s}$, and also occurring at time t. For example, let us suppose that $y_i(\mathbf{s}, t)$ is the biomass of some species at location $\mathbf{s}$ (e.g., s_1 is longitude and s_2 is latitude) at time t. Next, let us consider the nature of $f_i(\mathbf{x}_i(\mathbf{s}, t), \boldsymbol{\theta})$. Equation (10.1) says, simply, that $y_i(\mathbf{s}, t)$ depends in some functional way on other "explanatory" variables, contained in the vector $\mathbf{x}_i(\mathbf{s}, t)$, which itself is indexed by the spatial and temporal coordinates $\mathbf{s}$ and t; note that a random error term in Equation (10.1) is not explicitly given. The relationship between response and explanatory variables is controlled by parameters $\boldsymbol{\theta}$. Note further that the explanatory variables contained in the vector $\mathbf{x}_i(\mathbf{s}, t)$ may have different time and space indices. For example, $y_i(\mathbf{s}, t)$ might depend on $x_{ij}(\mathbf{u}, v)$ [the jth component of the vector $\mathbf{x}_i(\mathbf{s}, t)$] where $\mathbf{u}$ denotes some place in space other than $\mathbf{s}$, and v denotes some time at or prior to t. For the plant biomass example, the vector $\mathbf{x}_i(\mathbf{s}, t)$ might contain variables such as soil moisture, soil pH, soil nitrogen, soil carbon, the presence of herbivores, temperature at that particular location, the amount of sunlight at that location, or the numbers and types of neighboring plants, as well as some of the same variables in addition to others at another location at times past and present. It is obvious that this function $f_i(\mathbf{x}_i(\mathbf{s}, t), \boldsymbol{\theta})$ can be very complex. It is also worth considering the fundamental nature of $f_i(\mathbf{x}_i(\mathbf{s}, t), \boldsymbol{\theta})$; namely, whether it is random or deterministic. Are the x-variables themselves inherently random? That is, if we knew all of the explanatory variables and their functional relationships, could we predict $y_i(\mathbf{s}, t)$ perfectly? It does not really matter practically, because the ecological mechanisms are never known well enough to declare Equation (10.1) as capturing all of the variability in the responses. In the absence of total knowledge, we often add random components to a model to reflect our uncertainty or inability to model data perfectly.

10.1.2 Spatial Linear Model

Perfect prediction of the biomass of some species at location $\mathbf{s}$ at time t is not possible; therefore, we model $Y_i(\mathbf{s}, t)$ as a random quantity that is not perfectly predictable. There are many ways to incorporate random effects in $f_i(\mathbf{x}_i(\mathbf{s}, t), \boldsymbol{\theta})$. One might devise a fairly complex, nonlinear model of functional relationships that includes measured variables that are believed to be important. Some of these variables may themselves be considered random, so we can simulate responses. For example, take a forest-simulation model. The locations of the trees can be chosen randomly, and some of their growth parameters can be taken from a probability distribution with some mean and variance. Assuming a real forest follows the model, we can run the model for many years and deduce spatial properties of a forest from its final run. In general, assuming we know the parameters $\boldsymbol{\theta}$ (e.g., the mean and standard deviation of the density of trees and the mean and standard deviation of the forest growth rate for a given soil type, among others), we can simulate results from the model and determine spatial and temporal characteristics from the resulting output. This is deduction. In reality, we observe only the data, which are the outputs from some unknown physical process. We would like to assign a model to the physical process and use the observed data to say something about the model and its parameters. In statistics, this is called inference. How do we estimate the unknown parameters $\boldsymbol{\theta}$, from data, and incorporate the uncertainty of estimating a model's parameters and its predictions? Linear models in statistics are important because the models are tractable, and we can perform inference as well as deduction. That is the main topic of this chapter, although we will also discuss some newer approaches using hierarchical models.

Depending on objectives and data collection methods, the problem under study may involve data that are only spatial in nature; that is, the data may have been collected at one particular time, or the time component is removed through aggregation. In what is to follow, therefore, we shall focus attention on the spatial linear model:

$$Y_i(\mathbf{s}) = \beta_0 + \beta_1 x_{i1}(\mathbf{s}) + \cdots + \beta_p x_{ip}(\mathbf{s}) + Z_i(\mathbf{s}) \tag{10.2}$$

or

$$Y_i(\mathbf{s}) = [\mathbf{x}_i(\mathbf{s})]' \boldsymbol{\beta} + Z_i(\mathbf{s}) \tag{10.3}$$

where $x_{ij}(\mathbf{s})$ is a known, observed value of an explanatory variable that is believed to be important in determining the response, β_j is a parameter that provides a weighting for the observed $x_{ij}(\mathbf{s})$, and $Z_i(\mathbf{s})$ is a zero-mean, possibly autocorrelated, random error. Notice the subtle change in notation from θ to $\boldsymbol{\beta}$ in going from Equation (10.1) to Equation (10.2). We reserve θ for parameters of the true model, and use $\boldsymbol{\beta}$ for parameters of some model that we apply to data. Equation (10.2) can be thought of as

$$(\text{response}) = (\text{deterministic mean structure}) + (\text{random error variation}) \tag{10.4}$$

The usual assumption in linear models is that all error variation in Equation (10.4) is composed of independent random variables. This is a good approximation in designed field experiments where $x_{ij}(\mathbf{s})$ is an indicator (0 or 1) that the jth treatment was applied at location $\mathbf{s}$ (Hinkelmann and Kempthorne 1994, pg. 164). Ecological studies, however, are often observational. That is, we cannot randomize, and even if we do, we may want to condition on the design that we obtained and consider $\{Z_i(\mathbf{s})\}$ to be autocorrelated random quantities (Ver Hoef and Cressie 1993).

10.2 Autocorrelation in the Spatial Linear Model

What is a precise definition of *spatial autocorrelation*? In general, it means that a random variable may be correlated with itself when separated by some nonzero distance, and it has a similar interpretation in time series. The autocorrelation mathematically inherent in the random process $\{Z_i(\mathbf{s})\}$ is defined for all $\mathbf{s}$ and $\mathbf{u}$ to be

$$\rho_i(\mathbf{s}, \mathbf{u}) \equiv \frac{Cov[Z_i(\mathbf{s}), Z_i(\mathbf{u})]}{Var[Z_i(\mathbf{s})]^{1/2} Var[Z_i(\mathbf{u})]^{1/2}} \tag{10.5}$$

There are several things to notice about Equation (10.5). First, note that the index i is the same for both random variables, indicating correlation for the *same variable type*. For example, this could be the correlation between biomass at one place with biomass at another place. This is the reason for the "auto" part of the term. In statistics courses, we often first learn of the term *correlation* for the case of two *different types of random variables*, say biomass and soil moisture.

The second thing to notice about Equation (10.5) is that it is *not* the most general way to express spatial pattern, for which there can also be a contribution from the fixed effects part, $[\mathbf{x}_i(\mathbf{s})]' \boldsymbol{\beta}$ in Equation (10.3), which is called a "trend surface." For example, suppose that $Z_i(\mathbf{s})$ and $Z_i(\mathbf{u})$ have zero mean and zero correlation and $[\mathbf{x}_i(\mathbf{s})]'\boldsymbol{\beta} = s_1\boldsymbol{\beta}$ where s_1 is the spatial coordinate longitude. When $Y_i(\mathbf{s})$ and $Y_i(\mathbf{u})$ are nearby, they will tend to be more similar because of the presence of longitude as the fixed effect but, by Equation (10.5), $Y_i(\mathbf{s})$ and $Y_i(\mathbf{u})$ have $\rho_i(\mathbf{s}, \mathbf{u}) = 0$, and so they are *not* autocorrelated. We therefore prefer to reserve the word *autocorrelation* for properties of the random error components of the *model* that is applied to data, and use the words *spatial pattern* for the data themselves.

What, exactly, is autocorrelation? As seen in Equation (10.5), autocorrelation is the tendency for random variables to covary as a function of their locations in space. Although we can describe this statistically, what does it mean ecologically? It may be that autocorrelation, as a property of random quantities, actually exists in nature. On the other hand, let us consider again some "true" model given by a very complex relation (Eq. 10.1) with many explanatory variables with nonlinear relationships, and the much simpler (Eq. 10.2), which is linear and likely has only a small subset of the explanatory variables. Then, deviations from linearity and the effects of all of the explanatory variables that were not observed, and thus left out of the fixed mean structure, can show up in the error term in the spatial linear model, Equation (10.4). These explanatory variables often exhibit pattern themselves. As a result, the response variable may covary in ways that are not explained by the observed explanatory variables, which leads us to *model* the residuals as being autocorrelated. Hence, autocorrelation can be thought of as absorbing the spatial effects of nonlinear and unobserved explanatory variables through a stochastic error process.

10.2.1 Prediction and Estimation

The linear model is given by,

$$\mathbf{Y} = \mathbf{X}\boldsymbol{\beta} + \mathbf{Z}, \tag{10.6}$$

where $\mathbf{Y}$ is a vector of response variables (random), $\mathbf{X}$ is the design matrix (fixed) of observed covariates (e.g., indicators of treatments or values of covariates), $\boldsymbol{\beta}$ is a vector of unknown parameters, and $\mathbf{Z}$ is a vector of random variables. Equation (10.2) is a special case of Equation (10.6) where we make explicit use of the spatial indices to define the multivariate vector of data and model the random errors $\mathbf{Z}$. What are some of the usual inferences one wants to make with such a model? There are typically four:

1. The ecologist will often estimate β_i and interpret its effect on the response. Along with reporting the estimated value, a confidence interval expresses the uncertainty in the estimated value.

2. The ecologist may also be interested in estimating some linear function of $\boldsymbol{\beta}$, $g(\boldsymbol{\beta}) = \boldsymbol{\lambda}'\boldsymbol{\beta}$. For example, in simple linear regression, we may want to estimate the linear component at some new value $\mathbf{x}_0$ of the covariate, namely $g(\boldsymbol{\beta}) = \mathbf{x}_0'\boldsymbol{\beta} = \beta_0 + \beta_1 x_{01} + \beta_2 x_{02} + \ldots + \beta_p x_{0p}$. We are in the same

situation when making inferences for linear contrasts in designed experiments.

3. The ecologist might also want to "predict" a new value of the response variable $y(s_0)$ at some location s_0 where data have not been collected. Kriging is one such method of prediction, defined for a spatial linear model (e.g., see Ver Hoef 1993).

4. The ecologist might want to predict some function of the observed and unobserved response variables. For example, block kriging attempts to predict the average value of the response variable over a region.

These inferences can be broadly classified into two categories, prediction and estimation. We will use the word *estimation* when making inferences on *parameters* (unobservable fixed quantities) in the model, and *prediction* when making inferences on the *response variable* (potentially observable random quantity). Uses 1 and 2 are estimation problems (the former is just a special case of the latter, where λ is a vector of all 0's and one 1). Uses 3 and 4 are prediction problems (likewise, the former case of point prediction is a special case of the latter).

10.2.2 Spatial Regression and Kriging

Now, let us put to use the spatial indices contained in Equation (10.2). We shall assume that the errors $\mathbf{Z}$ in Equation (10.6) are spatially autocorrelated, as described by Equation (10.5). The autocorrelation will depend on further unknown parameters. We assume $E(\mathbf{Z}) = \mathbf{0}$ and $var(\mathbf{Z}) = \Sigma(\boldsymbol{\alpha})$, where we denote the dependence of the covariance matrix Σ on parameters $\boldsymbol{\alpha}$. The parameters control the strength and range of autocorrelation. For the purposes of estimation and prediction discussed earlier, $\boldsymbol{\alpha}$ contains nuisance parameters that also need to be estimated; however, the parameters in $\boldsymbol{\alpha}$, (e.g., the range of spatial autocorrelation) are sometimes of interest themselves. Spatial regression is largely concerned with estimation of $\boldsymbol{\beta}$ under these assumptions, and kriging is another name for prediction under these assumptions. Ordinary kriging occurs when we make the assumption that $\mathbf{X}\boldsymbol{\beta} = \mathbf{1}\mu$, where $\mathbf{1}$ is the vector of ones. This simply means that all random variables have a common mean, and has been called *mean stationarity*. Universal kriging occurs when we allow additional covariates in $\mathbf{X}\boldsymbol{\beta}$. We can make a further distinction that, if the matrix $\mathbf{X}$ is composed of the spatial coordinates themselves, we shall call it "kriging with trend."

10.2.3 Effects of Autocorrelation

The effects of autocorrelation are different for estimation versus prediction. Legendre (1993) asks, "spatial autocorrelation: trouble or new paradigm?" As we shall see, it depends. Consider the following simple "thought" experiment. Suppose that we have 10 normally distributed random variables located on a line (e.g., on a transect through space) with common mean

μ and common variance σ^2. Further, suppose that these 10 random variables have positive autocorrelation of 1, the maximum possible value. This means that $Y(1) = Y(2) = \ldots = Y(10)$. How would we simulate these data? Choose $Y(1)$ from the normal distribution and set all others equal to $Y(1)$. In contrast, let us consider the case where we have 10 normally distributed random variables located on a line with common mean μ and common variance σ^2, but they are all independent.

Now, suppose we want to *estimate* μ. Relative to the independence case, the case of maximum autocorrelation will have poor efficiency in estimating μ because it has only one independent observation. Now consider *prediction*. Suppose we had not observed $Y(5)$ and we wanted to predict it from the other data. Here, we will have perfect prediction for the maximum autocorrelation case, and this will be better than it is for the independence case. In general, positive autocorrelation allows more precise inference for prediction, but less precise inference for estimation.

10.2.4 Properties and Models for Autocorrelation

A central concept in statistical inference is that of replication. Without replication, it is difficult to estimate quantities and assess uncertainty. This is the reason for the classical assumption that data are produced independently with errors that have a common mean (usually 0) and common variance. What happens when they are not independent? The usual assumptions are that the errors still have a common mean, called *mean stationarity*, and that the autocovariance depends only on the spatial relationship between two variables (not their exact locations). That is,

$$C(\mathbf{h}) = cov(Z(\mathbf{s}), Z(\mathbf{s} + \mathbf{h})), \qquad (10.7)$$

where $C(\mathbf{h})$ is called the *autocovariance function*, and it depends only on $\mathbf{h}$, not $\mathbf{s}$. Thus, we get replication by having multiple pairs of data that have similar spatial relationships to each other. Mean stationarity and Equation (10.7), together, are called *second-order stationarity*. Then from Equation (10.5), autocorrelation becomes

$$\rho(\mathbf{h}) = C(\mathbf{h})/C(\mathbf{0}). \qquad (10.8)$$

Assumption (10.7) can be replaced by the more general

$$2\gamma(\mathbf{h}) = var(Z(\mathbf{s}) - Z(\mathbf{s} + \mathbf{h})) \qquad (10.9)$$

where $2\gamma(\mathbf{h})$ is called the *variogram* and $\gamma(\mathbf{h})$ is called the *semivariogram*. Mean stationarity and Equation (10.9), together, are called *intrinsic stationarity*. Notice that for second-order stationarity we have the relationship

$$\gamma(\mathbf{h}) = C(\mathbf{0}) - C(\mathbf{h}) \qquad (10.10)$$

so we can move freely between variograms, semivariograms, autocorrelation, and autocovariance functions. To increase replication even more, an additional assumption that is often made, if reasonable, is that of isotropy, where the autocovariance function and variogram depend only on distance (and not direction).

Definition (10.7) needs a model in order to explain exactly how the covariance changes with $\mathbf{h}$. An isotropic model that we use throughout this chapter is the spherical model:

$$C(||\mathbf{h}||) = \eta_n I[||\mathbf{h}|| = 0] + \eta_s \left[1 - \frac{3||\mathbf{h}||}{2\eta_r} + \frac{||\mathbf{h}||^3}{2\eta_r^3} \right] I[||\mathbf{h}|| \le \eta_r] \qquad (10.11)$$

where $||\mathbf{h}||$ denotes Euclidean distance and $I(A)$ denotes the indicator function of the event A. Equation (10.11) decreases toward zero as $||\mathbf{h}||$ increases, and the parameter η_n is called the *nugget*, η_s is called the *partial sill* (nugget + partial sill = sill), and η_r is called the *range*. Figure 10.1 shows the spherical autocovariance model (Eq. 10.11), along with the corresponding autocorrelation and semivariogram models.

10.2.5 *Techniques for Estimation and Prediction*

In this Chapter, we use three different estimation methods: (1) we assume $\{Z(\mathbf{s})\}$ in Equation (10.2) are all independent, zero-mean normal random variables, and then use ordinary least squares (call this the IND method) to estimate all parameters; (2) we assume $\{Z(\mathbf{s})\}$ are autocorrelated normally distributed errors with a spherical autocovariance model, Equation (10.11), and then use spatial maximum likelihood (SML method) (Mardia and Marshall 1984) to estimate all parameters; (3) we assume $\{Z(\mathbf{s})\}$ are autocorrelated normally distributed errors with a spherical autocovariance model, and then use spatial restricted maximum likelihood (SRML method) to estimate all parameters. Some explanation of the technical aspects of these three estimation methods is required.

For a general discussion of SML and SRML, see Cressie (1991, pp. 91–93). Restricted maximum likelihood was developed by Patterson and Thompson (1971, 1974); for computational details of SRML, see Zimmerman (1989). SML simply maximizes the multivariate normal likelihood, but produces biased estimates of the covariance parameters (Mardia and Marshall 1984). For a simple example, consider the model with n normal random variables with common mean μ and independent errors with variance σ^2. Then, the maximum likelihood estimator of the variance σ^2 is,

$$S^2 = \frac{\sum_i (Z(\mathbf{s}_i) - \overline{Z})^2}{n}$$

This is known to be biased for σ^2. The bias occurs because we use $\overline{Z}$ in the formula instead of the true value μ (which is unavailable). A restricted maximum likelihood estimator of σ^2 is $nS^2/(n-1)$, which is unbiased and is the estimator usually used. This theory can be extended to more complex cases for estimation of covariance-model parameters.

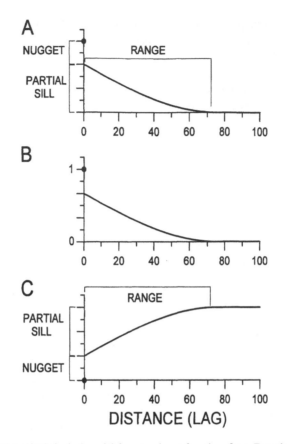

FIGURE 10.1. (A) Spherical model for covariance function, from Equation (10.11).
(B) Autocorrelation function that corresponds to graph A, from Equation (10.8).
(C) Semi-variogram model that corresponds to graph A, from Equation (10.10).

For making predictions, if all covariance parameters were known (e.g., the parameters of the spherical covariance model in Equation (10.11)), then we could use generalized least squares to make best linear unbiased predictions (BLUP); for further details for the spatial linear model, see Cressie (1991, pg. 163). Upon using estimated covariance parameters in the BLUP equations, the resulting prediction procedure is referred to as EBLUP (Zimmerman and Cressie 1992).

10.3 Examples

10.3.1 Trend Versus Autocorrelation

One problem faced when considering models with autocorrelation is that it may be difficult to determine whether any apparent trend in the data is the result of deterministic mean structure or spatial autocorrelation in the random error variation specified in Equation (10.4). This may seem surprising because the two sources of variability are fundamentally different (deterministic vs. random). The issue might best be illustrated with a simulation. For example, in Figure 10.2, two sets of data were produced for each panel.

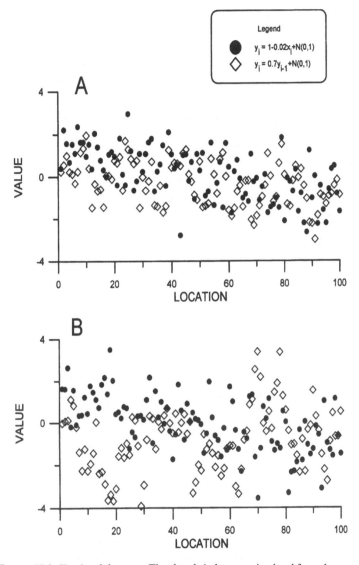

FIGURE 10.2. Simulated data sets. The closed circles were simulated from the regression model, $Y_i = \theta_0 + \theta_1 x_i + \epsilon_i$, where $\theta_0 = 1$, $\theta_1 = -0.02$, $x_i = i$, and $\epsilon_i \sim N(0,1)$; $i = 1, 2, \ldots, 100$. The open diamonds were simulated from a first-order autoregressive model, $Y_i = \theta_1 y_{i-1} + \epsilon_i$, where $\theta_1 = 0.7$. One realization from each model is given in panel A and panel B.

In Figure 10.2A, both simulated data sets exhibit a decreasing pattern. One data set (the solid circles), however, was produced by a model with a linearly decreasing mean and independent residual variation, and the other data set (open diamonds) was produced by a model with a constant mean (i.e., a model with mean stationarity) and autocorrelated errors. The patterns exhibited by both data sets are remarkably similar. In Figure 10.2B, another realization from each model was produced. Here, it is apparent that the patterns are quite different. Suppose now that we did not know the true models for each pattern in Figure 10.2A and B, and that we had to infer them from the data. The solid circles in Figure 10.2A and B both exhibit a decreasing pattern; therefore, we could be reasonably certain that the model that produced the circles was composed of fixed effects rather than auto-correlation. The open diamonds in Figure 10.2A and B both exhibit patterns in opposite directions; therefore, the model that produced them is obviously not one of fixed linear trend, and an autocorrelated model is more reason-able. Notice that it helps to have several realizations to distinguish between these models.

The trouble with real ecological data is that, very often, we have a data set that is only one realization from some unknown model. For example, suppose we had only the open diamonds in Figure 10.2A. How can we choose a model based on these data alone? Both models are plausible, and it is difficult to choose one over the other without more realizations. Another consideration is that real ecological data are rarely, if ever, expected to re-spond to longitude and latitude coordinates themselves. That is, the spatial coordinates serve as a proxy because they are highly correlated with other variables that really affect the response variable. Again, consider an example of species biomass. Suppose we notice an increasing trend in biomass, from south to north, within a state. The species is obviously not responding to latitude itself; rather, it is responding to the climatic conditions or soil characteristics that change gradually from south to north. Thus, it is possible to model spatial trend simply on the longitude and latitude coordinates. These are proxies for other explanatory variables that are known to trend along the coordinates; however, if possible, it is better to gather data on the variables that really affect the response variable and put them in the mean structure of the model.

This discussion demonstrates that when an ecologist observes spatial trend in data, it does not imply that the trend has to be removed with a surface of fixed effects—spatial autocorrelation is also capable of producing data that exhibit trend. The ecologist must instead think carefully about what effects might have caused the trend, if any, and if possible put those effects in the model. One way to think of this is as follows. Suppose we go back in time (say 100 years) and let the ecological processes start over. If things like dispersal, disease, climate, and the like, have random components, then we would not expect the species biomass to be exactly as it is today. Letting time start over again would produce a second realization. Would certain spatial patterns that we see today still be there (at similar locations) in this second realization? If we believe they would, we should put those fixed effects in the trend. We then suppose that all else can be absorbed as (possibly nonstationary) autocorrelation in the errors.

10.3.2 Building Models for Estimation and Prediction

We ran several simulations in order to reinforce some of the previous concepts and incorporate the ideas of trend and autocorrelation, as well as their effects on estimation and prediction. All simulations were done in one dimension with the spatial coordinates given by the integers, $s = 1, 2, \ldots 51$. Consider the following model:

$$Y(s) = \theta_0 + \theta_1 X_1(s) + \theta_2 X_2(s) + \theta_3 X_3(s) + \epsilon(s) \qquad (10.12)$$

where $\epsilon(s)$ is independent normal error with mean 0 and variance 1. We took $\theta_0 = 0$, $\theta_1 = 1$, $\theta_2 = 0$, and $\theta_3 = 1$. The random variables $X_i(s)$ were each obtained through of the following model:

$$X_i(s) = \theta_4[\sin(\theta_5 s) + R_i(s)]$$

where $\theta_4 = 3$, $\theta_5 = \frac{7\pi}{180}$, and $R_i(s)$ was autocorrelated and simulated from the multivariate normal distribution with zero mean and covariance matrix from the spherical model (Eq. 10.11):

$$Cov(R_i(s), R_i(u)) = \theta_6\left[1 - \frac{3|s-u|}{2\theta_7} + \frac{|s-u|^3}{2\theta_7^3}\right]I[|s-u| \le \theta_7] \qquad (10.13)$$

where we took $\theta_6 = 3$ and $\theta_7 = 40$ for all three random processes $X_i(s)$; $i = 1, 2, 3$. Further, $R_1(\bullet)$, $R_2(\bullet)$, and $R_3(\bullet)$ were mutually independent. Notice that $X_i(\bullet)$ is related to $X_j(\bullet)$ through the common sine wave.

We have described a "true" model (Eq. 10.12) from which the data are simulated. Now, suppose that $X_1(s)$ and $X_2(s)$ are variables that we can measure, along with the response $Y(s)$, and we choose a model for estimation given by

$$Y(s) = \beta_0 + \beta_1 X_1(s) + \beta_2 X_2(s) + Z(s) \qquad (10.14)$$

which is the spatial linear model (Eq. 10.2). This is a simplified example of what was discussed in the Introduction. The processes $\{X_i(\bullet)\}$ are themselves patterned and correlated with each other. We can observe only some of the explanatory variables ($X_1(s)$ and $X_2(s)$), and some of those may not be directly related to $Y(s)$ (e.g., notice that, because true $\theta_2 = 0$, $X_2(s)$ is not directly related to $Y(s)$). On the other hand, $Y(s)$ *is* directly related to $X_3(s)$, but it is one of those explanatory variables that is difficult and/or expensive to measure, so that we will not be able to observe it. In building our model, therefore, we have errors of included and omitted explanatory variables.

226   Jay M. Ver Hoef et al.

We ran a simulation of 1600 iterations from Equation (10.12) and we removed the simulated datum $Y(26)$. We therefore now have 50 observations for both estimation and prediction. Our goal was to estimate the parameters β_1 (true value is $\theta_1 = 1$) and β_2 (true value is $\theta_2 = 0$) of the linear model (Eq. 10.14) for each simulation and also to predict the missing value $Y(26)$. All simulations were done using PROC MIXED in SAS.

The results of the simulations, and the consequent estimation and prediction from them, are given in Table 10.1. The three estimation methods are given in the columns, and four summaries of their performance are given in the rows. Notice that, when estimating β_1 (the true value is $\theta_1 = 1$), the IND model had confidence intervals that were much too short because the 95% confidence interval only covered the true value 43.9% of the time. The SML and SRML methods were better, with SRML the best, but still having confidence intervals that were a bit too short (92.7%). There is no evidence of bias in any of the methods. The root mean-squared error (RMSE) is an indication of average "closeness" of the estimate of β_1 to the true value, with the smaller the RMSE the better. SML and SRML are clearly producing estimates that are much closer to the true value. Finally, the ratio of the average estimated standard error to the simulated RMSE is an assessment on the accuracy of the "uncertainty" estimate—we expect the ratio to be

TABLE 10.1. Simulation results.

	IND	SML	SRML
Estimation of β_1			
Coverage	0.4394	0.9044	0.9269
Bias	0.0043	−0.0011	−0.0005
RMSE	0.6710	0.2171	0.2131
Ratio	0.3097	0.8770	0.9349
Estimation of β_2			
Coverage	0.4688	0.9000	0.9231
Bias	0.0093	−0.0009	−0.0010
RMSE	0.6468	0.2155	0.2122
Ratio	0.3190	0.8785	0.9338
Prediction of $Y(26)$			
Coverage	0.9563	0.9344	0.9444
Bias	−0.0176	−0.0114	−0.0081
RMSPE	3.9532	2.1309	2.1168
Ratio	0.9968	0.9397	0.9639

Coverage indicates the proportion of times that the confidence interval or prediction interval contained the true value for 1600 simulations. Bias is the average difference between the estimated or predicted value and the true value. RMSE is the square root of the average-difference-squared between the estimated and the true value, and RMSPE is the square root of the average-difference-squared between the predicted and the true value. The last row for each parameter estimate is the ratio of the average estimated standard error (RMSE for estimation and RMSPE for prediction). The method that assumes independent residuals is denoted as IND; the method that assumes autocorrelated errors and is fitted using maximum likelihood is denoted by SML; and the method is denoted by SRML when autocorrelation parameters are fitted using restricted maximum likelihood.

very close to 1. Notice that estimation for the IND model produces standard errors that are too small, not reflecting the variability around the true estimate, so the ratio is substantially less than 1. This explains why the confidence intervals are too short. On the other hand, SML and SRML have ratios much closer to 1, with SRML the best.

The true value for β_2 is 0, and the results for its estimation, shown in Table 10.1, are much the same as for β_1; however, note the effect on building models. For the IND method, the true value of zero was covered by the confidence interval only 46.9% of the time. This means that 53.1% of the time, the P-value would indicate that the effect is significant, and we would want to include (incorrectly) the effect in the model. Thus, when the fitted model is misspecified, it often produces autocorrelation in the errors; the effect on model-building is that, if we assume classical assumptions of independence, we will include more effects than are really important (if the autocorrelation is positive). Notice from Table 10.1 that the confidence intervals using SML and SRML are more accurate, so that these two estimation methods will not make this mistake as often; however, the confidence intervals are still too short. This is not unexpected, and further corrections are possible (e.g., Harville 1985; Prasad and Rao 1990; Cressie 1992; Zimmerman and Cressie 1992; Ghosh and Rao 1994), although they are not implemented in SAS.

For prediction, notice that all three methods have prediction intervals with coverage in the 95% range (Table 10.1) and ratios near 1, so they are giving the appropriate reflection of the uncertainty of prediction. Again, all three methods are unbiased, but the SML and SRML methods are generally doing a much better job, as reflected in their smaller root mean-squared prediction errors (RMSPE).

In summary, this simulation (Table 10.1) clearly demonstrates the utility of assuming autocorrelated errors for a fitted spatial linear model (Eq. 10.14) that is misspecified for a more complicated true model (Eq. 10.12). We will often make initial errors of trying a fitted model that includes unimportant explanatory variables and omitting important ones. No models are correct, but we want a parsimonious model that gives understanding of relationships among variables and predictive ability, all in a valid statistical manner where we have properly assessed the uncertainty in our estimates and predictions. By assuming autocorrelated residuals, the model absorbs effects of unknown variables to give valid, more precise estimation and prediction. The SRML method seems to be the best choice. We can now practice these concepts on a real data set.

10.3.3 Whiptail Lizard Data

The orange-throated whiptail (*Cnemidophorus hyperythrus*) is a relatively small teiid lizard found in coastal southern California and much of Baja California (Stebbins 1985; Jennings and Hayes 1994; Fisher and Case 1997).

It generally occurs below 1000 m in elevation in scrubland habitats, including coastal sage scrub, chamise chaparral, and alluvial fan scrub. Adult lizards are generally active March through September (Bostic 1966c), with juveniles hatching in August and active through December (Bostic 1966b). Their diet consists primarily of insects of which termites are the dominant food (Bostic 1966a; Case 1979). This species has been considered a species of special concern in California for many years, although much of its current distribution and life history is poorly known (Hollander et al. 1994). *C. hyperythrus* has become a target species for conservation planning in the region due to the rapid urbanization of its habitats in southern California and its consideration as a sensitive species. The study by Hollander et al. (1994) used various geographical data sets in a hierarchical approach to determine how this species is distributed in space. They were able to identify biases in the existing distributional data for *C. hyperythrus* and gaps in our understanding of its habitat relationships. Chapter 3 described an ongoing study designed to correct these biases and fill these gaps.

Trapping stations for lizards were set up at 256 locations within 21 sites with 10–16 sampling periods evenly spaced over about a 2-year period. Each sampling period was 10 days long and traps were checked daily. Although this study collected data for all reptiles and amphibians captured at the stations, we only analyzed data for *C. hyperythrus* to illustrate the application of the spatial linear method. The lizard-count data were summed over time for each location. A total of 3,028 *C. hyperythrus* were captured and released. The variable of interest is the capture rate of *C. hyperythrus*, expressed as average number caught per trapping day. In order to use normal theory, the capture rate was log-transformed. Thirty-seven explanatory variables were collected at each location, along with *C. hyperythrus* abundance. These variables can be classed into five broad categories: vegetation layers, vegetation types, topographic position, soil types, and ant abundance.

We did some exploratory data analysis. Of the 256 locations, 107 of those, or about two fifths, had capture rates of zero. We envision the abundance of *C. hyperythrus* as a hierarchical process. First, there is an absence–presence (0–1) process that controls whether or not *C. hyperythrus* occurs at a location. Then, *given* this 0–1 process, another process controls abundance. The explanatory variables that affect each process may be quite different. For the purpose of demonstration, we shall concern ourselves with the process associated with abundance; therefore, we shall ignore the zeros and concentrate on locations where *C. hyperythrus* occurred, which totaled 149 locations within 15 sites.

First, we used the classical assumption of independent errors (IND method). We used stepwise regression in SAS, with a p-value to enter and a p-value to remove of 0.15. Of the 37 explanatory variables, seven were retained in the model, and six of them were significant at $p=0.05$. We checked the residuals for outliers and normality. One residual was an obvious outlier, so it was removed, leaving 148 locations. We again used stepwise regression and obtained the same seven variables in the model, with the same six variables having p-values < 0.05. We again checked the residuals and did not find any outliers, and the normal probability plot did not show lack of normality.

Next, we used a model with autocorrelation. We assumed a spherical model (Eq. 10.11) for the autocorrelation and used SRML in SAS PROC MIXED to estimate all parameters. Based on the simulation results given earlier, it appears that SRML is slightly better than SML, so we only used SRML for these data. Estimation for both SML and SRML can be quite slow and is quickened considerably when the covariance matrix has a block diagonal structure. We therefore began by assuming all sites were independent, but allowed spatial dependence within each site. The estimated range of spatial dependence, however, reached beyond some sites to nearby sites; therefore, we began to group sites that were nearby, and refit all parameters. We also assumed isotropy because the grouped sites had too few data individually to examine whether covariance changed with direction. For the final model, all of the locations, grouped sites, and the range of autocorrelation are shown in Figure 10.3. The fitted spherical covariance function, expressed as a semivariogram (Eq. 10.10), is shown in Figure 10.4. The empirical semivariogram (the method-of-moments estimator given by Cressie 1991, pg. 69) calculated on the residuals is also plotted as a diagnostic in Figure 10.4. Using SRML to estimate the autocorrelation parameters, the p-values (for the null hypothesis that the regression coefficient $= 0$) for most regression coefficients were larger because positive autocorrelation leads to fewer explanatory variables, as noted earlier. The final linear model with $p < 0.05$, using SRML, included only two of the six explanatory variables that were included using IND with $p < 0.05$. The final model was

$$Y(\mathbf{s}) = -3.9625 + 0.3217X_1(\mathbf{s}) + 0.9876X_2(\mathbf{s}) + Z(\mathbf{s})$$

where $X_1(\mathbf{s})$ is the abundance of *Crematogaster* ants at location indexed by $\mathbf{s}$, $X_2(\mathbf{s})$ is the logarithm of percent sandy soils, and $Z(\mathbf{s})$ is spatially autocorrelated, normally distributed random error with covariance

$$Cov(Z(\mathbf{s}), Z(\mathbf{u})) = 0.5091I[\mathbf{s} = \mathbf{u}] + 0.7259\left[1 - \frac{3\|\mathbf{s} - \mathbf{u}\|}{2(0.1972)} + \frac{\|\mathbf{s} - \mathbf{u}\|^3}{2(0.1972)^3}\right]$$
$$\times I[\|\mathbf{s} - \mathbf{u}\| \leq 0.1972]$$

The model indicates that the abundance of *C. hyperythrus* increases with more abundant *Crematogaster* ants and sandier soils. Notice that including autocorrelated errors causes us to retain fewer explanatory variables during the modeling process, which is consistent with the previously given simulation.

The parameters of the model are unobservable for estimation, so we rely on our simulation studies to understand the effects of autocorrelation on estimation better. For prediction, however, we have actually observed the

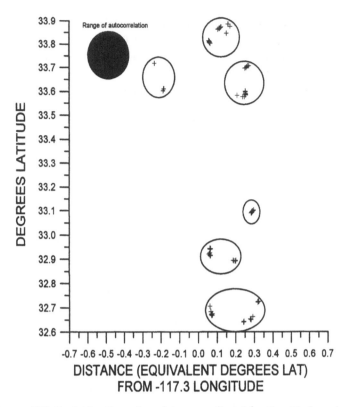

FIGURE 10.3. Spatial locations where data were collected for *Cnemidophorus hyper-ythrus*. The estimated range of autocorrelation is shown, along with the grouping of locations. Autocorrelation was assumed within groups, and groups were assumed independent.

realization of some of the spatial random variables; therefore, we did a cross-validation study. That is, we removed all 148 locations, one at a time, and then predicted their values after refitting with the IND method (with six explanatory variables) and the SRML model (with two explanatory variables). We then compared the predicted values with the true values. Based on the cross-validation predictions, the coverage of the 95% prediction intervals for the IND method was 97.3%, and the coverage for the SRML method was 93.2%; therefore, both methods gave good assessment of uncertainty. Neither had an obvious bias, which was 0.005 for both. There was, however, a dramatic difference in MSPE. The MSPE was 0.942 for the IND method, but only 0.649 for the SRML method. These results are consistent with our simulations previously discussed. Note that SRML provides better prediction (smaller MSPE) using fewer explanatory variables. This shows the power of utilizing autocorrelation for prediction. It uses information in both the explanatory variables and the response variable at neighboring locations.

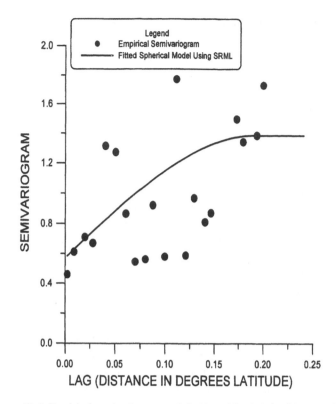

FIGURE 10.4. Empirical semivariogram and fitted model using the SRML method. The empirical semivariogram was computed on the residuals after subtracting the estimated fixed effects from the data.

The biological finding of these two factors—the abundance of *Crematogaster* ants and the logarithm of percent sandy soils—is consistent with known elements of the natural history of *C. hyperythrus*. Because they build their nests in well-drained friable soils (Bostic 1966c), the association with sandy soils is not unexpected. The diet of *C. hyperythrus* consists of 85–90% termites of the species *Reticulitermes hesperus*, which are native to the coastal and montane habitats throughout California (Pickens 1934, Bostic 1966a, Case 1979). These termites need wood for colony establishment; thus, they are often associated with dead plant material in nature, including trunks of bushes, sticks, and other surface debris (Pickens 1934). *Crematogaster* ants tend to use the same types of material for their colony establishment as do the termites, although they tend to forage away from their colony and are thus easy to record. The termites are only present on the surface briefly while swarming to form new colonies; otherwise, they are repelled by light and difficult to observe (Pickens 1934). Thus, the association between whiptail lizards and *Crematogaster* ants might reflect an association between *Crematogaster* and termites, which requires further verification.

10.4 Beyond the Spatial Linear Model

Ecologists often have data that are binary (0–1), counts, or positive and continuous. A common approach in these situations is to transform the data to near normality, as we did for the example with lizard data, although it is often more desirable and natural to work with data on their original scale. Geostatistics has produced methods such as disjunctive kriging (Matheron 1976; Armstrong and Matheron 1986a,b) and indicator kriging (Journel 1983); however, these methods are primarily concerned with prediction and they do not deal directly with fixed effects. See Gotway and Stroup (1997) and Diggle et al. (1998) for critiques and newer approaches, based on extensions of generalized linear models (Nelder and Wedderburn 1972; McCullagh and Nelder 1989). For example, it is possible using these extensions of generalized linear models to use a spatial logistic regression model for a 0–1 binary process that controls the occurrence of *C. hyperythrus*.

We would also like to come back to the generic space–time models (Eq. 10.1) introduced earlier. We sometimes obtain data and average over time to give a space-only data set; sometimes, we average over spatial locations to produce times-series data sets. Both of these fields of statistics employ models with autocorrelation (spatial or temporal). An area of active statistical research is in the area of space–time models. One attractive way to combine current understanding of ecological relationships into a statistical model is through hierarchical modeling. We shall describe an outline of a space–time hierarchical model for the lizard data to show the main features of this approach, although we shall not carry out the analysis.

Let us begin by assuming that we have the raw space–time data for *C. hyperythrus*, so we have the raw counts for each trapping bout (usually 10 days) with bouts spread throughout the year. Denote the counts per bout as $Y(\mathbf{s}, t)$, where $\mathbf{s}$ is the spatial location and t is the date of first day of the bout. We begin the hierarchical model with a zero-inflated Poisson regression model (see Lambert, 1992, for a version without space–time modeling):

$$Y(\mathbf{s}, t)|p(\mathbf{s}, t), \lambda(\mathbf{s}, t) \sim \begin{cases} 0, & \text{with probability } p(\mathbf{s}, t), \\ Poisson(\lambda(\mathbf{s}, t)), & \text{with probability } 1 - p(\mathbf{s}, t). \end{cases}$$

The preceding vertical-bar notation can be read as "$Y(\mathbf{s}, t)$, conditional on $p(\mathbf{s}, t)$ and $\lambda(\mathbf{s}, t)$." The counts are controlled by $\lambda(\mathbf{s}, t)$ and $p(\mathbf{s}, t)$, which are space–time "parameter" surfaces for Poisson and Bernoulli random variables, respectively. We believe that the parameter surfaces $\lambda(\mathbf{s}, t)$ and $p(\mathbf{s}, t)$ are affected by other variables, and the surfaces themselves might be autocorrelated, so we specify the second level in the hierarchy:

$$\log(\lambda(\mathbf{s}, t))|\boldsymbol{\beta}_\lambda, \mu_\lambda(\mathbf{s}), \tau_\lambda(t), \sigma_\lambda \sim N(\mathbf{x}_\lambda(\mathbf{s}, t)'\boldsymbol{\beta}_\lambda + \mu_\lambda(\mathbf{s}) + \tau_\lambda(t), \sigma_\lambda^2)$$

$$\text{logit}(p(\mathbf{s}, t))|\boldsymbol{\beta}_p, \mu_p(\mathbf{s}), \tau_p(t), \sigma_p \sim N(\mathbf{x}_p(\mathbf{s}, t)'\boldsymbol{\beta}_p + \mu_p(\mathbf{s}) + \tau_p(t), \sigma_p^2)$$

where the two normal distributions might be independent. Here, $\mathbf{x}_\lambda(\mathbf{s}, t)$ and $\mathbf{x}_p(\mathbf{s}, t)$ are vectors of covariates (explanatory variables) that can include other ecological variables, the number of capture days in each bout, and spatial and temporal trend coordinates (e.g., a linear trend in space and cosine functions of time); $\mu_\lambda(\mathbf{s})$ and $\mu_p(\mathbf{s})$ are generally stochastic autocorrelated spatial surfaces, and $\tau_\lambda(t)$ and $\tau_p(t)$ are stochastic time series. We gather up the parameters from this level into a vector: $\theta_1' = (\boldsymbol{\beta}_\lambda', \boldsymbol{\beta}_p', \sigma_\lambda^2, \sigma_p^2)$. At the third level of the hierarchy, we specify the spatial and temporal stochastic models, which might be:

$$\{\mu_\lambda(\mathbf{s})\} \sim N(\mathbf{0}, \Sigma(\boldsymbol{\alpha}_\lambda))$$

$$\{\mu_p(\mathbf{s})\} \sim N(\mathbf{0}, \Sigma(\boldsymbol{\alpha}_p))$$

$$\{\tau_\lambda(t)\} \sim N(\mathbf{0}, \Sigma(\boldsymbol{\eta}_\lambda))$$

$$\{\tau_p(t)\} \sim N(\mathbf{0}, \Sigma(\boldsymbol{\eta}_p))$$

where $\boldsymbol{\alpha}_\lambda$, $\boldsymbol{\alpha}_p$, $\boldsymbol{\eta}_\lambda$, and $\boldsymbol{\eta}_p$ are parameters that control the covariance matrix Σ for each process. We gather these up into the vector $\theta_2' = (\boldsymbol{\alpha}_\lambda', \boldsymbol{\alpha}_p', \boldsymbol{\eta}_\lambda', \boldsymbol{\eta}_p')$. For example, $\alpha_{\lambda,1}$ could be a parameter that controls the strength of autocorrelation, $\alpha_{\lambda,2}$ could be a parameter that controls the range of autocorrelation [e.g., as in the spherical covariance model (Eq. 10.11) without the nugget]. In the fourth level of the hierarchy, we specify prior models for all remaining parameters, $f(\theta_1, \theta_2)$, which might include vague priors such as $f(\boldsymbol{\beta}_\lambda, \boldsymbol{\beta}_p, \sigma_\lambda, \sigma_p, \boldsymbol{\alpha}_\lambda, \boldsymbol{\alpha}_p, \boldsymbol{\eta}_\lambda, \boldsymbol{\eta}_p) \propto \sigma_\lambda^{-2}\sigma_p^{-2}$, or fully parametric forms such as $f(\boldsymbol{\beta}_\lambda, \boldsymbol{\beta}_p, \sigma_\lambda, \sigma_p, \boldsymbol{\alpha}_\lambda, \boldsymbol{\alpha}_p, \boldsymbol{\eta}_\lambda, \boldsymbol{\eta}_p) = f(\boldsymbol{\beta}_\lambda)f(\boldsymbol{\beta}_p)f(\sigma_\lambda)f(\sigma_p)f(\boldsymbol{\alpha}_\lambda)f(\boldsymbol{\alpha}_p)f(\boldsymbol{\eta}_\lambda)f(\boldsymbol{\eta}_p)$, where $f(\beta_{ij})$ has a normal distribution with mean parameter β_{ij}^0 and variance ϕ_{ij}^2, $j = 1, 2, \ldots, q_i$, with q_i being the number of explanatory variables for i; $f(\sigma_i^2)$ is distributed as an inverse χ^2 with parameters n_i and v_i^2; $f(\alpha_{ij})$ has a gamma distribution with parameters $\alpha_{i,j}^0$ and $\xi_{i,j}$, $j = 1, 2, \ldots, c_i$, with c_i being the number of covariance parameters (typically two) for i; and $f(\eta_{ij})$ has a gamma distribution with parameters $\eta_{i,j}^0$ and ψ_{ij}, $j = 1, 2, \ldots, d_i$, with d_i being the number of covariance parameters (typically two) for i; $i = \lambda, p$. Of course, all hyperparameters need to be specified.

Inference for this model can be carried out using Markov chain Monte Carlo (MCMC) methods, which are a collection of stochastic simulation methods including the Gibbs sampler, the Metropolis algorithm, and the Hastings algorithm (Metropolis et al. 1953; Hastings 1970). These methods were used for Bayesian inference in a seminal paper by Geman and Geman (1984). The MCMC methods obtain samples from the posterior distribution, and are to Bayesian inference what the bootstrap is for obtaining sampling distributions in frequentist inference. See Tanner (1993) for a good introduction to MCMC methods, and Gilks et al. (1996) for an overview. The proposed hierarchical model given earlier is quite complicated, but similar models have been demonstrated to be tractable on quite large meteorological data sets (e.g., Wikle et al. 1998). The MCMC methodology produces simulations from the posterior distribution:

$$p(\lambda(\mathbf{s},t), p(\mathbf{s},t), \mu_\lambda(\mathbf{s}), \mu_p(\mathbf{s}), \tau_\lambda(t), \tau_p(t), \theta_1, \theta_2, |y(\mathbf{s},t))$$

which is obtained using repeated application of Bayes' theorem. Of most interest, we can make inferences on the parameter surfaces that control the counts ($\lambda(\mathbf{s},t)$ and $p(\mathbf{s},t)$), the spatial and temporal parts of the surfaces ($\mu_i(\mathbf{s})$ and $\tau_i(t)$), the parameters that control the observed covariates (β_i), the range and strength of autocorrelation in the spatial (α_i) and temporal (η_i) surfaces, and measurement error variances (σ_i^2); $i = \lambda, p$.

We can see that the hierarchical model outlined in this section goes a lot farther than the spatial linear model for modeling the complexities of ecological data. As our knowledge of relationships among variables increases, it is natural that Equation (10.1) becomes more and more complex. Hierarchical models can handle this complexity well, and inferential methods for these models are being developed. An advantage of the spatial linear model is that software (e.g., PROC MIXED in SAS) is available, making it more accessible, whereas a complicated hierarchical model requires custom programming.

10.5 Summary

We have concentrated on the spatial linear model, where we included the assumption of autocorrelated errors in the model to absorb the effects of patterned, unobserved factors related to the observed response variable. We have argued that no models are correct. Robust methods are needed to deal with ecological data, and there are two natural approaches. One is to use robust estimation methods; the other is to assume robust models. We have demonstrated that the spatial linear model is a robust model. We concur here with Box (1980): "For robust estimation of the parameters of interest we should modify the model which is at fault, rather than the method of estimation which is not." By modifying the assumption of independent errors to autocorrelated errors in the spatial linear model, we have a more robust model where we can get better estimation of unknown parameters and better prediction for unobserved variables. In addition, the uncertainty that we attach to these estimation and prediction problems is more valid than when we assume independent residuals, especially for estimation problems. Software for these models is readily available. The spatial linear model is developed for normally distributed data, and extensions to count and binary data have been developed but require custom software development. All ecological data are ultimately space–time data in that they are collected from some place at some time; therefore, space–time models will be an area of active research in the future. The natural complexity in ecological problems lends itself to hierarchical space–time models that, at the present time, can most effectively handle space–time data.

Acknowledgments. Financial support for this work was provided by Federal Aid in Wildlife Restoration to the Alaska Department of Fish and Game, by a cooperative agreement between the Alaska Department of Fish and Game and Iowa State University, and by cooperative agreement CR-822919-01-1 between the U.S. Environmental Protection Agency and Iowa State University.

References

Armstrong, M., and G. Matheron. 1986a. Disjunctive kriging revisited: part I. Mathematical Geology 18:711–28.

Armstrong, M., and G. Matheron. 1986b. Disjunctive kriging revisited: part II. Mathematical Geology 18:729–42.

Bostic, D.L. 1966a. Food and feeding behavior of the teiid lizard, *Cnemidophorus hyperythrus beldingi*. Herpetologica 22:23–31.

Bostic, D.L. 1966b. A preliminary report of reproduction in the teiid lizard, *Cnemidophorus hyperythrus beldingi*. Herpetologica 22:81–90.

Bostic, D.L. 1966c. Thermoregulation and hibernation of the lizard, *Cnemidophorus hyperythrus beldingi* (Sauria: Teiidae). Southwestern Naturalist 11:275–89.

Box, G.E.P. 1976. Science and statistics. Journal of the American Statistical Association 71:791–99.

Box, G.E.P. 1980. Sampling and Bayes' inference in scientific modelling and robustness (with Discussion). Journal of the Royal Statistical Society, Series A 143:383–430.

Case, T.J. 1979. Character displacement and coevolution in some *Cnemidophorus* lizards. Fortschritte Zoologica 25:235–82.

Cressie, N. 1991. Statistics for spatial data. John Wiley & Sons, New York.

Cressie, N. 1992. Smoothing regional maps using empirical Bayes predictors. Geographical Analysis 24:75–95.

Diggle, P.J., R.A. Moyeed, and J.A. Tawn. 1998. Model-based geostatistics (with Discussion). Applied Statistics 47:299–350.

Fisher, R.N., and T.J. Case. 1997. A field guide to the reptiles and amphibians of coastal southern California. USGS-BRD, Sacramento, CA.

Geman, S., and D. Geman. 1984. Stochastic relaxation, Gibbs distributions, and the Bayesian restoration of images. IEEE Transactions on Pattern Analysis and Machine Intelligence 6:721–41.

Ghosh, M., and J.N.K. Rao. 1994. Small area estimation: an appraisal. Statistical Science 9:55–93.

Gilks, W.R., S. Richardson, and D. Spiegelhalter, eds. 1996. Markov chain Monte Carlo in practice. Chapman and Hall, London.

Gotway, C.A., and W.W. Stroup. 1997. A general linear model approach to spatial data analysis and prediction. Journal of Agricultural, Biological, and Environmental Statistics 2:157–78.

Harville, D.A. 1985. Decomposition of prediction error. Journal of the American Statistical Association 80:132–38.

Hastings, W.K. 1970. Monte Carlo sampling methods using Markov chains and their applications. Biometrika 57:97–109.

Hinkelmann, K., and O. Kempthorne. 1994. Design and analysis of experiments, volume 1: introduction to experimental design. John Wiley & Sons, New York.

Hollander, A.D., F.W. Davis, and D.M. Stoms. 1994. Hierarchical representations of species distributions using maps, images and sighting data. Chapter 5 in R.I. Miller, ed. Mapping the diversity of nature. Chapman and Hall, London.

Jennings, M.R., and M.P. Hayes. 1994. Amphibian and reptile species of special concern in California. Final report to the California Department of Fish and Game, Inland Fisheries Division, Rancho Cordova, CA. Contract number 8023.

Journel, A.G. 1983. Nonparametric estimation of spatial distributions. Journal of the International Association for Mathematical Geology 15:445–68.

Lambert, D. 1992. Zero-inflated Poisson regression, with an application to defects in manufacturing. Technometrics 34:1–14.

Legendre, P. 1993. Spatial autocorrelation: trouble or new paradigm? Ecology 74:1659–73.

Mardia, K.V., and R.J. Marshall. 1984. Maximum likelihood estimation of models for residual covariance in spatial regression. Biometrika 71:135–46.

Matheron, G. 1976. A simple substitute for conditional expectation: the disjunctive kriging. Pages 221–36 in M. Guarascio, M. David, and C. Huijbregts, eds. Advanced geostatistics in the mining industry. Dordrecht, Reidel, The Netherlands.

McCullagh, P., and J.A. Nelder. 1989. Generalized linear models, second ed. Chapman and Hall, London.

Metropolis, N., A.W. Rosenbluth, M.N. Rosenbluth, A.H. Teller, and E. Teller. 1953. Equation of state calculations by fast computing machines. Journal of Chemical Physics 21:1087–92.

Nelder, J.A., and R.W.M. Wedderburn. 1972. Generalized linear models. Journal of the Royal Statistical Society, Series A 135:370–84.

Patterson, H.D., and R. Thompson. 1971. Recovery of interblock information when block sizes are unequal. Biometrika 58:545–54.

Patterson, H.D., and R. Thompson. 1974. Maximum likelihood estimation of components of variance. Pages 197–207 in Proceedings of the eighth international biometric conference, Biometric Society, Washington, D.C.

Pickens, A.L. 1934. The biology and economic significance of the western subterranean termite, *Reticulitermes hesperus*. Chapter 14 in C.A. Kofoid, S.F. Light, A.C. Horner, M. Randall, W.B. Hermes, and E.E. Bowe, eds. Termites and termite control. University of California Press, Berkeley, CA.

Prasad, N.G.N., and J.N.K. Rao. 1990. The estimation of mean squared errors of small-area estimators. Journal of the American Statistical Association 85:163–71.

Stebbins, R.C. 1985. A field guide to western reptiles and amphibians. Peterson field guide series. Houghton Mifflin Co., Boston.

Tanner, M. 1993. Tools for statistical inference. Methods for the exploration of posterior distributions and likelihood functions, second ed. Springer-Verlag, New York.

Ver Hoef, J.M., 1993. Universal kriging for ecological data. Pages 447–53 in M.F. Goodchild, B. Parks, and L.T. Steyaert, eds. Environmental modeling with GIS. Oxford University Press, New York.

Ver Hoef, J.M., and N. Cressie. 1993. Spatial statistics: analysis of field experiments. Pages 319–41 in S.M. Scheiner and J. Gurevitch, eds. Design and analysis of ecological experiments. Chapman and Hall, New York.

Wikle, C.K., L.M. Berliner, and N. Cressie. 1998. Hierarchical Bayesian space–time models. Environmental and Ecological Statistics 5:117–54.

Zimmerman, D. 1989. Computationally efficient restricted maximum likelihood estimation of generalized covariance functions. Mathematical Geology 21:655–72.

Zimmerman, D., and N. Cressie. 1992. Mean squared prediction error in the spatial linear model with estimated covariance parameters. Annals of the Institute of Statistical Mathematics 44:27–43.

11
Characterizing Uncertainty in Digital Elevation Models

ASHTON SHORTRIDGE

Many ecological modeling applications rely upon characterizations of the earth's surface. In some instances, the actual elevations are important (e.g., see Nisbet and Botkin 1993; de Swart et al. 1994; Liebhold et al. 1994). In others, secondary terrain attributes such as slope and aspect are also critical for the application (e.g., see Huber and Casler 1990; Nemani et al. 1993; Austin et al. 1996; Zhu et al. 1996; Russel et al. 1997). The term *digital elevation model* (DEM) refers to a variety of digital forms that characterize some portion of the earth's topography. These are used as proxies for the actual terrain surface in environmental modeling applications. The process of DEM production is one of abstraction, and is subject to error; therefore, a DEM does not perfectly match the real-world terrain it represents. The precise degree of this mismatch at every point is unknown, giving rise to uncertainty about the relationship between data and actual terrain.

This chapter identifies approaches to characterize DEM uncertainty and assesses its impact upon ecological applications. A brief overview of commonly employed data structures is provided, followed by a discussion of DEM uncertainty. Two categories of DEM-related uncertainty are identified. The first source of uncertainty results from differences between the form of the data model and the actual elevation surface. Uncertainty arises regarding elevations at locations not directly sampled in the DEM; this is referred to as *data model-based uncertainty*. The second is concerned with the fact that data production methods do not accurately capture elevations at specified *x, y* locations. This error is directly measurable and is referred to as *data-based uncertainty*.

Uncertainty in elevation data can be addressed if it can be characterized, either through field measurements or by the adoption of assumptions about the data and the terrain. An uncertainty model is employed to characterize uncertainty in a spatial dataset. Using this model, a researcher can produce a map of the most likely elevation surface, given the available information. For many applications, a better approach may be to propagate DEM uncertainty through the analysis to identify its impact upon the results of the application. This is accomplished by producing, via Monte Carlo simulation, a set of equiprobable realizations of the DEM. The ecological application is then run upon all realizations, producing a distribution of results.

Even though digital elevation models are commonly employed in ecological modeling, they are also representative of a more general class of spatial data that seeks to characterize continuous surfaces. The theory and methods discussed here are therefore applicable to many other forms of spatial data employed in ecological modeling. This chapter is intended to provide both a thorough background on DEMs and uncertainty, as well as to indicate general approaches to modeling uncertainty in data for continuous phenomena.

11.1 Production Methods and Data Structures

DEM sources vary widely. Some scientists perform their own field measurements and generate elevation models to their own specifications (e.g., see Dubayah and Rich 1993; de Swart et al. 1994). The increasing affordability and sophistication of global positioning system (GPS) technology has enabled the construction of project-specific DEMs, particularly for small study areas or for spot elevations at specific sites. Advantages of this approach include the researcher's ability to specify the spatial resolution and extent of the survey, design the sampling campaign, and oversee the production of the final elevation model. Accuracy of both the final DEM and the GPS data can be tested if the sampling campaign is carefully designed, leading to sophisticated uncertainty models (Oliver et al. 1989). Despite these advantages, environmental scientists often do not collect their own elevation data, particularly when the goal is to characterize the terrain surface over a more extensive region. Calibrating the GPS, particularly in rural or wilderness areas distant from elevation benchmarks, can be difficult, time consuming, and expensive. Collecting densely sampled elevation data in a careful and systematic manner is a lengthy process, particularly if it is combined with a plan to model data uncertainty. For watershed-scale areas or larger, building a quality DEM with a handheld GPS is impracticable. Finally, interpolation schemes to generate an elevation surface from the collected point data introduce uncertainty into the final product, and the choice of method can seem arbitrary (see Lam 1983 for a description of spatial interpolation methods).

Instead, most projects use general-purpose DEMs typically produced and/or distributed by government agencies like the U.S. Geological Survey (USGS) or the Ordinance Survey in the United Kingdom (e.g., see Huber and Casler 1990; Liebhold et al. 1994; Russel et al. 1997). These are currently produced in regular blocks at various resolutions and to various quality specifications. The USGS, for example, distributes several terrain data products at different resolutions and coverage ranges. Complete nationwide coverage is available from the 1:250,000-scale set. These are also called *1-degree DEMs*, which specifies the coverage area of each file. Elevations are stored on a regular 3 arc second grid. Distance between 3 arc second posts varies by direction and latitude but is roughly between 70 and 90 m in the conterminous United States. A second major format is the *7.5' DEM*. The spatial organization of this product corresponds to the 7.5' quadrangle system, and each file is named after its corresponding topographic map sheet. Elevations are stored in a regular grid with 30-m spacing, ostensibly at a scale of 1:24,000. Quality specifications for these DEMs vary and are dependent on the production method (see USGS 1995 for more about these DEM products).

Caution must be used when employing these elevation datasets. They were doubtless not designed with any particular ecological purpose in mind, and their resolution and accuracy might be entirely inappropriate for a given application. For example, Russel et al. (1997) found systematic "ripples" in the 7.5' DEM they used to derive wetness potential. These ripples propagated through their data processing and corrupted the end product, a wetland reclamation suitability site map. Current quality specifications for agency-produced DEMs are not adequate for specifying models of spatial data uncertainty because they use purely global accuracy measures. This topic will be treated more fully in a later section.

A third choice is to manually digitize elevation data from topographic map sources (e.g., see Gessler et al. 1993; Austin et al. 1996). This was frequently done in the United States before production DEMs became widely available (indeed, many USGS 7.5' DEMs are derived from nondigital contour maps), and remains a viable option for regions not included in inexpensive, readily available, high-resolution DEM data banks. Some researchers have preferred to produce elevation data manually in areas where existing DEMs were subject to excessive production artifact error (Garbrecht and Starks 1995).

Whether elevation data were collected by the scientist or downloaded from a digital library, the decision about how to characterize the surface for the purposes of analysis and GIS processing remains. This data modeling decision is typically limited to four choices (Weibel and Heller 1991): point model, digitized contour lines, irregular networks, and raster cells. The data may be treated simply as a set of locations with associated elevations. The point model may consist of a collection of scattered points, or of gridded locations in regular profiles, as from a USGS DEM. This point model makes no explicit assumptions about the nature of the surface between the specified locations. A second model is that of a set of digitized contour lines that fit the data. Although digital algorithms have been developed to work on this type of data (Moore et al. 1991), the model has been criticized for overspecifying the elevation surface at contour intervals and underspecifying areas falling between the intervals. Digital contours are essentially verbatim copies from the paper world of pen-based cartography; for terrain modeling purposes, other digital models should be more effective at characterizing terrain. Triangular irregular networks (TINs) store elevations at the vertexes of triangular elements. The terrain is then modeled as a surface composed of triangular facets (Weibel and Heller 1991). These facets are typically planar, so that the elevation surface is continuous with discontinuities in slope occurring at the facet boundaries. An advantage of the TIN model is that, because the facets can be of any size, sampling density may be increased to capture short-range variation in rugged terrain and reduced in less rugged areas (Kumler 1994). The most commonly used spatial data model employs raster cells to describe the elevation surface completely. This model partitions space into a rectangular matrix of identically sized, usually square cells. A single value is stored in each cell; all parts of the cell's space possess the same value. The raster model's popularity stems from its frequent implementation in geographic information systems, its utility for GIS analysis with other raster data layers, and its similarity to the gridded profiles of common production-level DEMs. It is worth noting that most GIS packages enable conversions between these models, so users should not feel constrained to one particular model for all processing and analysis.

For any of these models, two general sources of uncertainty may be specified: data-based uncertainty, due to the inaccuracy of measured elevations, and data model-based uncertainty, due to differences between the structural characteristics of the model and the landscape.

Data-based uncertainty is defined as the difference between the elevation of a location specified in the data set and the actual elevation at that location. This difference can in theory be measured by ground survey. DEM files produced by the USGS and other mapping agencies typically are assigned a root mean square error (RMSE) from a (usually small) set of locations for which the true elevation is known (USGS 1995). The actual elevations at these locations are compared with the DEM estimate; RMSE is derived from the sum of the squared differences. Agencies engaged in DEM production use this sort of measure for their data quality reporting, and the DEM accuracy literature describes this approach in detail (Shearer 1990; Bolstad and Stowe 1994; Monckton 1994). Global measures like RMSE are inadequate by themselves for analysis of uncertainty because they provide no information about local spatial structure. Indeed, map and data accuracy standards in general are not sufficient to characterize the spatial structure of uncertainty (Goodchild 1995; Unwin 1995).

Uncertainty arising from surface characterization depends very much upon the data model being used (see Goodchild 1992; for a discussion of the relationship between spatial data models and data structures). Two of the terrain data models described earlier do not explicitly represent a continuous terrain surface. In an array of points, no assumption is made about the elevation of intermediate locations. Digitized contours exhaustively capture all elevations at the contour intervals, but they do not specify elevations falling between contour intervals. Complete uncertainty exists regarding the elevation of any location not specified in either of these models. In practice, assumptions about intermediate values are frequently made because it is usual and often critical to model terrain as a continuous surface, with an elevation specified at every point. Contours and arrays of points are typically converted to data models that are used for surfaces, with defined values at every location (e.g., rasters and TINs).

A raster DEM assigns a single elevation to every location within a cell. There is often uncertainty about what this elevation represents. Is it the elevation of the center of the cell? Is it the height of the lower left corner of the cell? Is it the mean elevation value for the area within the cell? Surface elevation is discontinuous at cell boundaries. In contrast, the elevation surface is continuous across a TIN, although the slope surface is not continuous at facet edges if TIN facets are planar. Although elevations are specified for every location in these models, discrepancies can certainly arise between the real-world terrain and the structural characteristics of the model. The elevations at all vertexes of a TIN could conceivably be without error, yet the facets fail to capture actual terrain characteristics. Elevations for all cells in a raster DEM similarly could correctly characterize the mean cell elevation, but the fidelity of the model surface (flat-topped squares, like stacks of blocks) to the real world surface is very poor.

11.2 Data Model Uncertainty

This section is concerned with the effect of spatial resolution on parameters derived from elevation data. This is a problem related to data model uncertainty, as defined previously, because discrepancies at known locations are not relevant to the resolution effect. The impact of DEM raster cell resolution on secondary terrain variables like slope, aspect, and curvature has long been recognized and studied, particularly in hydrology, but also in soil science and geomorphology (Carter 1990; Chang and Tsai 1991; Zhang and Montgomery 1994; Hutchinson 1996; Gao 1997; Saulnier et al. 1997). Because slope is an elevation derivative frequently of interest to both hydrologists and ecologists, the following example has been developed to demonstrate this impact on slope algorithms.

This experiment employed the USGS 7.5' DEM Rancho Santa Fe, located north of San Diego, CA. The terrain for this region ranges from level to high relief. For this example the accuracy of the data will not be considered. The original DEM consisted of a cell resolution of 30 m. It was resampled to four larger resolutions: 60 m, 90 m, 150 m, and 300 m. All were resampled using each of the Arc/Info GIS' interpolation methods (nearest neighbor, bilinear, and cubic convolution) to identify what effect, if any, choice of interpolator had on the resulting slope grids. For each of the five DEMs, slope (in degrees) was calculated following the method presented in

TABLE 11.1. Characterization of degrees of slope for five cell resolutions and three interpolation methods.

Cell size (m)	Nearest neighbor range, mean, median (degrees)	Bilinear range, mean, median (degrees)	Cubic range, mean, median (degrees)
30	0.00–45.01, 9.74, 8.10		
60	0.00–38.48, 8.45, 6.99	0.00–38.11, 8.41, 6.92	0.00–39.66, 8.69, 7.18
90	0.00–34.30, 7.63, 6.19	0.00–34.30, 7.63, 6.19	0.00–34.30, 7.63, 6.19
150	0.00–30.26, 6.32, 5.07	0.00–30.26, 6.32, 5.07	0.00–30.26, 6.32, 5.07
300	0.05–22.15, 4.60, 3.64	0.14–22.33, 4.60, 3.64	0.13–22.40, 4.63, 3.67

Burrough 1986, which is that used in the Arc/Info geographic information system for gridded surfaces.

Table 11.1 presents information about the statistical distribution of the degrees slope map for each of the cell resolutions and interpolation methods. It should be clear that the first row of the table, the 30 m information, represents the original 7.5' data prior to any resampling. Two features stand out. First, choice of interpolation method does not appear to affect slope distribution for cell sizes greater than 60 m. Second, regardless of interpolator, average slope values and maximum slope values decline as cell size increases.

Figure 11.1 presents a set of boxplots to depict the effect of cell resolution on slope distribution graphically. This plot shows only the bilinear interpolation distributions. The most profound shift with increasing cell size is the

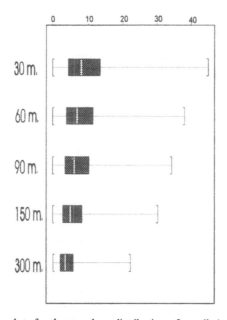

FIGURE 11.1. Boxplots for degrees slope distributions; five cell size resolutions for Rancho Santa Fe 7.5' raster DEM.

effect on larger slope values, as the distribution becomes increasingly fore-shortened on the high side. Similar trends have been noted in many different studies on many DEMs and many terrain types. An important conceptual point is that the resampled elevation values themselves may not be in error, but the calculated slope will change as cell size changes. Uncertainty in the true slope measure in this case is not due to error in the elevation values, the interpolation measure, or the slope algorithm; rather, it is due to the gap between the spatial data model and the real-world terrain. As cell size increases, this gap increases as well.

Some lessons can be drawn from this study. First and more obviously, data resolution can have a profound effect on terrain attributes, specifically slope distribution. Locations characterized by high slopes at fine resolution experience large reductions in slope as resolution coarsens. Researchers have attempted to use this sort of trend *to extrapolate* for terrain characterization at levels finer than the available data (Polidori et al. 1991), but this sort of approach is fraught with critical and largely untestable assumptions. A second point is that terrain data resolution should match the scale of the ecological process of interest. If the Mojave ringtailed chimera prefers slopes of greater than 10%, and the chimera is operating on scales smaller than 2 m, than using a 30-m resolution DEM to identify habitat is completely inappropriate, even if the 30-m data is perfectly accurate. There is no reason to expect much correlation between the slope maps of an area using data with 2-m versus 30-m resolution. Data model uncertainty issues clearly extend beyond this simple example. Alternative data structures like TINs, in which the only sampled locations are the triangle vertexes, and contour graphs, which are poorly defined at locations not on the contour interval, may be subject to different uncertainty characteristics. The impact of elevation uncertainty on alternative terrain attributes like slope and aspect will differ by data model as well.

11.3 Characterizing and Modeling DEM Uncertainty: The Simulation and Propagation Framework

Research into uncertainty modeling is related to the large body of work on spatial data accuracy assessment. The first portion of this section briefly reviews approaches to characterizing the accuracy of DEMs. The relationship between this work and uncertainty modeling is then discussed, and an overview of the modeling procedure follows.

A variety of directions have been taken to assess the accuracy of either a particular data production method or a specific elevation data file. The most straightforward method has been to compare accurate field elevation measures with DEM estimates. This has been accomplished in several studies (e.g., see Shearer 1990; Bolstad and Stowe 1994; Monckton 1994);

however, survey-quality terrain measurements are frequently not available for an area of interest. In such cases, investigators have three general options.

The first option is to make assumptions about intrinsic spatial qualities of the landscape and identify deviations from these assumptions as indications of error. For example, the potential for fractal properties of terrain surfaces has inspired research into the detection of artifacts in DEMs using a fractional Brownian motion model (Polidori et al. 1991; Brown and Bara 1994).

A second option is production-oriented. Likely artifacts resulting from the methods can be identified by studying problems in the methods used to generate digital elevation data. Robinson (1994) reviewed the generation of DEMs from contours. Common conversion routines suffer from a variety of problems relating to contour configuration. Earlier research compared several common contour interpolation algorithms to identify differences and problems (Clarke et al. 1982). Carter (1989) explored individual DEMs produced using different methods to identify artifacts. He found that particular error types tended to be symptomatic of particular production methods.

The third option is to compare independently derived terrain data for the same location. Isaacson and Ripple (1990) compared USGS 1:250,000 and 7.5' DEMs. They performed a linear regression on a sample of collocated points from the two datasets and determined that the best fit was not significantly different from a 1:1 line with an intercept at zero. The standard error for the regression was 31 m. See Guth (1992) for a critique of their methodology. Greater differences between DEM sets were found in slope and aspect. They found no sign of artifacts in their 1:250,000 DEM and concluded that the coarser DEM was adequate for their modeling purposes. Guth (1992) compared SPOT-derived 30-m data with a USGS 7.5' DEM. Distribution statistics were very comparable, but spatial patterns were identified for locations with large differences between the two data sets. Larger differences were found to be correlated with steep slopes and rugged terrain. Similar analysis was carried out for an area with two independently derived USGS 7.5' DEMs. Areas with large differences were again concentrated in high slope portions of the study area.

In the literature discussed earlier, accuracy characterization is concerned with description of error. In most of this literature, the link between error description and doing something with that description to improve the data's depiction of the world is not explicit. Users are warned that benching artifacts (i.e., artificially flat regions above and below contour lines) are common in DEMs derived from contour data, or that the average error is usually not zero (i.e., error is biased), or that elevation error is spatially autocorrelated and therefore not independent. How are ecological applications employing elevation data supposed to benefit from such warnings? Uncertainty modeling attempts to answer this question by using these descriptions of discrepancy in a statistically driven manner to develop improved data characterizations of terrain. The problem becomes one of identifying the impact of imperfect terrain characterization upon spatial applications using this imperfect data.

Two theoretical approaches for uncertainty modeling present themselves: analytical derivation and stochastic simulation/propagation. The first and more straightforward conceptually is to derive the uncertainty measure analytically. Hunter and Goodchild (1997) use the USGS level 1 DEM accuracy specification to characterize mathematically the RMSE of the orthogonal components of the gradient vector, which is a derived measure similar to slope. Their method relies on an estimation of the spatial autocorrelation of elevation errors, which is critical for any model of spatial uncertainty. Analytical solutions for uncertainty in more complex DEM applications like slope or aspect, however, are much more difficult. This problem has been addressed by adopting a second, more general approach, in which error is modeled stochastically, simulations are developed, and uncertainty is propagated through an analysis, producing a distribution of results which may be assessed statistically. This framework is presented in Figure 11.2 for characterization of uncertainty in DEMs and its effects on analysis. It has been explored by a number of researchers in geography and the earth sciences (e.g., see Heuvelink et al. 1989; Openshaw 1989; Goovaerts 1997; Hunter and Goodchild 1997; Heuvelink 1999). A more complete discussion of the general framework may be found in Chapter 9.

On the left in Figure 11.2 are a (possibly large) number of DEM realizations of a study region. Each of these realizations is a characterization of the true elevation surface; its statistical characteristics match those specified by a model of uncertainty for the DEM. In the center of the figure is an operation that results in some (possibly spatially distributed) outcome. This operation might be as simple as a single GIS command (identify the elevation at each of the 27 nesting sites; determine the viewshed area for the top of Mount Baldy), or it might be a complex series of algorithms (execute the sage owl habitat spatial model; develop a regression model to explain hemlock growth, using elevation as one explanatory variable) that incorporate cartographic modeling techniques (Berry 1993). For each realization,

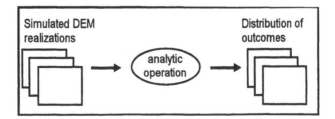

FIGURE 11.2. Uncertainty propagation using DEMs.

an answer is returned (e.g., an array of elevations, a value in square meters, a habitat map, a set of regression coefficients). Taken together, the outcomes from all of the realizations form a distribution that can be analyzed and described using graphs and statistics. Questions asked of these distributions can be subjected to statistical tests (provide a 95% confidence interval for the elevation of nest site D; find the likelihood that the viewshed is smaller than 25 km^2).

The method by which the DEM realizations are developed, the spatial uncertainty model, is of critical importance for this general propagation approach. Development of these models has been ongoing for some time, and they have received increasing attention in geography (Heuvelink et al. 1989; Fisher 1991; Lee et al. 1992; Englund 1993; Ehlschlaeger et al. 1997). Chapter 9 in this volume discusses the development of specific uncertainty models. This chapter will deal more generally with how elevation data may be handled with uncertainty models.

The following familiar circumstance forms the backdrop for this discussion on model development. Elevation data is required for a regional study. Although a gridded DEM is available for the region of interest, there is concern that infidelity between the actual terrain and this DEM will introduce uncertainty into the analysis. How can some limited information about the error in this DEM be used to build a model? A geostatistical approach to modeling DEM uncertainty begins by treating the elevation uncertainty surface as a random field. At each location on the surface, uncertainty is characterized by a probability distribution for all possible values at that location. Uncertainty model development hinges upon assumptions for characterizing the distribution of possible elevations (or elevation error) at locations where the actual elevation (or elevation error) is not known. In some instances, elevation may be known at a set of points within the study area (e.g., a GPS could have been used to sample several dozen locations within a study area). Conditional methods ensure that the surface model passes through these locations, "honoring" the ground truth data (for introductory discussions of geostatistics in general and conditional methods in particular, see Isaaks and Srivistava 1989; Goovaerts 1997). In other cases, true elevations may not be available within the area of interest, but error characteristics are known. They could be derived from data quality specifications, although these would have to include both aspatial characteristics like RMSE as well as measures for the spatial structure of error and any correlations between error and slope, or error and absolute elevation. On the other hand, they may be assumed to match those of nearby regions for which this information is available. Unconditional methods are useful in either circumstance, and are used to build error surfaces which are then added to the elevation data. Such methods are termed *unconditional* because no elevations are specifically honored; instead, elevations at all locations on the surface are perturbed (e.g., see Ehlschlaeger et al. 1997; Hunter and Goodchild 1997).

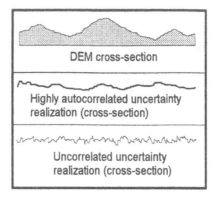

FIGURE 11.3. Spatial autocorrelation of the uncertainty surface.

A critical issue for either the conditional or unconditional approaches is proper specification of the spatial structure of the error surface. These models rely upon the notion that knowledge of either elevation or elevation error at a particular location informs the simulation of elevation or elevation error at nearby locations. Characterizing this correctly, typically with the semivariance or correlogram, is very important. Figure 11.3 compares two uncertainty realizations; one has very little spatial autocorrelation, whereas the other is highly spatially autocorrelated, along with the corresponding DEM cross-section. The resulting elevation surfaces would be quite different if this DEM cross-section were added to each of the uncertainty realizations in turn. Studies investigating DEM accuracy for a variety of data sets have found that elevation error is spatially autocorrelated, and incorporating this information in an error model is critical (Guth 1992; Monckton 1994; Ehlschlaeger et al. 1997).

The degree of uncertainty in slope and aspect calculations is highly dependent on the spatial structure of error in the DEM. Hunter and Goodchild (1997) determined that, as spatial autocorrelation of elevation error increased, the standard deviation of simulated slope differences decreased. Other applications using DEMs are also sensitive to spatial autocorrelation in the error structure, (e.g., viewshed calculation) (Fisher 1991), particularly when analysis is dependent upon areal estimates (the area in square meters above 600 m) or neighborhood measures (the upslope contributing area). Any uncertainty model must therefore account for the spatial dependence of the error surface.

11.4 Case Study: A Terrain-Based Model of Bigcone Spruce Habitat

By way of example a habitat model for the bigcone spruce (*Pseudotsuga macrocarpa*) is developed that uses elevation and elevation derivatives as its sole input. This model is for demonstration purposes only; the results may not be particularly good at identifying spruce habitat. The primary interest is in demonstrating how DEM uncertainty can be modeled, simulated, and propagated through an ecological application. For elevation data input we will use two sources: a portion of a USGS 1:250,000 quadrangle and a set of 250 high-quality elevation sample points, such as might be collected with a GPS. The study will demonstrate how the 250 points can be used to model elevation uncertainty and propagate uncertainty to the habitat model. Decisions about the implications of uncertainty in the final product are left for other chapters (see in particular Chap. 18).

The bigcone spruce is a conifer with foliage similar to the douglas fir. The species range extends from the Santa Barbara, CA, region south to San Diego County, and is found at higher elevations in the coastal ranges (Griffin and Critchfield 1972). Within Santa Barbara County, stands of bigcone spruce are found on steep north and west facing slopes in the mountains, especially at the heads of canyons in the interior (Smith 1976). This habitat description lends itself to a cartographic model using input raster GIS layers (Berry 1993), which is presented here in pseudomap algebra:

HABITAT = (ASPECT == (north OR west)) AND (SLOPE > 33) AND (ELEV > 1500)

where steep slope is somewhat arbitrarily chosen to be 33 degrees and higher elevation is chosen to be greater than 1500 m.

The study area is the USGS 7.5' quadrangle, Madulce Peak, CA, located in a rugged portion of northeastern Santa Barbara County. Elevation ranges from just over 800 m to 1988 m. Highest elevations are located in the northwestern portion of the quadrangle, with canyons draining to the southeast and north, as shown in Figure 11.4a. Two elevation data sets were obtained for Madulce Peak. The first is the USGS 7.5' DEM for the quadrangle, with a resolution of 30 m and 176,584 elevation postings. Elevation locations were transformed to the WGS72 horizontal datum to match the second data set. The second is the USGS 1:250,000 DEM Los Angeles-west, which represents the area with elevation postings at 3 arc second intervals. This DEM was projected to Universal Transverse Mercator, resulting in a square grid resolution of 85 m, with 22,205 cells containing elevation data; this data set will henceforth be called the 85-m DEM. Both DEMs are presented in Figure 11.4. Note the general agreement in depiction of the terrain but the relative lack of texture in the 85 m DEM. In fact, local disagreement between the two datasets is extremely large, as will be seen shortly.

This case study treats the 7.5' DEM as ground truth. This is clearly not true, but it does enable us to check the quality of the model. The bigcone spruce model is run on the ground truth terrain and compared with the habitat model developed from the 85 m DEM; the results are shown in

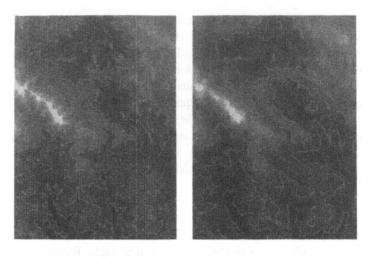

a. 30 m. DEM ("Ground Truth")   b. 85 m. DEM

FIGURE 11.4. Madulce Peak, California: 200-meter contours, 1000-m baseline. Lighter shades indicate higher elevations in all elevation maps in Figures 11.4–11.7. North is at the top for all maps in Figures 11.4–11.7.

Figure 11.5. The 85-m DEM-based model differs considerably from the ground truth model because the spatial resolution is much coarser and the DEM contains errors.

Instead of the entire ground truth, the study uses a set of 250 randomly sampled points across the quadrangle. These could be collected with a GPS (for this example they are sampled from the 7.5' DEM). Summary statistics reveal serious inaccuracies in the 85-m DEM: Mean error at the 250 sample locations is just 2.36 m, but the standard deviation is 36.7 m. At more than half the sample points, the DEM is in error by more than 20 m.

To account for the large discrepancy between the 85-m DEM and the actual terrain, the simulation/propagation approach discussed in the previous section is performed. A variogram model is constructed for the error data, and Gaussian simulation (a typical geostatistical simulation technique) is employed to generate realizations of the error surface, conditioned to the sample error points. The general procedure follows the stochastic simulation implementation discussion in (see Chap. 9, and Goovaerts 1997). Gstat, a public domain geostatistics package, is used to generate fifty error surface realizations (Pebesma and Wesseling 1998). Each surface is then subtracted from the 85 m DEM to produce a realization of the terrain of the Madulce quadrangle. Figure 11.6 presents one such error realization and its associated elevation realization. It is apparent that the realization is considerably rougher than the original 85 m DEM.

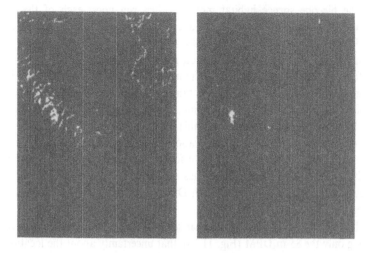

a. from 30 m. DEM ("Ground Truth")   b. from 85 m. DEM

FIGURE 11.5. Bigcone Spruce habitat model outcome (white indicates suitable). The 85-m DEM does not capture the range or complexity of the habitat identified by the ground truth DEM.

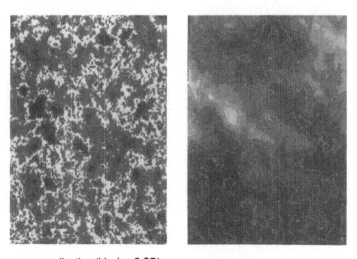

a. error realization (black > 2 SD)   b. elevation realization

FIGURE 11.6. One (of 50) simulations: 200-m contours, 800-m baseline. The elevation realization is generated by adding the error realization to the 85-m DEM. Terrain complexity matches that of the ground truth DEM.

The bigcone spruce habitat model is then run upon each of the 50 elevation realizations to produce 50 alternative habitat maps, one of which is depicted in Figure 11.7a. These 50 maps provide a probabilistic sense of what areas may be suitable for spruce, given the quality of the input data. Information from the 50 maps is summarized in the probability map in Figure 11.7b. Cell values in this map range from 0 (black), meaning that the cell is not identified in any realization, to 100 (white), indicating that all realizations identified the cell as suitable habitat. Note that this probabilistic map has identified many habitat locations in Figure 11.5a. The model has been partially successful at reproducing the complexity of the original habitat map—although the model has only had access to 12.7% as many elevation values as the ground truth DEM, and only 1.1% of those were accurate.

The map also suggests, via comparison with the habitat map produced using only the 85 m DEM (Fig. 11.5b), that uncertainty about the location of spruce habitat is quite high, given the input data. The white areas in Figure 11.7b were counted as habitat by at least 40% of the realizations; however, many of the gray-colored cells, including the vertical linear form in the northeastern portion of the map, were identified in only 15–20% of the realizations. One might conclude that the available terrain data is adequate for identifying the largest blocks of most suitable habitat for the bigcone

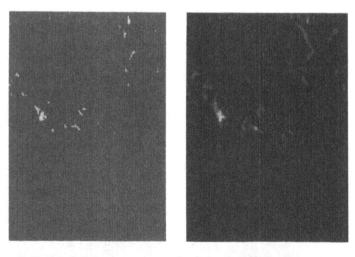

a. single habitat realization   b. stochastic habitat map

FIGURE 11.7. Characterizing DEM uncertainty in bigcone spruce habitat. White in (a) indicates suitable habitat identified in one realization. The map in (b) indicates the probability of each point being suitable for habitat. It was generated by summing across all 50 habitat realizations. Lighter regions indicate higher probabilities of suitable habitat.

spruce, but that uncertainty is too high for many marginal but possibly quite viable areas in the Madulce quadrangle.

11.5 Summary

Topography plays an important role in many environmental processes. Scientists studying land-based ecological systems often identify elevation derivatives like slope and aspect, as well as the elevation surface itself, as important descriptive or explanatory factors. Discretized characterizations of topography, DEMs, are used to represent the terrain. Discrepancies between DEMs and the real-world surfaces they represent introduce uncertainty into analyses which employ DEM data. This chapter classifies DEM-related uncertainty into two categories. The first source of uncertainty results from differences between the form of the spatial data model and the actual elevation surface. Uncertainty arises regarding elevations at locations not directly sampled in the DEM, referred to as data model-based uncertainty. An example application demonstrated the sensitivity of a terrain attribute, slope, to data model resolution.

A second source of uncertainty arises when data production methods do not accurately capture elevations at specified x, y locations. This error is directly measurable and is referred to in this chapter as data-based uncertainty. Uncertainty in elevation data can be addressed if it can be characterized, either through field measurements or by the adoption of assumptions about the data and the terrain. An uncertainty model is employed to characterize uncertainty in a spatial dataset. Given the available information and using a model of spatial uncertainty, a researcher can produce a map via a range of interpolation techniques, including a variety of kriging methods, of the most likely elevation surface.

A superior approach is often to propagate DEM uncertainty through the analysis to identify its impact upon the results of the application. This is accomplished by generating a set of realizations of the DEM via stochastic simulation from an uncertainty model. The ecological application is then run upon all realizations to produce a distribution of results. This procedure was demonstrated in the case study using a cartographic modeling approach to characterize bigcone spruce habitat. The model relied upon a DEM that matched the actual terrain quite poorly. Using scattered spot height data, however, an uncertainty model was developed. This model produced, via Gaussian simulation, a set of elevation realizations. By propagating uncertainty through the habitat model, a more complete picture emerged of the uncertainty in the application due to the quality of the input elevation data.

Several chapters in this book demonstrate approaches to modeling uncertainty that are applicable for projects using digital elevation data. In each case the ecologist is able to use higher-quality information to develop an uncertainty model for continuous data. Through subsequent simulation and propagation, the effect of elevation uncertainty on an application can be rigorously assessed, providing insight into the quality of the elevation data, as well as a closer bond between the ecological model and the real world processes it attempts to characterize.

Acknowledgments. This work was financially supported by the National Center for Geographic Information and Analysis, the National Center for Ecological Analysis and Synthesis, both in Santa Barbara, CA, and the National Imagery and Mapping Agency.

References

Austin, G.E., C.J. Thomas, D.C. Houston, and D.B.A. Thompson. 1994. Predicting the spatial distribution of buzzard Buteo buteo nesting sites using a geographical information system and remote sensing. Journal of Applied Ecology 33:1527–40.

Berry, J.K. 1993. Cartographic modeling: the analytical capabilities of GIS. Pages 58–74 in M.F. Goodchild, B.O. Parks, and L.T. Steyaert, eds. Environmental modeling with GIS. Oxford University Press, New York.

Bolstad, P.V., and T. Stowe. 1994. An evaluation of DEM accuracy: elevation, slope, and aspect. Photogrammetric Engineering and Remote Sensing 60(11):1327–32.

Brown, D.G., and T.G. Bara. 1994. Recognition and reduction of systematic error in elevation and derivative surfaces from 7.5-minute DEMs. Photogrammetric Engineering and Remote Sensing 60(2):189–94.

Burrough, P.A. 1986. Principles of geographic information systems for land resources assessment. Clarendon Press, Oxford, UK.

Carter, J.R. 1989. Relative errors identified in USGS gridded DEMs. Pages 255–65 in Proceedings of Autocarto 9, ASPRS-ACSM.

Carter, J.R. 1990. Some effects of spatial resolution in the calculation of slope using the spatial derivative. Pages 43–52 in Technical papers vol. 1, ACSM-ASPRS annual convention, Denver.

Chang, K., and B. Tsai. 1991. The effect of DEM resolution on slope and aspect mapping. Cartography and Geographic Information Systems 18(1):69–77.

Clarke, A.L, A. Gruen, and J.C. Loon. 1982. The application of contour data for generating high fidelity grid digital elevation models. Pages 213–22 in Proceedings of Autocarto 5, ASPRS-ACSM.

de Swart, E.O.A.M., A.G. van der Valk, K.J. Koehler, and A. Barendregt. 1994. Experimental evaluation of realized niche models for predicting responses of plant species to a change in environmental conditions. Journal of Vegetation Science 5:541–52.

Dubayah, R., and P. M. Rich. 1993. GIS-based solar radiation modeling. Pages 129–34 in M.F. Goodchild, B.O. Parks, and L.T. Steyaert, eds. Environmental Modeling with GIS. Oxford University Press, New York.

Ehlschlaeger, C.R., A.M. Shortridge, and M.F. Goodchild. 1997. Visualizing spatial data uncertainty using animation. Computers and Geosciences 23(4):387–95.

Englund, E.J. 1993. Spatial simulation: environmental applications. Pages 432–37 in M.F. Goodchild, B.O. Parks, and L.T. Steyaert, eds. Environmental modeling with GIS. Oxford University Press, New York.

Fisher, P.F. 1991. First experiments in viewshed uncertainty: the accuracy of the viewshed area. Photogrammetric Engineering and Remote Sensing 57(10):1321–27.

Gao, J. 1997. Resolution and accuracy of terrain representation by grid DEMs at a micro-scale. International Journal of Geographical Information Science 11(2): 199–212.

Garbrecht, J., and P. Starks. 1995. Note on the use of USGS level 1 7.5-minute DEM coverages for landscape drainage analyses. Photogrammetric Engineering and Remote Sensing 61(5):519–22.

Gessler, P.E., I.D. Moore, N.J. McKenzie, and P.J. Ryan. 1993. Soil-landscape modeling in southeastern Australia. Pages 53–58 in M.F. Goodchild, B.O. Parks, and L.T. Steyaert, eds. Environmental Modeling with GIS. Oxford University Press, New York.

Goodchild, M.F. 1992. Geographical data modeling. Computers and Geosciences 18(4):401–8.

Goodchild, M.F. 1995. Attribute accuracy. Pages 59–79 in S.C. Guptil, and J.L. Morrison, eds. Elements of Spatial Data Quality. Elsevier, London.

Goovaerts, P. 1997. Geostatistics for Natural Resources Evaluation. Oxford University Press, New York.

Griffin, J.R., and W.B. Critchfield. 1972. The distribution of forest trees in California. USDA Forest Service Research Paper PFW-82/1972.

Guth, P.L. 1992. Spatial analysis of DEM error. Pages 187–96 in ASPRS/ACSM/RT Technical Papers 2, Washington, DC.

Heuvelink, G.B., P.A. Burrough, and A. Stein. 1989. Propagation of errors in spatial modeling with GIS. International Journal of Geographical Information Systems 3(4):303–22.

Heuvelink, G.B.M. 1999. Propagation of error in spatial modeling with GIS. Pages 207–17 in P.A. Longley, M.F. Goodchild, D.J. Maguire, and D.W. Rhind, eds. Geographical information systems, volume 1: principles and technical issues, second ed. John Wiley & Sons, New York.

Huber, T.P., and K.E. Casler. 1990. Initial analysis of Landsat TM data for elk habitat mapping. International Journal of Remote Sensing 11(5):907–12.

Hunter, G.J., and M.F. Goodchild. 1997. Modeling the uncertainty of slope and aspect estimates derived from spatial databases. Geographical Analysis 29(1):35–49.

Hutchinson, M.F. 1996. A locally adaptive approach to the interpolation of digital elevation models. Third International Conference/Workshop on Integrating GIS and Environmental Modeling, Santa Fe. NCGIA, Santa Barbara, CA.

Isaacson, D.L., and W.J. Ripple. 1990. Comparison of 7.5-minute and 1 degree digital elevation models. Photogrammetric Engineering and Remote Sensing 56(11): 1523–27.

Isaaks, E.H., and R.M. Srivastava. 1989. Applied geostatistics. Oxford University Press, New York.

Kumler, M.P. 1994. An intensive comparison of triangulated irregular networks (TINs) and digital elevation models (DEMs). Cartographica Monograph 45 31(2): 1–99.

Lam, N-S. 1983. Spatial interpolation methods: a review. The American Cartographer 10(2):129–49.

Lee, J., P.K. Snyder, and P.F. Fisher. 1992. Modeling the effect of data errors on feature extraction from digital elevation models. Photogrammetric Engineering and Remote Sensing 58(10):1461–67.

Liebhold, A.M., G.A. Elmes, J.A. Halverson, and J. Quimby. 1994. Landscape characterization of forest susceptibility to gypsy moth defoliation. Forest Science 40(1):18–29.

Monckton, C.G. 1994. An investigation into the spatial structure of error in digital elevation data. Pages 201–10 in M.F. Worboys, ed. Innovations in GIS 1. Taylor & Francis, London.

Moore, I.D., R.B. Grayson, and A.R. Ladson. 1991. Digital terrain modelling: a review of hydrological, geomorphological, and biological applications. Hydrological Processes, 5:3–30.

Nisbet, R.A., and D.B. Botkin. 1993. Integrating a forest growth model with a geographic information system. Pages 265–69 in M.F. Goodchild, B.O. Parks, and L.T. Steyaert, eds. Environmental modeling with GIS. Oxford University Press, New York.

Nemani, R., S.W. Running, L.E. Band, and D.L. Peterson. 1993. Regional hydroecological simulation system: an illustration of the integration of ecosystem models in a GIS. Pages 296–304 in M.F. Goodchild, B.O. Parks, and L.T. Steyaert, eds. Environmental Modeling with GIS. Oxford University Press, New York.

Oliver, M., R. Webster, and J. Gerrard. 1989. Geostatistics in physical geography. Part II: applications. Trans. Inst. British Geography 14:270–86.

Openshaw, S. 1989. Learning to live with errors in spatial databases. Pages 263–76 in M. Goodchild, and S. Gopal, eds. The accuracy of spatial databases. Taylor & Francis, London.

Pebesma, E.J., and C.G. Wesseling. 1998. Gstat, a program for geostatistical modelling, prediction and simulation. Computers and Geosciences 24(1):17–31.

Polidori, L., J. Chorowicz, and R. Guillande. 1991. Description of terrain as a fractal surface, and application to digital elevation model quality assessment. Photogrammetric Engineering and Remote Sensing 48(2):18–23.

Robinson, G.J. 1994. The accuracy of digital elevation models derived from digitised contour data. Photogrammetric Record 14(83):805–14.

Russel, G.D., C.P. Hawkins, and M.P. O'Neill. 1997. The role of GIS in selecting sites for riparian restoration based on hydrology and land use. Restoration Ecology 5(4s):56–68.

Saulnier, G.M., C. Obled, and K. Beven. 1997. Analytical compensation between DTM grid resolution and effective values of saturated hydraulic conductivity within the TOPMODEL framework. Hydrological Processes 11:1331–46.

Shearer, J.W. 1990. Accuracy of digital terrain models. Pages 315–36 in G. Petrie, and T.J.M. Kennie, eds. Terrain modelling in surveying and civil engineering. Thomas Telford, London.

Smith, C.F. 1976. A Flora of the Santa Barbara Region, California. Santa Barbara Museum of Natural History, Santa Barbara, CA.

Unwin, D.J. 1995. Geographical information systems and the problem of "error and uncertainty". Progress in Human Geography 19(4):549–58.

U.S. GS (U.S. Geological Survey). 1995. National mapping program technical instructions. Standards for digital elevation models. U.S. Dept. Interior, Washington, DC.

Weibel, R., and M. Heller. 1991. Digital terain modelling. Pages 269–97 in D.J. Maguire, M.F. Goodchild, and D.W. Rhind, eds. Geographical information systems: principles and applications (1). Longman, London.

Zhang, W., and D.R. Montgomery. 1994. Digital elevation model grid size, landscape representation, and hydrologic simulations. Water Resources Research 30(4):1019–28.

Zhu, A.X., L.E. Band, B. Dutton, and T.J. Nimlos. 1996. Automated soil inference under fuzzy logic. Ecological Modelling 90:123–45.

12
An Overview of Uncertainty
in Optical Remotely Sensed Data
for Ecological Applications

MARK A. FRIEDL, KENNETH C. MCGWIRE, and DOUGLAS K. MCIVER

Remote sensing has become a widely used tool in ecology. Examples of ecological applications that use remote sensing include species conservation efforts such as GAP analysis (Scott et al. 1993), land cover and land use change monitoring (Skole and Tucker 1993; DeFries and Townshend 1994), and estimation of ecosystem carbon assimilation rates and net primary production (Prince 1991). At biome to global scales, it has also been demonstrated that the utility of remote sensing for monitoring ecosystem dynamics at time scales is commensurate with global change processes (Braswell et al. 1997; Myneni et al. 1997). Developments in remote sensing technologies, including airborne radar, video imaging systems, and satellite instruments with high spatial and spectral resolution show substantial promise for ecological studies. Further, even though this chapter focuses on optical remote sensing, new technologies (e.g., radar, laser altimeter, and lidar systems, which provide detailed information regarding topography and vegetation structure in three dimensions) suggest that the use of remote sensing by ecologists is likely to increase in the future.

Conventional uses of remote sensing in ecology fall into two main application domains. First, and most commonly, remote sensing has been used to map discrete classes of vegetation and species habitat using pattern recognition techniques applied to multispectral data acquired in the visible, near-infrared, and mid-infrared wavelength regions (Richards 1993). These applications have used both supervised and unsupervised classification algorithms with the goal of mapping specific vegetation or habitat classes that depend on the ecological problem of interest. Second, remote sensing is being increasingly used to map land surface biophysical properties using a wide array of both active and passive remote sensing instruments through inversion of physical models that describe radiative transfer within plant canopies in the visible, near infrared, and microwave wavelength regions (here we use the term "*inversion*" in a very general sense; the estimation of an unknown variable, e.g., LAI, from indirect measurements, e.g., remote sensing). Key ecological variables that have been examined include the leaf area index (LAI), biomass, canopy cover, and three dimensional (3-D) structure of vegetation (for a review see Davis and Roberts 2000).

An important attribute of remotely sensed data that is often overlooked in ecological applications is that the data represent indirect, and therefore imperfect, measures of the ecological attributes of interest. In the case of thematic maps derived from remote sensing (e.g., vegetation maps), classification algorithms assume that the classes of interest can be discriminated on the basis of their spectral reflectance properties, which is an assumption that is often somewhat weak. In the case of maps produced by inversion modeling, the accuracy of inverted vegetation properties strongly depends on whether or not sufficient remote sensing observations are available to constrain the inversion procedure (Goel 1987). In either case, the quality of a map derived from remote sensing is dependent on a variety of issues related to both the nature of the data and the processing algorithm applied to those data.

This chapter will provide an overview of the use of optical remote sensing for ecological applications, focusing specifically on the nature and sources of uncertainty and error inherent to this rather complex source of data. This chapter will also provide insight and guidance to scientific literature regarding the fundamental nature of remote sensing data and its uses and limitations for ecological applications. The philosophy implicit in these goals is that like any source of data, it is incumbent on the user to understand the nature and magnitude of uncertainty in remotely sensed data, and by extension, to understand how such uncertainty may propagate through ecological analyses that use maps derived from such data.

To achieve these goals, the chapter is divided into four main sections. We first review the nature of remote sensing as a data source for ecological problems. This section seeks to identify and illustrate the key limitations and sources of error inherent to both the data themselves and the processing algorithms that are conventionally used to extract information from these data. To this end, key references are provided where the interested reader can find more detailed information. The second section of this chapter discusses sources of error associated with standard techniques for extracting information from remotely sensed data. The third section describes standard methods for assessing the quality of map products derived from remote sensing, emphasizing the assessment of thematic map quality as well as continuous surface maps derived from inversion modeling. The final section of the chapter examines the interaction between the spatial resolution of remotely sensed data and the scale of ground scene properties because this relationship largely determines the utility of a remote sensing data set for a specific problem.

12.1 The Nature of Remotely Sensed Data

For the purposes of this chapter, we distinguish between three primary sources of error: (1) errors introduced through the image acquisition process;

(2) errors produced by the application of data processing techniques to raw remote sensing data; and (3) errors associated with interactions between instrument resolution and the scale of ecological processes on the ground. In this section and the next we will briefly describe the nature of errors belonging to sources (1) and (2). Section 12.4 specifically considers issues of sensor resolution in more detail.

It is important to note that the definition of error and uncertainty is somewhat imprecise. Here we consider sources of error of a fairly precise nature. We specifically consider errors arising from engineering aspects of the remote sensing process; however, in any ecological application that uses remote sensing, a substantial source of uncertainty is always introduced because ecological variables are often vaguely defined. A classic example of this problem is the use of strict class definitions to describe vegetation stands, which are by their nature heterogeneous and whose class transitions are gradual, not discrete. We will not explicitly consider these sources of error in this chapter; however, they are important to many remote sensing-based analyses, and are therefore implicit in much of the discussion here.

12.1.1 Image Acquisition

This section provides an overview of the remote sensing process, focusing on sources of error and uncertainty in unprocessed remotely sensed data that result purely from the remote sensing data acquisition process. Note that a detailed description of the various types and sources of remote sensing data is beyond the scope of this chapter. For an excellent discussion of remote sensing data sources and types the reader is referred to Kramer (1996).

Figure 12.1 presents a flow diagram illustrating the main steps involved in the acquisition of remotely sensed data. This framework identifies three distinct components to the remote sensing process (Strahler et al. 1986): the ground scene, the atmosphere, and the sensor. Implicit in this process are a variety of issues and details that affect the quality of map-based products derived from remote sensing. Duggin and Robinove (1990), for example, identify 11 key assumptions implicit in this process. We will consider sources of error associated with each of the three components identified by Strahler et al. (1986).

Scene Properties. Sources of error associated with the image acquisition process include a suite of effects related to interactions among the scene properties (i.e., the attributes of the land area of interest), the atmosphere, and the instrument used to record the data. With respect to the ground scene, two main sources of error are dominant. First, topographic variability within scenes strongly influences the apparent surface reflectance measured by remote sensing instruments (Fig. 12.2). To reduce topographic effects, sophisticated correction algorithms have been developed using

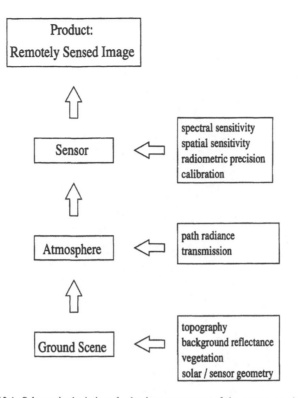

FIGURE 12.1. Schematic depicting the basic components of the remote sensing data acquisition process. Different types of error and uncertainty are introduced to remotely sensed data that are a function of each component.

digital elevation models (DEMs) to adjust for the influence of topography on both incident and reflected radiances (Civco 1989; Proy et al. 1989; Dubayah 1992; Richter 1997; Sandmeier and Itten 1997). Such corrections are generally not necessary for scenes where topography is subdued; however, in regions characterized by substantial relief, topographic effects will introduce artifacts to remotely sensed data that will be propagated in classification and inversion modeling procedures. Algorithms to correct for these effects are unfortunately limited by the quality of DEMs that are currently available (Hunter and Goodchild 1997).

Second, land surface directional reflectance properties can also exert strong control on land surface reflectance properties (Deering 1989). The directional dependence of remote sensing measurements is described by the surface bidirectional reflectance distribution function (BRDF), which defines the variation in spectral reflectance of a surface as a function of illumination and sensor view geometries (Fig. 12.3). A classic example of this effect is given by the so-called hot-spot effect that occurs when the sensor views the surface from the retrosolar position (i.e., the sun is directly behind the

FIGURE 12.2. Landsat Thematic Mapper image of an area adjacent to Irvine, California, illustrating the effect of topography on remotely sensed imagery. The upper right corner of the scene includes substantial variation in brightness induced by topography. Imagery courtesy of PCI Geomatics, Richmond Hill, Ontario, Canada.

sensor) (Jupp and Strahler 1991). In this situation, shadows cast by surface structures (e.g., tree crowns) are hidden from the sensor, and the surface therefore appears much brighter from this viewing perspective than from other angles. A variety of models have been developed to account for this effect (Walthall et al. 1985; Li and Strahler 1986; Myneni et al. 1992; Rahman et al. 1993); however, considerable expertise is required in order to use these models in practice. Whether or not such effects will significantly impact analysis results will strongly depend on the specific analysis being conducted. Awareness of potential variation in remotely sensed data induced by directional reflectance effects can help to minimize BRDF artifacts introduced into a remote sensing-based analysis and will provide guidance regarding whether or not corrections to the data need to be made.

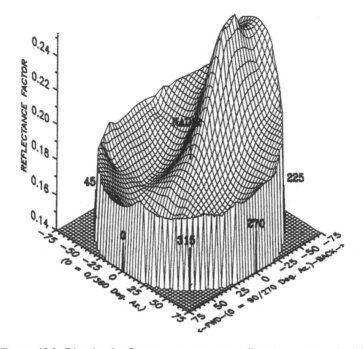

FIGURE 12.3. Directional reflectance measurements collected over a grassland in the near-infrared wavelengths. The BRDF describes the variation in surface reflectance as a function of viewing and illumination geometry. Note the distinctive peak in reflectance corresponding to the hot-spot. Reprinted from Deering (1989) by permission of John Wiley & Sons.

Atmospheric Effects. Spatial and temporal variation in atmospheric properties can also introduce significant artifacts to radiances measured from aircraft and satellites (Kaufman 1989). In the visible wavelength region, scattering processes dominate radiative transfer within the atmosphere, whereas absorption dominates in the near and mid-infrared regions. As a result, atmospheric effects can introduce bias to biophysical quantities inverted from either canopy reflectance models (Myneni and Asrar 1993) or semi-empirical models that are based on simulated relationships between spectral vegetation indices and vegetation canopy properties (Myneni 1994). In addition, radiation scattered from adjacent land areas into the sensor field of view can introduce blurring of scene properties via the so-called adjacency effect; therefore, because atmospheric properties influence estimates of surface upwelling radiances, results from inversion or classification algorithms will also be influenced by atmospheric effects (Kerekes and Landgrebe 1989; Myneni 1994). These effects can be especially problematic when multidate images, which are subject to variation in atmospheric effects, are used in an analysis.

Two main types of procedures may be applied to correct for atmospheric effects. The first approach is empirical in nature and is referred to as "dark pixel subtraction" (Chavez 1996). The basic idea behind this technique is that by selecting targets within imagery that have very low reflectance (e.g., water), the magnitude of the atmospheric effect within an image can be quantified and removed. The second approach uses much more elaborate methods based on atmospheric radiative transfer models (e.g., Tanre et al. 1990). These techniques are attractive because they are physically realistic and general. In order for radiative transfer models to provide good corrections, however, detailed information regarding atmospheric optical properties (e.g., vertical profiles of humidity, aerosols, temperature, and pressure) must be available at the time of data acquisition (Kaufman 1989). For problems involving multidate imagery Hall et al. (1991) have proposed a correction procedure that normalizes data across dates based on surfaces within the imagery that are assumed to be pseudoinvariant with respect to their reflectivity. In doing so, this procedure attempts to remove date-specific differences in imagery that are related to variation in atmospheric optics and solar geometry, and which are independent of the scene properties of interest.

Sensor Engineering. Whereas sensor engineering is often ignored by the end user, the spectral, spatial, and radiometric properties of remote sensing instruments exert strong influence on the properties and quality of image data. The sensor spatial point spread function (PSF; e.g., see Markham 1985) controls the sensor instantaneous field of view (IFOV) on the ground, and produces an effect known as *regularization*, where continuously varying ground features are sampled on a systematic basis to form an image. Interaction between the sensor PSF and the scale and variability of ground scene properties has been shown to be a key control on image information content (Jupp et al. 1988; Jupp et al. 1989). This topic will be discussed in detail in Section 12.4.

In a similar fashion, because the spectral domain has traditionally been the main dimension used to extract ecological information from remotely sensed data, sensor spectral response is a key attribute of a remote sensing instrument (Markham and Barker 1985; Richards 1993). So-called hyperspectral instruments have become available with very high spectral resolution, which may prove to be useful for a wide range of applications in ecology ranging from mapping and identification of species-level vegetation (Martin 1998) to inversion of canopy chemistry (Wessman et al. 1988; Martin and Aber 1998).

Sensor spectral and radiometric resolution and calibration are therefore key controls on the nature and quality of image data (Price 1987; Price 1988). In this framework, *radiometric resolution* refers to the precision of the data recorded by a remote sensing instrument, whereas *calibration* refers to the specific gain and offset of the instrument. These issues are particularly relevant where absolute (not relative) radiometric values are important (e.g., inversion modeling using canopy radiative transfer models or semi-empirical methods—see later). Goward et al. (1991) provide an excellent discussion of these issues in the context of NDVI measurements from the Advanced Very High Resolution Radiometer (AVHRR), which is being widely used for studies of global land cover and net primary production. Radiometric calibration and precision are also critically important for studies that require high precision measurements of directional reflectance (e.g., Braswell et al. 1996; Privette et al. 1996), or for studies examining long-term trends in eco-system properties from remote sensing (Braswell et al. 1997; Myneni et al. 1997).

Finally, it is important to note that even though each of the issues de-scribed above have been discussed in isolation, the integrated effect of these factors has proven to be very difficult to study. Kerekes and Landgrebe (1989, 1991) and Friedl et al. (1995, 1997), for example, have used simu-lation models to examine how interactions among sensor engineering, atmospheric effects, and scene properties influence thematic maps and inversion products derived from remote sensing. Systematic assessment of the additive or multiplicative effects of these factors in combination with each other using real data (as opposed to simulated data) has unfortunately proven to be problematic because the problem is underdetermined in the absence of extensive and detailed information related to each of the components acquired during the image acquisition process.

12.2 Data Processing

Errors associated with data processing in remote sensing are introduced via algorithms applied to digital data once acquired by a remote sensing device (Richards 1993). We identify two primary classes of errors that fall within this general domain. We will refer to the first class of errors as preprocessing errors that influence the raw image data. These errors are introduced to remote sensing data by data processing algorithms designed to correct image data *geometrically* to a specified coordinate system, and *radiometric errors* that are introduced to remotely sensed data when image data are adjusted to account for sensor calibration effects, or, alternatively, when topographic or atmospheric effects are poorly corrected. As a consequence geometric errors result in locational uncertainty, whereas radiometric errors introduce error to the remotely sensed data themselves. The second class of errors we will discuss result when classification or inversion models are applied to image data. These errors arise due the combination of error in the source data (i.e., radiometric errors) and errors introduced by the inversion or classification algorithms applied to those data.

12.2.1 Preprocessing

Errors arising from preprocessing are produced by well-defined sources and are well understood. The first class arises from the need to coregister image data to a map or image-based coordinate system, thereby allowing overlay of the image data with other map or image data. *Geometric rectification* techniques apply geometric transformations [e.g., rotation, translation, and resampling algorithms (Schowengerdt 1997)] to compensate for the fact that the 3-D surface of the earth must be projected into two dimensions in order to overlay image data with other digital data sources (e.g., other remote sensing or GIS data). As a consequence of those transformations, both positional errors and errors in the raw digital data are introduced (see Richards 1993 and Schowengerdt 1997 for detailed discussions of these issues).

In addition to geometric corrections, *radiometric corrections* are often applied to image data that can also introduce error to data products derived from remote sensing. Such corrections are often applied to account for topographic or atmospheric effects in image data, as described earlier. Radiometric corrections are more commonly applied to remove errors associated with changes in sensor calibration (i.e., drift in the sensor gain and offset) or systematic noise associated with poor cross-calibration among detectors that can produce image striping. Schowengerdt (1997) provides an excellent discussion of the wide array of techniques that have been developed to correct for these sources of error.

12.2.2 Classification

The preprocessing errors discussed earlier introduce radiometric and positional noise to remote sensing image data. In contrast, errors generated by classification algorithms introduce errors to map layers with inferred thematic values that may or may not be systematic. Two general classes of classification algorithms are used conventionally to extract and map thematic information from remotely sensed data: (1) unsupervised classification algorithms, and (2) supervised classification algorithms. Unsupervised approaches are data-driven techniques that use clustering algorithms to discriminate or identify relatively distinct spectral groupings within an image data set (Fig. 12.4). Common examples of this type of algorithm include k-means and hierarchical clustering techniques (Schowengerdt 1997). Because the algorithms do not require a priori definition of the classes to be classified, posterior labeling of spectral clusters is required, which may introduce further errors (McGwire 1992).

Supervised classification techniques, on the other hand, use training samples of spectral data for classes defined a priori to estimate decision surfaces in spectral space that discriminate among the different classes of interest. Common examples of this type of algorithm include maximum likelihood classification techniques (Strahler 1980), k-nearest neighbor algorithms

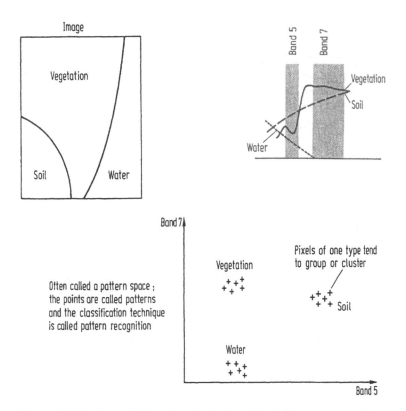

FIGURE 12.4. A schematic illustrating the conceptual basis for pattern recognition-based classification techniques (reproduced with permission from Richards 1993). Clusters of data that are similar in spectral space are assumed to belong to the same class.

(Schowengerdt 1997), and, more recently, neural networks (Gopal and Woodcock 1996) and decision trees (Friedl and Brodley 1997). The last three techniques have the distinct advantage over maximum likelihood algorithms in that they are nonparametric and make no assumptions regarding the distribution of input data.

In either case, two fundamental questions influence the accuracy of map products derived from the application of classification algorithms to image data. First, a fundamental distinction is made between discrimination and classification; specifically, *discrimination* refers to the process of differentiating between distinct groups of data within spectral space. Classification, on the other hand, refers to the assignment of class labels to image data that belong to previously unknown classes. Thus, unsupervised classification techniques are really best described as discrimination methods and do not assign class labels without human interaction. Supervised techniques, on the other hand, attempt to assign class labels to observations that may or may not be truly discriminable in feature space. Thus, a key determinant controlling how well (or accurately) a classification algorithm performs is the degree to which the available features (i.e., spectral channels or derived measures such as image texture) are capable of discriminating among the classes of interest. Indeed, hybrid-based methods that exploit the strengths of both supervised and unsupervised methods have been tested as a means to exploit the strengths of both paradigms (Schowengerdt 1997; Stoms et al. 1998).

Second, we distinguish between classifications (and classification assessments) that are performed at the site or stand level versus those that are performed on a pixel-by-pixel basis. Image classifications are traditionally performed in the latter fashion. For example, using a supervised approach, a classification model is estimated based upon a subsample of the entire image (i.e., a set of training data), which is then applied on a per-pixel basis to the entire image data set to produce a map; however, researchers have adapted such so-called per-pixel classification algorithms for application at scales that are more natural with respect to the classes being mapped on the ground (i.e., at the scale of individual vegetation patches or stands, whatever the user considers the mapping unit of interest). This approach provides a mechanism for applying image classification algorithms at scales appropriate to the mapping problem at hand. There is a trade-off involved, however, in that finite levels of within-stand heterogeneity and error must also be tolerated. In other words, the presence of multiple scales of variation are the norm in nature, even at the scale of individual vegetation stands. This variation is also manifested in remotely sensed data. Within this framework, the notion of fuzzy labels and secondary labels has received attention as a more realistic alternative to the assignment of a single class value at each pixel (e.g., see Chap. 15, as well as Woodcock and Gopal 1994).

12.2.3 Inversion Models

We have so far mainly focused on the nature and sources of errors in thematic map products derived from remote sensing; however, as we have already indicated, maps depicting continuously varying parameters related to vegetation structure and density are increasingly being derived from remote sensing data sources. The approach used to perform this type of inversion can vary widely depending on the specific problem at hand, but the specific inversion strategy used generally depends on the wavelength region of the remote sensing data used.

Three general approaches have been used in the visible and near-infrared region. First, purely empirical methods have been used to estimate regression models relating biophysical parameters such as LAI and biomass to remotely sensed measurements (Asrar et al. 1986; Friedl et al. 1994). These methods have the advantage that the uncertainty associated with the estimated regression model is well defined using such standard metrics as the variance explained (R^2) and the root mean squared error (RMSE) between measurements and the model predictions. Purely empirical models unfortunately have the disadvantage of being site specific and require detailed ground data for model calibration purposes (Curran and Hay 1986).

Second, semi-empirical models have been used to develop inversion strategies that use modeled relationships between canopy biophysical variables and spectral reflectance or vegetation indices (SVIs) (Sellers et al. 1992). This family of models uses indices based on linear combinations or ratios of reflectance measurements to derive relatively robust and invertible linear or pseudolinear relationships between a specific SVI and biophysical variable of interest, principally LAI and the fraction of absorbed photosynthetically active radiation (FPAR) (Goward and Huemmrich 1992). A wide array of factors, however influence the overall accuracy of inverted parameters derived in this fashion including sensor precision and calibration, atmospheric effects, and surface directional reflectance (Goward et al. 1991; Myneni 1994).

Third, canopy reflectance models have been used in association with multiangular measurements to estimate canopy structure and density (Strahler 1997). These algorithms use remote sensing observations acquired from different view (and perhaps solar) geometries in association with numerical inversion methods that minimize the cumulative error between model predictions and observations. Because forward-mode canopy reflectance models (i.e., a model used to predict surface reflectance given solar–sensor geometry and surface properties) typically require numerous parameters, the problem is often underdetermined, and therefore difficult to solve. Further, because the information content of the remote sensing observations for this problem depends in part on the angular distribution of the observations, the accuracy of model inversions also depends on the number and quality of the multiangular observations available. Despite these problems, research has demonstrated the viability of this approach using available operational remote sensing instruments (Goel 1987; Braswell et al. 1996; Privette et al. 1996).

It is important to note that the errors introduced to maps produced by such remote sensing inversion techniques are of an entirely different nature and origin than thematic errors present in classification maps; specifically, map products derived from numerical inversion of canopy reflectance or scattering models include varying levels of error associated with biophysical parameters inverted from the remotely sensed data. Unlike the global classification or per-class accuracy statistics reported for thematic map products (see Sec. 12.3), errors introduced via the inversion process are more relevant at the pixel level (i.e., the inverted value at each pixel has an associated accuracy and precision). These errors depend on a wide array of factors, including the number and quality of observations available at a given pixel in combination with the quality of the forward model and its "invertibility" (i.e., the use of numerical procedures to estimate surface variables from observations using a forward model).

12.3 Assessment and Quantification of Errors in Thematic Maps

Quantification and assessment of classification accuracy for thematic maps derived from remote sensing is typically performed using a set of independent validation or "ground truth" data. When using a supervised classification method, validation data is withheld from the data used to train the classification algorithm. The predicted class values for the validation data are then compared with their known values using a confusion matrix or contingency table (Story and Congalton 1986; Campbell 1996). This matrix maps the true class values from the validation set along the columns against the predicted classes from the classifier in the rows. Correctly classified pixels therefore fall along the diagonal of the matrix, whereas pixels that have been misclassified, or confused with another class fall in the off-diagonal cells. The confusion matrix therefore provides information on both the overall classification accuracy as well as the errors among specific pairs of classes.

Table 12.1 presents a sample confusion matrix. This matrix was computed using test data from a classification of landcover based upon multitemporal AVHRR data from Central America (Muchoney et al. 2000) using the classification scheme defined by the International Geosphere-Biosphere Program (IGBP) (Belward 1996). The overall classification accuracy (based on

TABLE 12.1. Confusion matrix showing the number of test cases assigned to each of 17 classes by a supervised decision tree classification algorithm from an independent test data set.

	1	2	3	4	5	6	7	8	9	10	11	12	13	14	15	16	17
1	229	22		4	14		1	1		2	2	15	1	4		1	
2	28	616		8	14	2	2	1	2	1	11	23	2	7			
3			3														
4	1	3		40	2		1			2	1	4	1				1
5	16	21		3	114			2	2		3	19	1	2			
6	5	5	1			6				4	1	2		1		1	
7	6	1		2	2	1	40	2		1		9	1				1
8	6	3		3			1	71		2	8	8	1				
9	1	5						2	5	1	1	1					
10	4	3		5	3		5	5		119	4	23		2			
11	11	15		1	6	1	2	3	1	1	95	11	1	4			2
12	13	25		6	6		6	7	1	11	13	285	3	2			
13	1	3					1	1			1	4	40				
14	2	11		1	4			2	2	1	2	3		49			
15																	
16		1	1	1	1		1									2	
17							1						2	2			148

Columns represent the true class values, whereas rows represent the values assigned by the classification algorithm. Correctly classified cases fall on the diagonal. For example, for class 1, 229 cases were correctly classified, 22 cases were incorrectly assigned a class label of 2, four cases were assigned a label of 4, and so on.

an independent sample of test data) is 75%. At the same time, the dispersion of off-diagonal entries demonstrates that substantial confusion exists between many of the classes.

Obtaining high-quality validation data is generally difficult and costly and is often constrained by practical limitations (Edwards et al. 1988; Congalton 1991). Further, depending upon the source, they are generally subject to many of the same positional and classification errors as the remotely sensed data being evaluated (Congalton 1991). Errors in the validation data, therefore, may either inflate or reduce the estimated accuracy of the classification.

The central objective in compiling validation data is that it contain minimal errors, be representative of the data being assessed, and should also consist of a large enough sample to provide statistically significant results for each class in the classification (Hay 1979; Congalton 1991). Within this framework, numerous sampling techniques have been suggested to obtain representative samples for validation, including random, systematic, clustered, and stratified sample designs (Congalton 1988a; Stehman 1992). Using an appropriate sampling scheme is particularly important because classification errors often exhibit a high degree of spatial autocorrelation (Congalton 1988b). Spatial autocorrelation among the classification errors, therefore, can bias estimates of accuracy if a spatially representative and independent sample is not obtained.

When supervised classification techniques are used, there is an additional need for adequate data for both training and validation purposes; specifically, to obtain an accurate evaluation of the classification, the validation set must be independent of the data used to train the classifier. Thus, a portion of the available "truth" data must be withheld from the training phase in order to assess performance. This trade-off between training and testing data sets is particularly important for nonparametric classifiers that can require large training sets to produce accurate results. As we mentioned earlier, the validation set must also be large enough to produce statistically significant results. Note that there are a number of studies in the remote sensing literature where the reported classification accuracy is based on back-classification of training data. This procedure can strongly bias results toward an overoptimistic assessment.

Another important point to note is that the assessment of classification accuracy is normally based upon random samples of pixels; however, because most land cover and vegetation classifications are attempting to map features that occur in patches, stands, or regions on the ground, pixel-based accuracy assessments may not be the most appropriate scale of analysis or accuracy assessment. In other words, because significant spatial correlation may be present among pixels within validation sites, results from accuracy assessments based on random splits of data pooled across sites will often provide spurious estimates of the classification accuracy. A more valid approach, therefore, is to partition available data for training and testing at the stand or site level (Friedl et al. 2000). That is, randomly sample entire stands from the training data to be used as test data rather than randomly sampling pixels within stands. This type of approach does not provide information regarding variation at scales below the stand level; however, it does provide an assessment of the map accuracy at a scale that is commensurate with the types of sampling designs used to collect so-called ground-truth data.

Once a validation set has been obtained, numerous measures of accuracy may be used to quantify the error in thematic maps produced by pattern recognition classification algorithms (Stehman 1997b). The simplest of these measures is the overall classification accuracy (also known as the percent correctly classified, PCC), which is obtained by dividing the sum of the diagonal elements of the confusion matrix by the total number of pixels in the test set. More information about the classification accuracy can be obtained by examining the off-diagonal elements of the confusion matrix.

Finally, a number of methods have been developed to account for the fact that some pixels may be correctly labeled by a classification algorithm due solely to random chance, thereby overestimating the overall classification accuracy. The Kappa (κ) statistic (Cohen 1960; Congalton and Mead 1983; Congalton et al. 1983; Rosenfield and Fitzpartrick-Lins 1986) is used most commonly to evaluate the accuracy of classification, although other possibilities exist (Ma and Redmond 1995; Stehman 1997b). The κ statistic accounts for both the off-diagonal elements of the error matrix and the possibility of chance agreements:

$$\hat{\kappa} = \frac{N \sum_{i=1}^{r} x_{ii} - \sum_{i=1}^{r}(x_{i+} \times x_{+i})}{N^2 - \sum_{i=1}^{r}(x_{i+} \times x_{+i})} \tag{12.1}$$

where N is the number of observations, r is the number of rows in the matrix, x_{i+} and x_{+i} are the marginal totals of row i and column i, respectively, and x_{ii} is the number of observations in column i and row i. For a given confusion matrix, a $\hat{\kappa}$ value of 1.0 represents perfect accuracy, 0 represents no better than chance agreement, and a negative value less than chance agreement. $\hat{\kappa}$ tends to be conservative, normally providing lower values than those provided by the overall classification accuracy. Because the distribution of $\hat{\kappa}$ is known, significance levels may be assigned to results obtained using this test statistic. In this context, it is important to note that different sampling schemes have different estimates or standard errors for $\hat{\kappa}$ (Stehman 1995). Stehman (1996, 1997a) provide details in this regard for stratified and cluster sampling, respectively.

12.4 Interaction Between Processing Models and Data

12.4.1 High Versus Low Resolution Data

One of the most fundamental sources of uncertainty in remotely sensed data is caused by interaction between the scale of variation within the ground scene and the spatial resolution of the sensor. To be specific, the spatial resolution of remote sensing data places limits on the types of information that may be extracted from remotely sensed imagery and has a significant influence on the spatial uncertainty of remote sensing data sources in ecological analyses. A remote sensing data source is often selected to satisfy many requirements, including the frequency at which data are acquired and the spectral resolution of the instrument. In particular, the relationship between the IFOV of the selected sensor system and the spatial variability in ecological variables on the ground will strongly influence the types of analyses that may be performed. This includes whether or not the desired parameter is discernible, the precision of parameter estimates, and the degree to which the dynamic range of the parameter is captured. This section will identify key issues related to spatial scale with respect to both categorical and continuous land cover parameters. With respect to classification, this will include a discussion of the relevance of the classification taxonomy to the goal of regionalization and the potential for accurate identification of classes from reflectance or emittance characteristics. For continuous biophysical parameters, spatial heterogeneity occurs at a number of scales from stomata to leaf to canopy. The selection of an appropriate scale of analysis therefore depends on the correlation scales that are manifested by the parameter.

Multiple scales of variation are represented in remotely sensed imagery, with the spatial resolution of the sensor and the extent of the image bounding the relevant scales of study (Woodcock and Harward 1992). As demonstrated by Townshend and Justice (1988) with AVHRR data covering several different geographic regions, spatial variations over a wide range of scales contribute to the overall variance in scenes. The information content of an image is a function of the complexity of the ground scene and the spatial, spectral, radiometric, and temporal resolution of the sensor. Woodcock and Strahler (1987), for example, demonstrated that image variance is directly related to the size of ground objects relative to the spatial resolution of the sensor, and that this relationship is manifested in measures of both local and total image variance.

Scale-dependent effects arising from pixel size may be considered in terms of so-called H and L resolution (Woodcock and Strahler 1987). *H resolution* refers to the situation where the sensor IFOV on the ground is finer than the size of the features on the ground to be resolved. Under these circumstances, the frequency distribution of remotely sensed values tends to be multimodal in character, representing either target or background. In the *L resolution* case, measurements are coarser than the size of objects to be resolved. This results in remotely sensed measurements that tend to have a more continuous frequency distribution, representing a gradient in sensor response corresponding to variations of target versus background within the sensor IFOV. Within this framework, semivariograms have been used to examine the spatial variation in remotely sensed data as a function of ground scene elements and remote sensor parameters (Curran 1988; Woodcock et al. 1988a; Woodcock et al. 1988b). Theoretical research using simplified scene models has indicated direct ties between scene components and variogram form (Jupp et al. 1988; Jupp et al. 1989).

The spatial response function of the sensor often influences the pixel size at which data are initially resampled by the source data provider (e.g., see Markham 1985). It is important to recognize, however, that this is not always the case. For example, many of the analyses performed with products derived from the AVHRR sensor use a 1.0-km pixel size; however, the actual IFOV of the AVHRR sensor ranges from 1.1 at nadir to more than 4.0 km depending on how far off nadir the view geometry of a pixel is. This situation raises serious problems with respect to the consistency of representation for features on the ground, both within and between images. In addition, variation in pixel size as a function of view geometry compromises the validity of many statistical assessment methods (e.g., accuracy assessment based on the assumption of independent validation data) because adjacent pixels will not necessarily be independent samples. Image processing operations may also degrade the effective resolution of the data, regardless of the pixel size. For example, data may be spatially filtered to reduce noise (Schowengerdt 1997).

12.4.2 Measurement Precision

The sensor spatial PSF and IFOV interact with the underlying scales of variability on the ground to determine the degree of precision that may be represented by remotely sensed parameters. As a result, discrete remotely sensed observations (pixels) represent area-averages of land surface properties within the sensor IFOV, smoothing away information that is expressed at finer scales. The resulting reduction in the dynamic range of the measured values as a function of sensor IFOV can be important in cases where the phenomenon of interest displays nonlinear response to the remotely sensed observations (e.g., radiation attenuation within plant canopies as a function of leaf area index; see Friedl 1997).

Pixel size also influences the information content of thematic data derived from images. Taxonomic systems for classifying vegetation communities are often based on the relative dominance of selected species, and classes within a taxonomy will often be defined by proportions. For example, a woodland class might be defined as all areas possessing between 25 and 50% tree cover. As we discussed earlier, in situations where the pixel size is much smaller than the size of a tree (H resolution), the frequency distribution of pixel values within a scene will be bimodal. That is, pixels will tend to be composed of either "tree" or "understory." In isolation, neither tree nor understory relate directly to the concept of "woodland." Thus, attempts to map woodland from such H-resolution data can be problematic on a per-pixel basis. On the other hand, if the pixel size is roughly the same as the crown size of trees within the scene (i.e., intermediate between H and L resolution), the spectral reflectance may not provide a clear indication of either class because pixels would then encompass mostly tree, mostly understory, or a mixture of both. As a result a woodland class would possess substantial spectral overlap with classes such as "forest" and "grassland" and would not be well characterized using conventional summary statistics (e.g., the sample mean and variance).

If a resolution is used that is coarse enough to encompass relatively consistent proportions of tree and background, a woodland class might start to emerge as a stable and distinct distribution along the spectral continuum from grassland to forest. As the spatial resolution is coarsened, however, the spectral separability of these three classes might again become undistinguishable. Thus, the selection of a taxonomic framework has direct implications for the required scale of analysis.

A solution to this problem is to treat pixels as being composed of mixtures of classes (Adams et al. 1986; Huete 1986). Such "mixture modeling" techniques treat the spectral response of a pixel as a linear combination of subpixel surface types. If the spectral properties of each of the subpixel classes is known (so-called end-members), then unmixing techniques can be applied to estimate the subpixel proportions of each component class. For example, DeFries et al. (2000) used this technique to map the distribution of vegetation properties at global scales based on unmixing of AVHRR spectral data. This approach is unfortunately really only viable if good estimates of the end-member spectral reflectance curves are available. As we discuss in the next section, the reliability of such end-member reflectance curves decreases as the area of interest increases. Other limitations on spectral unmixing methods include nonlinear mixing of end-member reflectances and nonunique solutions arising from spectra that are not sufficiently distinct in spectral space.

12.4.3 Signature Extension and Extrapolation

A central goal in many ecological applications that use remotely sensed data is to isolate those reflectance characteristics that are most closely related to the phenomenon under study (Fig. 12.5). This type of analysis will then lead to transformations of the image data that will provide the most precise and unbiased representations of that phenomenon. To do this, training samples are taken from the image to characterize the spectral signature of the phenomenon of interest.

Even if a spectral transformation is available that isolates a particular feature (e.g., Kauth and Thomas 1976), however, the issue of adequately defining the inherent variability of that feature is still unresolved. If we consider a particular plant species in isolation, the morphology of that species often is related to a number of environmental characteristics (e.g., nutrient availability, water availability, etc.) In particular, morphological changes can occur over space in response to environmental gradients, which in turn

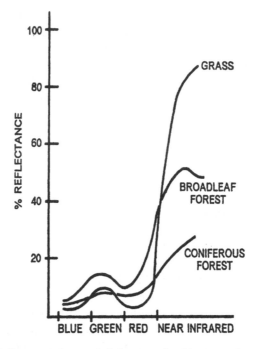

FIGURE 12.5. Representative spectral signatures for three vegetation types. Differences in the reflectance of vegetation across wavelengths is the primary mechanism that has been used to map vegetation using pattern recognition techniques (reproduced with permission from Campbell 1996).

will impose spatial variability on the measured spectral response. Further, morphological characteristics that change over time will be expressed in space at different scales (e.g., geographic variation in phenology produced by regional differences in climate) (Schwartz 1994).

The problem of unexplained geographic variability in spectral signatures leads to the well-known problem of "signature extension." This problem refers to the fact that training samples for spectral classes may be limited in terms of their representativeness across time and space. In particular, training data extracted from a specific image generally represent a composite of spectrally different subclass components. For example, corn fields may contain different cultivars or may be planted on different dates. The species composition of a "yellow pine forest" will similarly vary from place to place. If representative samples are selected for a class that do not match the true range of values for that class on the ground, then the probability distribution function used to characterize the spectral signature will be biased, or at least incomplete. As a consequence, observations with spectral values at the tails of the true distribution may be misclassified, even if the pixels are spectrally unique. Note that this issue is also problematic for nonparametric classification algorithms such as neural networks or decision trees because the training sample does not include exemplars for the entire range of values in the underlying population.

Depending on the scales of variation in spectral response relative to the pixel size of the imagery, spectral variability over space may be expressed as random noise, as spatially autocorrelated fields of pixels, or as trends across the entire study area (nonstationarity) (Bartlett et al. 1988). *Nonstationarity* is a gradual or discrete shift in the mean and/or variance of spectral features across a study area. As a consequence, depending on the spatial scales of variation, the effectiveness of traditional statistical methods for classification may be greatly reduced by spatial autocorrelation and/or nonstationarity in surface reflectance. For example, spatial autocorrelation in spectral response violates the assumption of sample independence that underlies many classification techniques. Training samples are often obtained by delineating a region on an image. If this region is not significantly larger than the spatial extent of spatial autocorrelation effects, then the sample variance will underestimate that of the population. Several authors have shown how this effect reduces classification accuracy (Craig 1979; Campbell 1981; Labovitz 1984). The use of nonspatial statistical methods to map features whose reflectance contains spatial structure will have this structure passed into the map as error (see McGwire et al. 1993 and Chap. 14, this volume). The effect of nonstationarity on image classifications will be similar to that for spatial autocorrelation, but is likely to be even more pronounced.

12.5 Summary

Remote sensing has long been promoted as an excellent source of data for ecological studies. This promise is increasingly being realized; however, the effective use of remote sensing data for ecological applications requires a thorough understanding of both the strengths and weaknesses of this data source. In this regard, the most significant strength of remote sensing is clearly the ability to observe ecological phenomena over large areas in a fashion that is perhaps not possible using field-based methods. At the same, this strength also represents a key limitation in that remotely sensed data represent indirect measurements of the parameters of interest, and may not provide adequate spatial detail.

In this chapter we have attempted to survey the nature and sources of uncertainty in optical remotely sensed data for ecological applications. In this context, we identified three main sources of error: those attributable to the image acquisition process, those introduced via data-processing techniques, and those related to interaction between the scale of variation on the ground and the spatial sampling imposed by the sensor spatial resolution. In many regards, the general topics addressed in this chapter may be perceived to be a Pandora's box; indeed, the discussion presented here is not comprehensive.

This is not to say that ecologists should therefore abandon the use of remote sensing. Substantial information and value is present in remotely sensed data for a wide range of ecological applications. If it is used appropriately, remote sensing can provide a wealth of information that is otherwise very time consuming, expensive, and perhaps impossible for an ecologist to collect in the field. Further, the ability to acquire repeated observations over extended time periods provides ecologists with the ability to monitor temporal variations in ecosystem properties and processes over large areas. The key to successful use of remote sensing lies in careful experimental designs in which remotely sensed data is collected in coordination with field data. Given the complexity of this data source most ecologists attempting to use remote sensing will benefit from close collaboration with scientists and engineers who specialize in remote sensing. We hope that this chapter provides both context and a resource that will lead to wider collaborations of this nature, and, by extension, to wider and more effective use of remote sensing in ecology.

Acknowledgments. Partial support for this work under grant NAG5-7218 from NASA's Terrestrial Ecology Program is gratefully acknowledged.

References

Adams, J., M. Smith, and P. Johnson. 1986. Spectral mixture modeling: a new analysis of rock and soil types at the Viking lander 1 site. Journal of Geophysical Research 91(B8):8098–12.

Asrar, G., A.E. Kanemasu, G. Miller, and R. Weiser. 1986. Light interception and leaf area estimates from measurements of grass canopy reflectance. IEEE Transactions on Geoscience and Remote Sensing GE-24:76–81.

Bartlett, D., M. Hardisky, R. Johnson, M. Gross, and J. Hartman. 1988. Continental scale variability in vegetation reflectance and its relationship to canopy morphology. International Journal of Remote Sensing 9:1223–41.

Belward, A. 1996. The IGBP-DIS Global 1 km land cover data set DISCOVER: proposal and implementation plans. IGBP-DIS Working Paper, IGBP–DIS Office, Meteo–France, 42 Av. G. Coriolis, F–31057, Toulouse, France.

Braswell, B., D. Schimel, J. Privette, B. Moore, W. Emery, E. Sulzman, et al. 1996. Extracting ecological and biophysical information from AVHRR oprical data: an integrated algorithm based on inverse modeling. Journal of Geophysical Research 101(D18):23335–48.

Braswell, B., D. Schimel, E. Linder, and B. Moore. 1997. The response of global terrestrial ecosystems to interannual temperature variability. Science 278:870–72.

Campbell, J. 1981. Spatial correlation effects upon accuracy of supervised classification of land cover. Photogrammetric Engineering and Remote Sensing 47:355–64.

Campbell, J.B. 1996. Introduction to remote sensing, second ed. Guilford, New York.

Chavez, P. 1996. Image-based atmospheric corrections revisited and improved. Photogrammetric Engineering and Remote Sensing 62(9):1025–36.

Civco, D. 1989. Topographic normalization of Landsat Thematic Mapper digital imagery. Photogrammetric Engineering and Remote Sensing (55):1303–09.

Cohen, J. 1960. A coefficient of agreement for nominal scales. Educational and Psychological Measurement 20:37–46.

Congalton, R.G., and R.A. Mead. 1983. Quantitative method to test for consistency and correctness of photointerpretation. Photogrammetric Engineering and Remote Sensing 49:69–74.

Congalton, R.G., R.G. Oderwald, and R.A. Mead. 1983. Assessing Landsat classification accuracy using discrete, multivariate analysis statistical techniques. Photogrammetric Engineering and Remote Sensing 49:1671–78.

Congalton, R.G. 1988a. A comparision of sampling schemes used in generating error matrices for assessing the accuracy of maps generated from remotely sensed data. Photogrammetric Engineering and Remote Sensing 54:593–600.

Congalton, R.G. 1988b. Using spatial autocorrelation analysis to explore the errors in maps generated from remotely sensed data. Photogrammetric Engineering and Remote Sensing 54:587–92.

Congalton, R.G. 1991. A review of assessing the accuracy of classifications of remotely sensed data. Remote Sensing of Environment 37:35–46.

Craig, R. 1979. Autocorrelation in Landsat data. Pages 1517–24 in Proceedings of the thirteenth international symposia on remote sensing of environment. Environmental Research Institute of Michigan, Ann Arbor.

Curran, P., and A. Hay. 1986. The importance of measurement error for certain procedures in remote sensing at optical wavelengths. Photogrammetric Engineering and Remote Sensing 52:229–41.

Curran, P. 1988. The semi-variogram in remote sensing: an introduction. Remote Sensing of Environment 24:493–507.

Deering, D. 1989. Field measurements of bidirectional reflectance. Pages 14–65 in G. Asrar, ed. Theory and applications of optical remote sensing. John Wiley & Sons, New York.

DeFries, R., and J. Townshend. 1994. NDVI-derived land cover classifications at a global scale. International Journal of Remote Sensing 5:3567–86.

DeFries, R., M. Hansen, and J. Townshend. 2000. Global continuous fields of vegetation characteristics: a linear mixture model applied to multi-year 8 km AVHRR data. International Journal of Remote Sensing 21:1389–414.

Davis, F., and D. Roberts. 2000. Stand structure in terrestrial ecosystems. Pages 7–30 in O. Sala, R.B. Jackson, H.A. Mooney, and R.W. Howarth, eds. Methods in Ecosystem Science (chapter in review). Springer-Verlag, New York.

Dubayah, R. 1992. Estimating net solar radiation using landsat thematic mapper and digital elevation data. Water Resources Research 28:2469–84.

Duggin, M., and C. Robinove. 1990. Assumptions implicit in remote sensing data acquisition and analysis. International Journal of Remote Sensing 11(10):1669–94.

Edwards, T.C., G.G. Moisen, and D.R. Cutler. 1988. Assessing map accuracy in a remotely sensed ecoregion-scale cover map. Remote Sensing of Environment 63:73–83.

Friedl, M., J. Michaclsen, F. Davis, and D. Schimel. 1994. Estimating grassland biomass and leaf area index using ground and satellite data. International Journal of Remote Sensing 15(7):1401–20.

Friedl, M., F. Davis, J. Michaelsen, and M. Moritz. 1995. Scaling and uncertainty in the relationship between the NDVI and land surface biophysical variables: an analysis using a scene simulation model and data from FIFE. Remote Sensing of Environment 54:233–46.

Friedl, M. 1997. Examining the effects of sensor resolution and sub-pixel heterogeneity on spectral vegetation indices: implications for biophysical modeling. Pages 113–40 in D. Quattrochi, and M. Goodchild, eds. Scale in Remote Sensing and GIS. Lewis, Boca Raton, FL.

Friedl, M., and C. Brodley. 1997. Decision tree classification of land cover from remotely sensed data. Remote Sensing of Environment 61:399–409.

Friedl, M., C. Woodcock, S. Gopal, D. Muchoney, A.H. Strahler, and C. Barker-Schaaf. 2000. A note on procedures used for accuracy assessment in land cover maps derived from AVHRR data. International Journal of Remote Sensing 21:1073–77.

Goel, N. 1987. Models of vegetation canopy reflectance and their use in estimation of biophysical parameters from reflectance data. Remote Sensing Reviews 3:1–212.

Gopal, S., and C. Woodcock. 1996. Remote sensing of forest change using artificial neural networks. IEEE Transactions on Geoscience and Remote Sensing 34(2):398–404.

Goward, S., B. Markham, D. Dye, W. Dulaney, and J. Yang. 1991. Normalized difference vegetation index measurements from the advanced very high resolution radiometer. Remote Sensing of Environment 35:257–77.

Goward, S., and K. Huemmrich. 1992. Vegetation canopy PAR absorptance and the normalized difference vegetation index: an assessment using the SAIL model. Remote Sensing of Environment 39:119–40.

Hall, F., D. Strebel, J. Nickeson, and S. Goetz. 1991. Radiometric rectification: toward a common radiometric response among multidate, multisensor images. Remote Sensing of Environment 35:11–27.

Hay, A. 1979. Sampling designs to test land-use map accuracy. Photogrammetric Engineering and Remote Sensing 45:529–33.

Huete, A. 1986. Separation of soil-plant spectral mixtures by factor analysis. Remote Sensing of Environment 19:237–51.

Hunter, G., and M.F. Goodchild. 1997. Modeling the uncertainty in slope and aspect estimates from spatial databases. Geographical Analysis 29:35–49.

Jupp, D., A. Strahler, and C. Woodcock. 1988. Autocorrelation and regularization in digital images I. Basic theory. IEEE Transactions in Geoscience and Remote Sensing 26:463–73.

Jupp, D., A. Strahler, and C. Woodcock. 1989. Autocorrelation and regularization in digital images. II. Simple Image Models. IEEE Transactions in Geoscience and Remote Sensing 27:247–56.

Jupp, D., and A. Strahler. 1991. A hotspot model for leaf canopies. Remote Sensing of Environment 38:193–210.

Kaufman, Y. 1989. The atmospheric effect in remote sensing and its corrections. Pages 336–428 in G. Asrar, ed. Theory and applications of optical remote sensing. John Wiley & Sons, New York.

Kauth, R., and G. Thomas. 1976. The tasselled cap—a graphic description of the spectral-temporal development of crops as seen by Landsat. Proc. symp. on machine processing of remotely sensed data 41–51.

Kerekes, J., and D. Landgrebe. 1989. Simulation of optical remote sensing systems. IEEE Transactions on Geoscience and Remote Sensing GE-27(6):762–71.

Kerekes, J., and D. Landgrebe. 1991. An analytical model of earth-observational remote sensing systems. IEEE Transactions on Systems, Man, and Cybernetics 21(1):125–33.

Kramer, H. 1996. Observation of the earth and its environment, third ed. Springer, New York.

Labovitz, M. 1984. The influence of autocorrelation in signature extraction—an example from the Cotter Basin, Montana. International Journal of Remote Sensing 50:315–32.

Li, X., and A. Strahler. 1986. Geometric-optical bidirectional reflectance modeling of a conifer forest canopy. IEEE Transactions on Geoscience and Remote Sensing GE-24:281–93.

Ma, Z., and R.L. Redmond. 1995. Tau coefficient for accuracy assessment of classification of remote sensing data. Photogrammetric Engineering and Remote Sensing 61:435–39.

Markham, B. 1985. The Landsat sensor spatial responses. IEEE Transactions on Geoscience and Remote Sensing GE-23(6):864–75.

Markham, B., and J. Barker. 1985. Spectral characterization of the LANDSAT thematic mapper sensors. International Journal of Remote Sensing 6(5):697–716.

Martin, M. 1998. Determining forest species composition using high spectral resolution remote sensing data. Remote Sensing of Environment 65:255–66.

Martin, M., and J. Aber. 1998. High spectral resolution remote sensing of forest canopy lignin, nitrogen and ecosystem processes. Ecological Applications 7:431–43.

McGwire, K. 1992. Analyst variability in labeling of unsupervised classifications. Photogrammetric Engineering and Remote Sensing 58(12):1673–77.

McGwire, K., M. Friedl, and J. Estes. 1993. Spatial structure, sampling design and scale in remotely sensed imagery of a California savanna woodland. International Journal of Remote Sensing 14(11):2137–64.

Muchoney, D., J. Borak, H. Chi, M. Friedl, J. Hodges, N. Morrow, et al. 2000. Application of the MODIS global supervised classification model to vegetation and land cover mapping of Central America. International Journal of Remote Sensing. 21:1115–38.

Myneni, R., G. Asrar, and F. Hall. 1992. A three-dimensional radiative transfer method for optical remote sensing of vegetated land surfaces. Remote Sensing of Environment 41:105–21.

Myneni, R., and G. Asrar. 1993. Simulation of space measurements of vegetation canopy bidirectional reflectance factors. Remote Sensing Reviews 7:19–41.

Myneni, R. 1994. Atmospheric effects and spectral vegetation indices. Remote Sensing of Environment 47:390–402.

Myneni, R., C. Keeling, C. Tucker, G. Asrar, and R. Nemani. 1997. Increased plant growth in the northern high latitudes from 1981 to 1991. Nature 386:698–702.

Price, J. 1987. Calibration of satellite radiometers and the calibration of vegetation indices. Remote Sensing of Environment 21:15–27.

Price, J. 1988. An update on visible and near infrared calibration of satellite instruments. Remote Sensing of Environment 24:419–22.

Prince, S. 1991. Satellite remote sensing of primary production: comparison of results for Sahelian grasslands 1981–1988. International Journal of Remote Sensing 12: 1301–11.

Privette, J., W. Emery, and D. Schimel. 1996. Inversion of a vegetation reflectance model with NOAA AVHRR Data. Remote Sensing of Environment 58:187–200.

Proy, C., D. Tanre, and P. Deschamps. 1989. Evaluation of topographic effects in remotely sensed data. Remote Sensing of Environment 30:21–32.

Rahman, H., M. Verstraete, and B. Pinty. 1993. Coupled surface-atmosphere reflectance (CSAR) model, 1. Model description and inversion on synthetic data. Journal of Geophysical Research 98:20779–89.

Richards, J. 1993. Remote sensing digital image analysis: an introduction, second ed. Springer-Verlag, New York.

Richter, R. 1997. Correction of atmospheric and topographic effects for high spatial resolution satellite imagery. International Journal of Remote Sensing 18:1099–111.

Rosenfield, G.H., and K. Fitzpatrick-Lins. 1986. A coefficient of agreement as a measure of thematic classification accuracy. Photogrammetric Engineering and Remote Sensing 52:223–27.

Sandmeier, S., and K. Itten. 1997. A physically-based model to correct atmospheric and illumination effects in optical satellite data of rugged terrain. IEEE Transactions on Geoscience and Remote Sensing 35:708–17.

Schowengerdt, R. 1997. Remote sensing: models and methods for image processing. Academic Press, New York.

Schwartz, M. 1994. Monitoring global change with phenology: the case of the spring green wave. International Journal of Biometeorology 38(1):18–22.

Scott, J., F. Davis, B. Csuti, R. Noss, B. Butterfield, C. Groves, et al. 1993. GAP analysis: a geographic approach to protection of biological diversity. Wildlife Monographs 123:1–41.

Sellers, P., F. Hall, G. Asrar, D. Strebel, and R. Murphy. 1992. An overview of the First International Satellite Land Surface Climatology Project (ISLSCP) Field Experiment (FIFE). Journal of Geophysical Research 97(D17):18345–71.

Skole, D., and C. Tucker. 1993. Tropical deforestation and habitat fragmentation in the Amazon: satellite data from 1978 to 1988. Science 260:1905–01.

Stehman, S.V. 1992. Comparision of systematic and random sampling for estimating the accuracy of maps generated from remotely sensed data. Photogrammetric Engineering and Remote Sensing 58:1343–50.

Stehman, S.V. 1995. Thematic map accuracy assessment from the perspective of finite population sampling. International Journal of Remote Sensing 16:589–93.

Stehman, S.V. 1996. Estimating the kappa coefficient and its variance under stratified random sampling. Photogrammetric Engineering and Remote Sensing 62:401–7.

Stehman, S.V. 1997a. Estimating standard errors of accuracy assessment statistics under cluster sampling. Remote Sensing of Environment 60:258–69.

Stehman, S.V. 1997b. Selecting and interpreting measures of thematic classification accuracy. Remote Sensing of Environment 62:77–89.

Stoms, D.M., M.J. Bueno, F.W. Davis, K.M. Cassidy, K.L. Driese, and J.S. Kagan. 1998. Map-guided classification of regional land-cover with multi-temporal AVHRR data. Photogrammetric Engineering and Remote Sensing 64(8):831–38.

Story, M., and R.G. Congalton. 1986. Accuracy assessment: a user's perspective. Photogrammetric Engineering and Remote Sensing 52:397–99.

Strahler, A. 1980. The use of prior probabilities on maximum likelihood classification of remotely sensed data. Remote Sensing of Environment 10:135–63.

Strahler, A., C. Woodcock, and J. Smith. 1986. On the nature of models in remote sensing. Remote Sensing of Environment 20:121–39.

Strahler, A. 1997. Vegetation canopy reflectance modeling—recent developments and remote sensing perspectives. Remote Sensing Reviews 15:179–94.

Tanre, D., C. Deroo, M. Herman, J. Morcrette, J. Perbos, and P. Deschamps. 1990. Description of a computer code to simulate the satellite signal in the solar spectrum. International Journal of Remote Sensing 11:659–68.

Townshend, J., and C. Justice. 1988. Selecting the spatial resolution of satellite sensors for global monitoring of land transformations. International Journal of Remote Sensing 9:187–236.

Walthall, C., J. Norman, J. Welles, G. Campbell, and B. Blad. 1985. Simple equation to approximate the bidirectional reflectance from vegetation canopies and bare soil surfaces. Applied Optics 24:383–87.

Wessman, C., J. Aber, D. Peterson, and J. Melillo. 1988. Remote sensing of canopy chemistry and nitrogen cycling in temperate forest ecosystems. Nature 335:154–56.

Woodcock, C., A. Strahler, and D. Jupp. 1988a. The use of variograms in remote sensing: I Scene models and simulated images. Remote Sensing of Environment 25:323–48.

Woodcock, C., A. Strahler, and D. Jupp. 1988b. The use of variograms in remote sensing: II. Real digital images. Remote Sensing of Environment 25:349–79.

Woodcock, C., and A.H. Strahler. 1987. The factor of scale in remote sensing. Remote Sensing of Environment 21:311–32.

Woodcock, C., and V. Harward. 1992. Nested-hierarchical scene models and image segmentation. International Journal of Remote Sensing 13:3167–87.

Woodcock, C., and S. Gopal. 1994. Theory and methods for accuracy assessment of thematic maps using fuzzy sets. Photogrammetric Engineering and Remote Sensing 60:181–88.

13
Modeling Forest Net Primary Productivity with Reduced Uncertainty by Remote Sensing of Cover Type and Leaf Area Index

Steven E. Franklin

Process-based ecosystem models have emerged as a powerful new tool in forest management with applications at multiple scales (Landsberg and Gower 1997; Waring and Running 1998; see also Running et al. 1989; Running 1990; Peterson and Waring 1994; Ruimy et al. 1994; Green et al. 1996; Hunt et al. 1996; Milner et al. 1996; McNulty et al. 1997; Coops 1999; Landsberg and Coops 1999). Resource managers can use ecosystem models to describe the state of a forest at a point in time relative to a range of potential management treatments, and to generate projections of future growth and stand development. As commercial forestry approaches the sustainable limit of resource use in a wide range of ecological settings, the value of these process models as new tools and an information source for managers in a wide variety of applications, including wildlife habitat mapping, biodiversity monitoring, and forest growth assessment, is increasingly clear. For example, the models can be used to estimate stand or site net primary production (NPP) when the necessary information on species, soils, topography, and climate are available. Improved ecosystem process models in the future may replace empirical stand growth and yield models (Landsberg and Coops 1999).

Model performance can be improved with remotely sensed estimates of key parameters (e.g., forest cover type and leaf area index, LAI) (Running et al. 1989; Bonan 1993; Nemani et al. 1993; Peterson 1997). These remote sensing inputs are exemplars of the two main application domains discussed in Chapter 12. For example, forest cover type can be represented by discrete classes of species composition, diameter and height, or density, mapped from digital remote sensing data using supervised or unsupervised classification algorithms. Forest LAI is an important structural attribute of forest ecosystems because of its potential to be a measure of energy, gas, and water exchanges, and is representative of a large number of continuously varying biophysical properties that may be obtained from digital remote sensing data through various modeling techniques.

This chapter will present a review and an example of methods used to integrate remotely sensed forest cover type and LAI input into ecosystem process models in the estimation of forest productivity. The first section

284

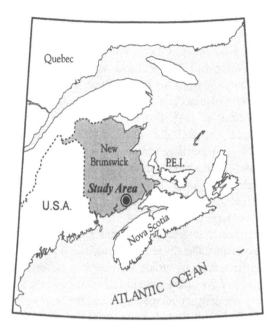

FIGURE 13.1. Location of the study area in Fundy Model Forest, southeast New Brunswick, Canada.

introduces some of the large area satellite remote sensing applications in forestry, and considers issues in ecosystem modeling related to the remote sensing data characteristics. Image classification and estimation techniques are described in the second section, followed by two sections illustrating forest cover type classification and LAI estimation in the northern deciduous, mixed-wood forest area of eastern Canada. The examples are drawn from forest modeling work in the Acadian Forest Region in southeastern New Brunswick (Fig. 13.1) (Franklin et al. 1997a,b). Uncertainties in cover type classification results are shown that are related to the type of training data and the methods of processing such data; uncertainties in remote estimation of stand LAI are shown that are related to the species composition of each forest stand; and, finally, uncertainties in estimates of NPP are shown for several stands that are related to the input variables and the assumptions in the model used to estimate productivity. This aspect of the forest ecosystem process model sensitivity is illustrated in Section 13.6. The chapter concludes with an assessment of the challenges in reducing uncertainty in process-based ecosystem modeling with remote sensing.

13.1 Remote Sensing Applications

Forest assessment by satellite remote sensing may range from an analysis or classification of forest cover type (Beaubien 1994; Franklin 1994; Mickelson et al. 1998), stand age (Nilsen and Peterson 1994), timber volume (Trotter et al. 1997), forest structure (Spanner et al. 1990, 1994; Cohen et al. 1995; Chen and Cihlar 1996; Fassnact et al. 1997; Woodcock et al. 1997), stand growth characteristics (Ahern et al. 1991; Franklin and Luther 1995), canopy insect defoliation (Muchoney and Haack 1994; Franklin et al. 1995; Ekstrand 1996; Luther et al. 1997; Chalifoux et al. 1998), other forest decline or disturbance phenomena (Rock et al. 1988; Brockhaus et al. 1993), and general forest (leaf) physiological, nutrient, and chemical status (Running et al. 1989; Curran et al. 1997). All of these remote assessments are made possible by the relationships between multispectral reflectance and scattering and absorption by vegetation (Curran 1980), and the numerous advances in image processing and computing that permit the confident extension of these relationships over large areas.

   A normative research method is the dominant approach in remote sensing because of the cost and difficulty in establishing causal relationships between multispectral reflectance and surface features such as forest canopies. An empirical approach relies on establishing relationships in the forest region of interest between remotely sensed reflectance and the biophysical parameters of interest. One critical relationship, most useful to ecologists, foresters, and professional resource managers in their resource management planning and implementation, is between multispectral reflectance and forest cover type. This relationship could be summarized using a key forest inventory class variable (e.g., species composition or stand density). A second critical relationship is between multispectral reflectance and stand structure, which could be summarized using a key process variable such as LAI. These relationships are generally expressed, respectively, as (1) a multispectral classification of forest cover types and, (2) a vegetation index related to forest structure. A plethora of techniques exist to establish, calibrate, and validate the index relationships or the classification decision rules, and then to extend them to the larger, surrounding, less-well-known areas. A few key decisions required in this critical process of translating an image into a model input are discussed in this chapter and more generally in the presentations in Chapters 12, 14, and 15.

13.2 Characteristics of Remote Sensing Data

What is the best way to derive the required information from remote sensing data? Decisions are initially required on (1) the type of data to be used, (2) the type of image processing required, and (3) the output formats and performance measures needed to assess the results. For example, forestry measurements from (low spatial resolution) satellite sensors have typically involved per-pixel classifiers or simple regression-based inversion that can be checked against independent tests sites (if available), or against training data sets. Because the low spatial resolution covers a large area in each pixel measurement, mixed-pixel spectral information is the main element contributing to image classification and estimation. Multivariate statistical techniques are the dominant image processing model. On the other hand, aerial platforms can generate high spatial resolution imagery with meter-to-submeter spatial resolution (King 1995; Guindon 1997), which are also now available from space (Glackin 1998). For these data, image segmentation techniques may be required because many classification algorithms cannot handle the increased local variance produced by the small pixels (relative to object size). Radar sensors such as Radarsat are generating imagery with very different spectral, radiometric, and spatial characteristics, and with very different utility in ecology (Kasiscke et al. 1994; Waring et al. 1995; Hyyppä et al. 1996; Weishempel et al. 1997). The first requirement in the use of remote sensing in ecological studies is an understanding of the appropriate image source (aerial, satellite) and electromagnetic region (optical, thermal, microwave).

13.2.1 The Effect of Image Spatial Resolution

At the heart of ecosystem process models is an area on the ground that will be assumed to be homogeneous for the purpose of the model (Bonan 1993; Waring and Running 1998; Landsberg and Coops 1999). Several sources of information can be used to define this homogeneous area, and all have implications for the modeling results. Most forest areas under active (or even passive) management have an inventory map based on aerial photointerpretation and field sampling—such maps are often available in a GIS polygon format. Photointerpreted polygonal information is typically collected to specifications set out by forest management agencies. One specification of particular importance [i.e., the minimum polygon size (usually four hectares)] is necessary to keep the number of stands under control. The result is that interpreters are forced to incorporate extra variability within polygons. The main problems are that users of this information have to assume that the boundaries are exact and that they enclose homogeneous or "acceptably heterogeneous areas" for their purpose (Robinove 1981). Using this GIS polygon information in an ecosystem model can be based on the understanding that the variability within the polygon is not important for the model functioning, or that no other reliable source of information exists to refine or replace the GIS polygon information.

High spatial resolution remote sensing data can yield precise information on individual trees and small objects—the challenge here is to assemble these observations into objects that can then be aggregated into a forest inventory label. Low spatial resolution remote sensing data can yield precise information on arbitrary-sized ground areas (pixels)—the challenge here is to predict correctly the class of the individual pixels and then assemble these class observations into a forest inventory label. At the moment no one approach is optimal for generating forest inventory or biophysical information from aerial or satellite imagery; the high spatial resolution imagery are limited by the high variance and spatial detail provided by fine resolution, and the low spatial resolution imagery are limited by the aggregation of individual objects and the lack of dynamic range provided by the coarse resolution! Practical limits to extracting forest information are also imposed by our finite capability to generate the required output from the available input data using computer image processing methods.

Figure 13.2 (see color insert) shows the possibility of increasing uncertainty associated with decreasing spatial resolution and, paradoxically, the increasing uncertainty associated with increasing spatial resolution, in the assignment of forest cover type labels to forest stands depicted in imagery. At the lowest level of spatial resolution the entire polygonal area outlined (usually from aerial photographs) is attributed a single cover-type label, resulting in a large degree of uncertainty for the polygon and for subareas of the polygon. The result is the familiar maxim that "every point in the polygon can represent the polygon equally well"—but only because all other information has been discarded (i.e., the aerial photographs used to determine the mapping unit or polygon). At the next level (going to the right-hand side of Fig. 13.2) the same polygon is overlayed on a Landsat TM image with 30-m spatial resolution. Uncertainty for the polygon label as a whole can be reduced because each pixel can be considered a discrete area of the polygon, and can be assigned a specific cover-type label derived from an image classification. The desired polygon label may be constructed in many ways. The simplest approach could be to use the classification of each pixel using a decision rule based on training data, conduct a summation of pixel class membership within the polygonal boundary, and then assign the polygonal label as the modal class. The sources of uncertainty at this level can be traced to the class structure imposed on the image data, the methods of assigning pixels to classes (e.g., the degree of confidence in the training data, or the extent to which training areas were "purified" using surface observations or additional aerial photointerpretation), or perhaps even the postprocessing that is often used to reduce class speckle.

The next level in Figure 13.2 shows the same polygon overlayed on an aerial image acquired from 10 km altitude above ground by the Compact Airborne Spectrographic Imager sensor, resulting in still greater spectral resolution (more and narrower spectral bands) and spatial resolution (pixel size is approximately 2.5 m). Uncertainty can again be reduced for the polygon as a whole because of the availability of within-polygon information that could be used to refine the original polygon label or to replace that label with a more accurate summary of the attributes of the polygon. For example, image texture could be used to create higher accuracy in the classification decision rules (Wulder et al. 1998).

Finally, at the highest level in this hierarchy is shown a portion of the same area in very high spatial resolution (submeter pixels) digital frame camera imagery acquired over the forest at less than 1 km altitude above ground (Fig. 13.2). At this level the individual trees can be segmented and a stand description based on feature identification (Gougeon 1995a,b; Gerylo et al. 1998), rather than with a grouping of larger areas already averaged within the sensor field of view. Segmentation is a spatial aggregation or area classification technique (Lobo 1997). The area of the image in Figure 13.2 with the highest spatial resolution is a fraction of the original polygon, so a large degree of uncertainty arises in assigning a label to the entire area based on the individual frames. With this level of spatial detail the computer processing is approaching the decisions that photointerpreters must make. The data summary techniques must be complex in order to build a large area description that can be used with confidence. The GIS labels at the lowest level of the figure might be used to guide and strengthen the individual frame descriptions. In addition, there are many radiometric and geometric problems in mapping applications with such high spatial resolution imagery.

13.2.2 Ecosystem Modeling and Remote Sensing

Process-based ecosystem models, such as FOREST-BGC (Running 1994), BIOME-BGC (Running and Hunt 1993), BGC++ (Hunt et al. 1999), and 3PGS (Coops 1999), are emerging as one approach to providing managers with information on forest productivity. Simulations can be used to test hypotheses on stand growth and structure (Hunt et al. 1999) and are an effective way of providing estimates of important variables that are difficult to measure directly (Peterson 1997; Waring and Running 1998). Remote sensing can be used to generate initial conditions (e.g., cover type) and driving variables (e.g., LAI) for such models, and to validate model output (Peterson and Waring 1994). One obvious opportunity in which uncertainty can be introduced in process modeling is through provision of these model input and driving variables.

BIOME-BGC is the model selected for simulations in this chapter (Running and Hunt 1993). It is a mechanistic ecosystem model derived from the earlier conifer forest ecosystem model called FOREST-BGC, but now designed to generalize ecosystem biogeochemical and hydrological cycles across a wide range of life forms and climate. The evolution and context of this model, and the current data requirements and assumptions, are summarized in Running and Hunt (1993), Running (1990, 1994), and Coughlan and Dungan (1997), and applications of the model can be found in a few of the papers cited earlier as well as in Waring and Running (1998).

The model requires climate station records (e.g., air temperature, radiation, precipitation, humidity, atmospheric CO_2), and GIS site data (e.g., soil texture, coarse fragment content, and depth) to estimate soil water holding capacity for use in a daily water balance. Because of the scarcity of reliable soil information, a digital elevation model (DEM) can be used to estimate depth by assuming a relationship between position of the stand on the slope and soil development (Moore et al. 1993; Zheng et al. 1996). Modeled carbon dynamics include daily canopy net photosynthesis and maintenance respiration, annual photosynthate allocation, tissue growth, growth respiration, litterfall, and decomposition. Modeled N processes are annual mineralization, deposition, uptake and allocation to canopy, and losses. Annual allocation of C and N is made to leaf, stem, and coarse and fine roots.

To reduce computational constraints a "look-up table approach" was used for all simulations in this chapter. The table is comprised of 3840 NPP entries based on model runs for 40 LAI increments (0–20 in 0.5 steps) by four ecoregions by four soil units by six cover types.

13.3 Extraction of Information from Remotely Sensed Data

Remote sensing is commonly considered a tool or a technique in ecology and other cognate disciplines (Chap. 12, and Franklin and Woodcock 1997). This is appropriate when remote sensing data are used in an early, exploratory phase, and the questions revolve around the suitability of the data for a given application, the extent to which remote sensing data can supplement or replace traditional field observations, the degree to which remote sensing data can be relied on to provide information not available in any other way, and so on. Remote sensing is increasingly considered a normative method in many disciplines, including ecology, because it provides the general framework or research structure that determines the kind of result sought—in this case, a large area, regional, or locally dynamic model based on synoptic assessment of key process variables. As Waring and Running (1998) have pointed out: "Any serious quantitative study of a landscape must begin with some type of remote sensing; there is no other way to obtain consistent measurements across large areas."

In many studies remote sensing potential is realized, as a data collection tool, a series of digital techniques including image processing, modeling, and cartographic design, and a research method. In those applications where remote sensing is considered a research method, it is recognized as the best available way to approach spatially explicit problems in the natural environment. Questions then revolve around how best to extract information from the imagery, how to combine or integrate these data with other available information, how to optimize landscape parameterization in subsequent models, how to assess results and calibrate other models, and so on.

It is still true that our ability to analyze remote sensing data is far outstripped by the availability and complexity of the data (Graetz 1990). The tool of remote sensing generally remains blunt and unwieldy, while simultaneously yielding enormous quantities of undifferentiated data and unrealized potential. The georadiometric problems with aerial and satellite remotely sensed data remain formidable (Teillet 1986; Itten and Meyer 1993; Gemmel 1998; Gu and Gillespie 1998). Beyond the question of image preparation and preprocessing, techniques of image analysis applied to remote sensing images can be based on an enormous range of algorithms. To simplify the discussion these algorithms can be generally categorized by the type of data that are required to make the classification of cover types or the estimation of LAI successful. One emerging idea, supported by multiple field-based and remote sensing studies in different ecological settings, is that even though there may be a reasonable complementarity in the different image analysis approaches, a prime focus for analysis in the future will need to include spectral or hyperspectral data, plus the various forms of spatial information from extracted data features (Weishempel et al. 1997; Frohn 1998). The best examples are drawn from satellite-based forest classifications (e.g., Cohen and Spies 1992; Jakubauskas 1997) and in aerial estimation of biophysical variables such as LAI (e.g., Wulder et al. 1998).

13.3.1 Forest Cover-Type Classification

Cover-type information available for managed forests is typically stored in a regional or local GIS database. For example, in Canada the cover types are available in GIS polygons assembled in mapping units by province. These polygonal databases are typically based on earlier forest inventory work derived from medium-scale, black and white metric aerial photointerpretation, stand cruising, and permanent plot compilations (Leckie and Gillis 1995; Lowell and Edwards 1996). This method of organizing the landscape may be appropriate for many management practices or treatments, but may not be an optimal source of information on cover types for modeling and growth estimation. GIS polygonal databases of this type would also contain only marginal information on the leaf area or structure of each stand.

Satellite remote sensing studies typically substitute the aerial photointerpreter's minimum mapping unit with an arbitrary pixel dimension that coincides with the particular image resolution used in the mapping project (Cohen et al. 1996). The aerial photointerpreter's judgment on the composition of the polygon is replaced by a decision rule that is image based. An increasingly common approach in aerial remote sensing studies is to build minimum mapping units with segmentation or aggregation algorithms that are also designed to replace aerial photointerpretation of polygon boundaries (Eldridge and Edwards 1993; Fournier et al. 1995; Gougeon 1995a,b); however, relatively coarse satellite image resolution has thus far prevented wider applicability for regional studies (Woodcock and Harward 1992; Lobo 1997).

In general, there are four major steps in multispectral image classification (see Chap. 12): (1) definition of the classification scheme (Franklin and Woodcock 1997), (2) training area selection (McCaffrey and Franklin 1993),

(3) decision rule choice and application (Thomas et al. 1987; Peddle et al. 1994; Foody 1996; Foody and Arora 1996), and (4) accuracy assessment (Hammond and Verbyla, 1996; Stehman and Czaplewski 1998).

13.3.2 Forest Leaf Area Index Estimation

Classification procedures may be preceded or followed by the development of specific regression equations relating spectral response to stand structure or LAI using field data on young, mature, and overmature forest stands. However, certain forest conditions are problematic in estimation of LAI by conventional remote sensing image analysis of multispectral data (Nemani et al. 1993; Spanner et al. 1994; Jasinski 1996; Peddle et al. 1999) and in classification of forest type by coarse resolution (e.g., Landsat) and fine resolution (e.g., aerial digital frame camera data) remote sensing. For example, spectral vegetation indexes such as the normalized difference vegetation index (NDVI) relate to forest LAI, but such relationships differ for broadleaf and needleleaf species because of different reflectances in the near infrared portion of the spectrum. Hardwood and mixed-wood stands with variable amounts of understory and a range of crown sizes are very common globally, but they are largely ignored in remote sensing studies of LAI, which often focus on pure stands of conifers with full crown closure. Attempts to account for understory contributions have been reported (Nemani et al. 1993) using different band combinations, including the shortwave infrared channels of the TM sensor.

Bonan (1993) found that relationships between NDVI and LAI depend on species and stand structure and that cover-type variations should be accounted for in subsequent model calculations of photosynthesis and other ecosystem processes. Cohen et al. (1996) suggest that other indexes that use more of the original spectral information (than the NDVI) (e.g., the tasseled cap transformation) can outperform simple vegetation indexes but may require more detailed interpretation because the coefficients are scene specific.

13.4 Forest Inventory Classification (New Brunswick Fundy Model Forest)

An example of forest inventory classification using a 1992 Landsat Thematic Mapper image is taken from the Fundy Model Forest, a 400,000 ha working forest in the Acadian Forest Region (Rowe 1972) of southeastern New Brunswick. Field sampling with probability proportional to size (5–10 prism plots per stand) was done in 128 stands during 1994 using a basal area factor 2 prism. Diameter at breast height (DBH) and species were recorded for all trees, and several dominant trees were selected for total height, crown length measurements, and increment coring. Sapwood width was measured on each core and used to estimate sapwood cross-sectional area at breast height. Relationships between DBH and sapwood cross-sectional area on these trees were used to calculate the total sapwood cross-sectional area for all trees. Equations relating leaf area per tree to sapwood cross-sectional area at breast height were used to estimate LAI for each plot and each stand. At 17 plots (20 × 20 m), destructive sampling was used to develop allometric relationships between sapwood cross-sectional area and leaf area index for commercially important species for comparison to remotely sensed LAI.

The class structure for each stand surveyed was based on the New Brunswick Department of Natural Resources Integrated Land Classification System (Table 13.1). Not all the possible variations in class training, decision rule development, and accuracy assessment are shown here; but one of the more obvious procedures that can strongly influence modeling results is described, with a brief analysis of the effect on the resulting classification or estimation maps. Differences in training a classification can have an enormous effect on the resulting classification accuracy and the utility of the final products as input to the ecosystem model. Two commonly applied training area procedures are described here. Table 13.2 contains the classification accuracy for three general cover types using (1) the mean TM spectral reflectance inside the GIS polygons to discriminate the GIS label (i.e., softwood, hardwood, mixed-wood) (Table 13.2, column 1), and (2)

TABLE 13.1. Forest cover types and species descriptions for representative stands in the Fundy Model Forest GIS database.

Class label	Dominant species or mix
Hardwoods	
TH Tolerant hardwoods	Maple, beech, birch
IH Intolerant hardwoods	Aspen
MH Mixed hardwoods	All deciduous species
Softwoods	
Sp	Spruce
Bf	Balsam Fir
Jp	Jack Pine
MC	All conifer species
Mixed-wood (examples)	
SpTH	Spruce-tolerant hardwoods
SpIH	Spruce-intolerant hardwoods
THSp	Tolerant hardwoods-spruce
IHSp	Intolerant hardwoods-spruce
JpTH	Pine-tolerant hardwoods
JpIH	Pine-intolerant hardwoods
THJp	Tolerant hardwoods-pine
IHjP	Intolerant hardwoods-pine
BfTH	Fir-tolerant hardwoods
BfIH	Fir-intolerant hardwoods
THBf	Tolerant hardwoods-fir
IHBf	Intolerant hardwoods-fir

294   Steven E. Franklin

TABLE 13.2. Classification accuracy of 128 forest stands in three general cover-type classes in New Brunswick using two separate classification procedures based on different methods of collecting training data.

Class	Classification Procedure[*] 1	2
	Accuracy (%)	
Softwood	71	90
Hardwood	100	96
Mixedwood	52	88
Mean	74	91

[*]Procedure 1 used the mean TM values in the GIS polygon as training data, and every pixel in the polygon was considered representative of that polygon and that cover type; Procedure 2 used 1853 individual TM pixels selected as training data from within a sample of polygons based on field surveys to determine the representativeness of each pixel for that polygon and that cover type.

individual pixel accuracies using an independent sample of pixels (Table 13.2, column 2).

13.4.1 Use of GIS Polygons to Collect Training Data

The first classification attempt used all the pixels within a sample of GIS polygons as training data. A number of GIS polygons representing the various classes were used to generate signatures to be applied to all the pixels in the imagery, and the accuracy was checked by examining the polygon label relative to the class of the mean of TM spectral response for that polygon. The overall classification accuracy was 74% in these 128 stands, with the softwood class 71% correct, the hardwood class 100% correct, and the mixed-wood class 52% correct. These results are reasonable and consistent with earlier classificatory work in similar boreal and northern pure and mixed forests and elsewhere (Mickelson et al. 1998; Wulder 1998). The mixed-wood class errors are primarily a function of the species composition in the original stands. They are readily explained as confusion with the softwood and hardwood classes that have the more distinct signatures. An independent check of the dominant species in each stand in the cruise that was misclassified revealed that many of these stands are quite patchy, with different combinations of species dominating different portions of the polygons. Another simple test of the similarity of signatures in such stands is to examine the probabilities of class membership for the individual samples; thus, in virtually every case of misclassification of mixed-wood stands, the second highest probability of class membership was with the mixed-wood class.

In some cases the GIS data were clearly in error and the classification using TM data appeared to generate a more appropriate label to apply to the stand. These results, however, are based on the mean TM values in the stand outlined in a GIS, which Ghitter et al. (1995), among others, have shown will rarely provide distinct signatures suitable for training satellite remote sensing data classifiers. It is important to note that this tabulation assumes the GIS polygon label is 100% correct for each and every pixel within the stand boundary, which is an assumption that is obviously highly questionable in light of the Acadian Region's diverse and complex mixtures of forest conditions. A more likely interpretation is that the remote sensing depiction of stand heterogeneity is realistic compared with the GIS assumption that any point in the polygon can represent the polygon equally well. An independent check of stand heterogeneity in the cruise data again revealed that the remote sensing classification is in fact a reasonable representation of stand heterogeneity, and that individual pixels within each GIS polygon could, and in fact do, belong to the other general classes.

13.4.2 Use of Individual Pixel Samples as Training Data

The second classification attempt (Table 13.2, column 2) used individual training pixels selected without reference to GIS polygons, but in a typical supervised approach whereby the full range of class conditions was sampled. The accuracy was tested against a similar, independent sample of pixels. The results show that the highest classification accuracies for individual pixels were achieved with this independent, non–GIS-related classification method. The assumption that stands mapped using aerial photointerpretation for forest inventory objectives clearly could serve the purpose of training an image classifier is not supported in these tests. This variability in stand conditions—LAI and cover type—should not be ignored in the modeling of productivity.

Figure 13.3 (see color insert) contains four illustrative examples of TM pixel variability within GIS polygons that are assumed to be homogeneous forest stands/cover type. These images graphically illustrate the difficulty of using polygonal data in training a classifier. They also point to the more interesting difficulty in modeling NPP or growth characteristics for such heterogeneous landscape units. In Figure 13.3a a stand labeled SPBF softwood is revealed to contain no more than half the pixels as softwood class; overall, the GIS polygon is actually predominantly mixed-wood pixels with two relatively small, dense areas of pure conifer (dark brown). In Figure 13.3b a stand labeled in the GIS as IHSP Mixed-wood is shown to contain approximately 60% of the pixels (area) that are actually mixed-wood (light brown), and the remaining 40% of the pixels appear to be pure conifer pixels (dark brown). An area of three blue pixels appeared as open water surrounded by riparian deciduous shrub cover (beige, orange, green). In Figure 13.3c another IHSP Mixed-wood stand actually contains a small pond of open water (blue), and appeared to be almost equally divided between pure conifer pixels (brown) and pure hardwood pixels (orange). Finally, in Figure 13.3d, a stand labeled THIH Hardwood is confirmed with the TM data as exhibiting a continuous, unbroken deciduous canopy cover; however, the position of the boundary in such a homogeneous area was considered arbitrary. This relatively uniform spectral response may mask information on the relative distribution of LAI (and hence NPP) between strata in the stand. An interesting possibility in such areas would be to use ancillary data or other remote observations (perhaps different bands, image texture, indices) to attempt to partition NPP between the canopy and the understory layers to increase the value of NPP estimation in management decisions.

The resulting spatial arrangement of pixel classification appears to represent true class membership within the stand boundaries. It certainly provides a much better source of information that can be obtained from a single, homogeneous GIS polygon label in many applications, including modeling of growth and productivity, and in the calculation of landscape metrics (Wickham et al. 1997). Computational constraints on modeling may continue to apply (Running et al. 1989; Coughlan and Dungan 1997), but even if that were the case in this application and in others, a three-class summary of the polygonal heterogeneity may be more useful than assuming a homogeneous cover type.

13.5 Estimation of LAI (New Brunswick Fundy Model Forest)

A test of the relationships between LAI and multispectral reflectance was conducted using the data acquired using destructive sampling in the 17 plots (20×20 m), and over the larger region, in the 128 stands that were sampled for sapwood area using increment cores. A maximum LAI for each stand was estimated using the ecosystem model based on climate and soil constraints in each of four ecoregions. An example of the correlation between forest LAI and remotely sensed NDVI is shown in Table 13.3.

TABLE 13.3. The empirical relationships between forest stand LAI in 17 plots (destructively sampled to obtain field LAI estimates) and TM reflectance (NDVI measures) within each cover type and overall in the New Brunswick Acadian Forest region.

Strata	Simple R	Multiple R
Softwood ($n = 6$)	0.93	0.98
Hardwood ($n = 5$)	0.13	0.25
Mixedwood ($n = 6$)	0.66	0.82
Overall ($n = 17$)	0.15	0.28

[*]All correlations significant at 0.01; simple R is the relationship between TM NDVI and field measured LAI; multiple R is the relationship between TM NDVI plus stems/ha from the GIS database and field measured LAI.

The overall relationship between LAI and NDVI across all sampled cover types in the 17 plots was weak but statistically significant. The correlation coefficient was 0.15. This is a weaker relationship between LAI and NDVI than that described by Franklin and Luther (1995) in a simpler (structurally) monospecies forest in Newfoundland. They reported an R^2 value of 0.29 and a standard error of 2.31 in LAI prediction for 36 stands of balsam fir. In Oregon, Spanner et al. (1994) report higher correlation coefficients, but similar standard errors in a wide range of forest conditions (e.g., species and structure) for a variety of sensors. Their sample extended across a large gradient of LAI from juniper woodlands to coastal hemlock stands. It is clear, however, that the overall relationship between LAI and NDVI is weak and not substantively useful across the sampled New Brunswick stand types and species combinations.

The 17 plot locations were grouped into softwood, hardwood, and mixed-wood cover types to stratify the analysis of relationships between LAI and remotely sensed data. The NDVI calculated from the TM data was highly correlated to LAI in softwood stands, but much less correlated in hardwood and mixedwood stands (Table 13.3). Our correlation coefficient (0.93) for conifer stands was similar to that observed by Chen and Cihlar (1996), and much better than that found in the earlier work in Newfoundland which included a wider range of stand age, density, and growth characteristics (Franklin and Luther 1995). The lower correlations for hardwood and mixed-wood stands illustrate the difficulty in LAI estimation by remote sensing. There are specific differences in leaf thickness, shape, and distribution within crowns that influence reflectance. There may also be differences among cover types in stand structure, for instance in the amount of understory vegetation, that also influence LAI–NDVI relationships.

Multiple linear regressions using stem density and NDVI substantially improved correlations with LAI for hardwood and mixedwood stands for the measured plots in comparison to the simple correlations (Table 13.3). These results suggest that LAI might accurately be estimated for large areas with TM data by using forest inventory data to (1) stratify by cover type, and (2) provide estimates of stem density. Fassnact et al. (1997) found that cover-type strata were necessary for accurate estimation of LAI in Wisconsin, and recommended that such estimates be related to stand density. In another study, Spanner et al. (1990) found that estimates of LAI improved in areas where crown closure was greater than 90% compared with areas with more open canopies, less dense stocking, and with understory exposed to the sensor. White et al. (1997) found that accurate estimates of LAI using satellite data that were correlated with field estimates were possible if all sources of error were identified from the ground up, including allometric or field errors due to nonrandom self-shading effects and topographic effects. One obvious approach would be to classify the image data to determine cover type (with or without reference to a GIS polygon label derived in some other way), and then map this cover-type distribution together with any available supplemental information (e.g., stem density counts) to develop the LAI relationships for that forest strata.

Over the larger sample of 128 stands differences in LAI estimation based on the field measurement of sapwood, the remote sensing vegetation index and the ecosystem model are significant (Fig. 13.4). For example, the mean LAI in the softwood stands was 10.48 based on the measured sapwood allometric equation, but 11.10 using the remote sensing LAI–NDVI relationship. The model assumption of maximum LAI for these sites based on soils and climate was 7.5. Although the softwood LAI estimates are reasonably comparable, it appears that the remotely sensed LAI in stands with significant deciduous species is underestimated relative to the field measurements.

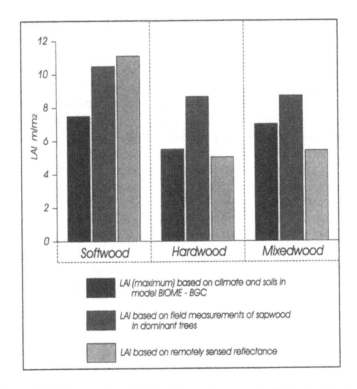

FIGURE 13.4. A comparison of LAI estimates for a sample of 128 stands in three general cover-types in New Brunswick, obtained through three different methods. Field-based estimates of LAI were based on sapwood measurements in each stand and the LAI allometric equations by species derived by destructive sampling in 17 plots; remotely sensed estimates of LAI are based on NDVI–LAI regression equations derived in these same 17 plots and applied to each pixel within the 128 stand polygons; BIOME-BGC process model estimates were based on assumptions of the maximum LAI that could be expected on sites with these soils and climate conditions.

Actual LAI is higher than modeled LAI for most sites with the exception of the mixedwood stands.

13.6 Ecosystem Model Sensitivity

Figure 13.5 shows the difference between NPP calculated based on the assumption of a maximum LAI for each site (based on climate and soils), and NPP calculated using the actual LAI derived from the Landsat imagery for each site based on image classification within the GIS cover-type label. In the model, the aspen and spruce parameters are from Hunt and Running (1992). The simulations are for the Sussex climate station on the Harcourt soil unit (red mudstone, weathered compact till, sandy loam), which represents some of the best sites in the region.

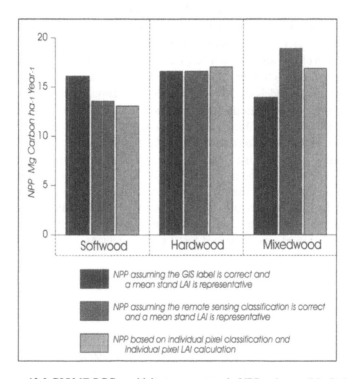

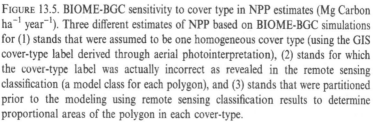

FIGURE 13.5. BIOME-BGC sensitivity to cover type in NPP estimates (Mg Carbon ha^{-1} year^{-1}). Three different estimates of NPP based on BIOME-BGC simulations for (1) stands that were assumed to be one homogeneous cover type (using the GIS cover-type label derived through aerial photointerpretation), (2) stands for which the cover-type label was actually incorrect as revealed in the remote sensing classification (a model class for each polygon), and (3) stands that were partitioned prior to the modeling using remote sensing classification results to determine proportional areas of the polygon in each cover-type.

Overall, the Model Forest stand productivity appears to be relatively close to the maximum in all types except the tolerant hardwoods (maple), where NPP is approximately 75% of the potential on all sites. A spatial analysis of these areas that are shown to be well below productivity revealed cover types and individual stands that can be considered for different management treatments. For example, simulations show that NPP was consistently below potential in two jack pine plantations and as expected in the tolerant hardwood plots. In the two jack pine plantations actual productivity averaged only about 60% of the potential productivity as estimated by the model because of the reduced (actual) LAI compared with the maximum (potential) LAI for these sites. One possible interpretation of these differences in NPP simulations is that jack pine is not an appropriate species for these two sites because it cannot take advantage of the water and nutrients that are available.

In this example, NPP estimates can vary by as much as 50% on selected sites for the same site conditions if the incorrect LAI equation is used for any given stand. In other words, if the GIS label is incorrect or not representative (as an average) of the actual species mixture in the stand, the subsequent estimates of LAI and NPP can be quite inaccurate and there will be increasing uncertainty as to their utility in forest management applications. Current wood supply models could be making these assumptions, and there is no simple way to deal with that error and the attendant uncertainty within the existing GIS data. Reduced uncertainty can result from the application of remote sensing, as shown in these simulations; specifically, uncertainty is reduced as a result of the use of image analysis methods in the classification of forest cover type, and in the subsequent estimation of forest stand LAI. A final consideration in the reduction in uncertainty is in the assessment of spatially distributed results (checking stands in the field that appear to be anomalous in the model), and the comparison of model runs based on remotely sensed versus assumed forest stand conditions.

13.7 Summary

Process-based ecosystem models are a useful new tool and an information source for forest resource managers. With available climate, soils, topography, and species data, the models can provide estimates of local and regional-scale net primary productivity. This approach has the potential to supplement or replace traditional empirical stand growth and yield models (Landsberg and Coops 1999) because the models can generate this information over large areas at multiple scales (Waring and Running 1998). Uncertainty in model NPP estimates can be reduced with remote sensing input in the form of cover-type classifications and LAI estimates. Cover-type information may be obtained from existing GIS databases (usually constructed using aerial photointerpretation methods), but these will rarely provide the level of detail and accuracy required, nor provide information on other aspects of the ecosystem important in modeling (e.g., LAI).

The remote sensing data and methods required to convert multispectral reflectance to a model input are not overly complex, but a few key decisions must be made to ensure acceptable accuracy. First, the effect of image wavelength range and spatial resolution must be understood because the ecosystem models require a homogeneous—or acceptably heterogeneous for the purpose of the model—area on the ground as a starting point for numerous process calculations. Second, the image-processing strategy must be tied to the resolution and scale of the remote sensing observations. The challenge with high spatial resolution remote sensing imagery (e.g., that available from aerial platforms and some next-generation satellite platforms) is to find a way to assemble the individual components of the scene into a usable inventory-like label; promising approaches include the use of feature-based image processing (Gerylo et al. 1998) and texture derivatives (Wulder et al. 1998). The challenge with the low spatial resolution remote sensing imagery is to find enough detail in the individual classification of pixels that they can be combined in a simple way to provide a meaningful inventory-like label. Promising developments that increase the utility of relatively coarse resolution satellite imagery include the application of canopy reflectance or geometrical/optical models (Woodcock et al. 1997), as well as spectral unmixing approaches (Wulder 1998; Peddle et al. 1999).

Remotely sensed information extraction can benefit through analysis with reference to the available GIS data (e.g., forest inventory data can be used to stratify individual pixels prior to the development of regression equations to predict stand LAI; and cover-type labels can be confirmed or improved with reference to individual pixel classifications that can characterize within-polygon homogeneity). In the simulations using the ecosystem process model BIOME-BGC in the Acadian Region of New Brunswick, estimates of NPP could vary by as much as 50% depending on the source of information for the cover type and the LAI estimates. This variability in output represents a large source of uncertainty in the use and application of process ecosystem models in forest management.

A few of the practical outcomes of this modeling work include the identification of two plantations operating significantly below potential because of species incompatibility with site conditions, the understanding that even disturbed sites in the hardwood stands can have LAI close to the maximum site potential, and the idea that other factors such as soil quality and nutrient fluxes may be limiting productivity.

Acknowledgments. The Natural Science and Engineering Research Council of Canada, the Fundy Model Forest, and the Canadian Forest Service provided financial support. I am grateful to Dr. M.B. Lavigne (Canadian

Forest Service—Atlantic), Dr. E.R. Hunt, Jr. (University of Wyoming), and Dr. M.A. Wulder (Canadian Forest Service—Pacific) for their many contributions to this research. Much appreciated technical help was provided by M.J. Deuling (University of Calgary).

References

Ahern, F., T. Erdle, D.A. MacLean, and I.D. Kneppeck. 1991. A quantitative relationship between Landsat TM spectral response and forest growth rates. International Journal of Remote Sensing 12:387–400.

Beaubien, J. 1994. Landsat TM images of forests: from enhancements to classification. Canadian Journal of Remote Sensing 20:17–26.

Bonan, G. 1993. Importance of LAI and forest type when estimating photosynthesis in boreal forests. Remote Sensing of Environment 43:303–14.

Brockhaus, J.A., S. Khorram, R. Bruck, and M.V. Campbell. 1993. Characterization of defoliation conditions within a boreal montane forest ecosystem. Geocarto International 8:35–42.

Chalifoux, S., F. Cavayas, and J.T. Gray. 1998. Map-guided approach for the automatic detection on Landsat TM images of forest stands damaged by the spruce budworm. Photogrammetric Engineering and Remote Sensing 64:629–35.

Chen, J., and J. Cihlar. 1996. Retrieving leaf area index of boreal conifer forests using Landsat TM images. Remote Sensing of Environment 55:153–62.

Cohen, W., and T. Spies. 1992. Estimating structural attributes of Douglas-fir/ western hemlock forest stands from Landsat and SPOT imagery. Remote Sensing of Environment 41:1–17.

Cohen, W., T. Spies, and M. Fiorella. 1995. Estimating the age and structure of forests in a multi-ownership landscape of western Oregon, USA. International Journal of Remote Sensing 16:721–46.

Cohen, W., J.D. Kushla, W.J. Ripple, and S.L. Garman. 1996. An introduction to digital methods in remote sensing of forested ecosystems: focus on the Pacific northwest, USA. Environmental Management 20:421–35.

Coops, N.C. 1999. Linking multiresolution satellite derived estimates of canopy photosynthetic capacity and metereological data to assess forest productivity in a *Pinus radiata* (D. Don) stand. Photogrammetric Engineering and Remote Sensing 65:1149–65.

Coughlan, J.C., and J.L. Dungan. 1997. Combining remote sensing and forest ecosystem modeling: an example using the Regional HydroEcological Simulation System (RHESSys). Pages 135–58 in H.L. Gholz, K. Nakane, and H. Shimoda, eds. The use of remote sensing in the modeling of forest productivity. Kluwer Academic Publishers, Boston.

Curran, P. 1980. Multispectral remote sensing of vegetation amount. Progress in Physical Geography 4:315–41.

Curran, P., J.A. Kupiec, and G.M. Smith. 1997. Remote sensing the biochemical composition of a slash pine canopy. IEEE Transactions on Geoscience and Remote Sensing 35:415–20.

Ekstrand, S. 1996. Landsat TM based forest damage assessment: correction for topographic effects. Photogrammetric Engineering and Remote Sensing 62:151–61.

Eldridge, N.R., and G. Edwards. 1993. Acquiring localized forest inventory information: extraction from high resolution airborne digital images. Pages 443–48 in Proceedings, Thirteenth Canadian Symposium on Remote Sensing. Canadian Aeronautics and Space Institute, Ottawa, Canada.

Fassnacht, K.S., S.T. Gower, M.D. MacKenzie, E. Nordheim, and T.M. Lillesand. 1997. Estimating the leaf area index of north central Wisconsin forests using the Landsat Thematic Mapper. Remote Sensing of Environment 61:229–45.

Foody, G.M. 1996. Approaches for the production and evaluation of fuzzy land cover classifications from remotely sensed data. International Journal of Remote Sensing 17:1317–40.

Foody, G.M., and M.K. Arora. 1996. Incorporating mixed pixels in the training, allocation and testing stages of supervised classifications. Pattern Recognition Letters 17:1389–98.

Fournier, R., G. Edwards, and N. Eldridge. 1995. A catalogue of potential spatial discriminators for high spatial resolution digital images of individual tree crowns. Canadian Journal of Remote Sensing 21:285–98.

Franklin, J., and C.E. Woodcock. 1997. Multiscale vegetation data for the mountains of southern California: spatial and categorical resolution. Pages 141–68 in D.A. Quattrochi, and M.F. Goodchild, eds. Scaling in remote sensing and GIS. CRC Press, Boca Raton, FL.

Franklin, S.E. 1994. Discrimination of subalpine forest species and canopy density using digital CASI, SPOT PLA and Landsat TM data. Photogrammetric Engineering and Remote Sensing 60:1233–41.

Franklin, S.E., and J.E. Luther. 1995. Satellite remote sensing of balsam fir forest structure, growth and cumulative defoliation. Canadian Journal of Remote Sensing 21:400–11.

Franklin, S.E., R.H. Waring, R. McCreight, W.B. Cohen, and M. Fiorella. 1995. Aerial and satellite sensor detection and classification of western spruce budworm defoliation in a subalpine forest. Canadian Journal of Remote Sensing 21:299–308.

Franklin, S.E., M.B. Lavigne, M.J. Deuling, M.A. Wulder, and E.R. Hunt, Jr. 1997a. Landsat TM-derived forest cover type for use in ecosystem models of net primary production. Canadian Journal of Remote Sensing 23:91–99.

Franklin, S.E., M.B. Lavigne, M.J. Deuling, M.A. Wulder, and E.R. Hunt, Jr. 1997b. Estimation of forest leaf area index using remote sensing and GIS data for modeling net primary production. International Journal of Remote Sensing 18: 3459–71.

Frohn, R.C. 1998. Remote sensing for landscape ecology. CRC Press, Boca Raton, FL.

Gemmel, F. 1998. An investigation of terrain effects on the inversion of a forest reflectance model. Remote Sensing of Environment 65:155–69.

Gerylo, G., R.J. Hall, S.E. Franklin, A. Roberts, and E.J. Milton. 1998. Hierarchical image classification and extraction of forest species composition and crown closure from airborne multispectral images. Canadian Journal of Remote Sensing 24:219–32.

Ghitter, G.S., R.J. Hall, and S.E. Franklin. 1995. Variability of Landsat Thematic Mapper data in boreal deciduous and mixedwood stands with conifer understory. International Journal of Remote Sensing 16:2989–3002.

Glackin, D.L. 1998. International space-based remote sensing overview. Canadian Journal of Remote Sensing 24:307–14.

Gougeon, F.A. 1995a. Comparison of possible multispectral classification schemes for tree crowns individually delineated on high spatial resolution MEIS images. Canadian Journal of Remote Sensing 21:1–9.

Gougeon, F.A. 1995b. A crown-following approach to the automatic delineation of individual tree crowns in high spatial resolution aerial images. Canadian Journal of Remote Sensing 21:274–84.

Graetz, R.D. 1990. Remote sensing of terrestrial ecosystem structure: an ecologist's pragmatic view. Pages 5–30 in R.J. Hobbs, and H.A. Mooney, eds. Remote sensing of Biosphere Functioning. Springer-Verlag, New York.

Green, R.M., N.S. Lucas, P.J. Curran, and G.M. Foody. 1996. Coupling remotely sensed data to an ecosystem simulation model—an example involving a coniferous plantation in upland Wales. Global Ecology and Biogeography Letters 5: 192–205.

Gu, D., and A. Gillespie. 1998. Topographic normalization of Landsat TM images of forest based on subpixel sun-canopy-sensor geometry. Remote Sensing of Environment 64:166–75.

Guindon, B. 1996. Computer-based aerial image understanding: a review and assessment of its application to planimetric information extraction from very high resolution satellite images. Canadian Journal of Remote Sensing 23:38–47.

Hammond, T.O., and D.L. Verbyla. 1996. Optimistic bias in classification accuracy assessment. International Journal of Remote Sensing 17:1261–66.

Hunt, E.R., Jr., and S.W. Running. 1992. Simulated dry matter yields for aspen and spruce stands in the North American boreal forest. Canadian Journal of Remote Sensing 18:126–33.

Hunt, E.R., Jr., S.C. Piper, R. Nemani, C.D. Keeling, R.D. Otto, and S.W. Running. 1996. Global net carbon exchange and intra-annual atmospheric CO_2 concentrations predicted by an ecosystem process model and three-dimensional atmospheric transport model. Global Biogeochemical Cycles 10:431–56.

Hunt, E.R., Jr., M.B. Lavigne, and S.E. Franklin. 1999. Factors controlling the decline of growth efficiency and net primary production for balsam fir in Newfoundland. Ecological Modeling 122:151–64.

Hyyppä, J., J. Pulliainen, M. Hallikainene, and A. Saatsi. 1997. Radar-derived standwise forest inventory. IEEE Transactions on Geoscience and Remote Sensing 35:392–404.

Itten, K.I., and P. Meyer. 1993. Geometric and radiometric correction of TM data of mountainous forested areas. IEEE Transactions on Geoscience and Remote Sensing 31:764–70.

Jakubauskas, M.E. 1997. Effects of forest succession on texture in Landsat Thematic Mapper imagery. Canadian Journal of Remote Sensing 23:257–63.

Jasinski, M. 1996. Estimation of subpixel vegetation density of natural regions using satellite multispectral imagery. IEEE Transactions on Geoscience and Remote Sensing 34:804–13.

Kasischke, E., L. Bourgeau-Chavez, N. Christensen, and E. Haney. 1994. Observations on the sensitivity of ERS-1 SAR image intensity to changes in aboveground biomass in young loblolly pine forests. International Journal of Remote Sensing 15:3–16.

King, D. 1995. Airborne multispectral digital camera and video sensors: a critical review of system designs and applications. Canadian Journal of Remote Sensing 21:245–74.

Landsberg, J., and S.T. Gower. 1997. Applications of physiological ecology to forest production. Academic Press, San Diego, CA.

Landsberg, J., and N.C. Coops. 1999. Modeling forest productivity across large areas and long periods. Natural Resource Modeling 12:1–28.

Leckie, D.G., and M.D. Gillis. 1995. Forest inventory in Canada with an emphasis on map production. The Forestry Chronicle 71:74–88.

Lobo, A. 1997. Image segmentation and discriminant analysis for the identification of land cover units in ecology. IEEE Transactions on Geoscience and Remote Sensing 35:1136–45.

Lowell, K.E., and G. Edwards. 1996. Modeling the heterogeneity and change of natural forests. Geomatica 50:425–40.

Luther, J.E., S.E. Franklin, J. Hudak, and J.P. Meades. 1997. Forecasting the susceptibility and vulnerability of balsam fir forests to insect defoliation with satellite remote sensing. Remote Sensing of Environment 59:77–91.

McCaffrey, T.M., and S.E. Franklin. 1993. Automated training site selection for large-area remote sensing image analysis. Computers and Geosciences 19:1413–28.

McNulty, S.G., J.M. Vose, and W.T. Swank. 1997. Scaling predicted pine forest hydrology and productivity across the southern United States. Pages 187–209 in D.A. Quattrochi, and M.F. Goodchild, eds. Scale in remote sensing and GIS. CRC Press, Boca Raton, FL.

Mickelson, J.G., D.L. Civco, and J.A. Silander, Jr. 1998. Delineating forest canopy species in the northeastern United States using multitemporal TM imagery. Photogrammetric Engineering and Remote Sensing 64:891–904.

Milner, K., S.W. Running, and D.W. Coble. 1996. A biophysical soil-site model for estimating potential productivity of forested landscapes. Canadian Journal of Forest Research 26:1174–86.

Moore, I.D., P.E. Gessler, G.A. Nielson, and G.A. Peterson. 1993. Soil attribute prediction using terrain analysis. Soil Science Society of America Journal 57:443–52.

Muchoney, D.M., and B.N. Haack. 1994. Change detection for monitoring forest defoliation. Photogrammetric Engineering and Remote Sensing 60:1243–51.

Nemani, R., L. Pierce, S. Running, and L. Band. 1993. Forest ecosystem processes at the watershed scale: sensitivity to remotely sensed Leaf Area Index estimates. International Journal of Remote Sensing 14:2519–34.

Nilsen, T., and U. Peterson. 1994. Age dependence of forest reflectance: analysis of main driving factors. Remote Sensing of Environment 48:319–33.

Peddle, D.R., G. Foody, A. Zhang, S.E. Franklin, and E.F. LeDrew. 1994. Multi-source image classification. II: an empirical comparison of the evidential reasoning and neural network approaches. Canadian Journal of Remote Sensing 20:396–407.

Peddle, D., F.G. Hall, and E.F. LeDrew. 1999. Spectral mixture analysis and geometric-optical reflectance modeling of a boreal forest biophysical structure. Remote Sensing of Environment 67:288–97.

Peterson, D.L., and R.H. Waring. 1994. Overview of the Oregon Transect Ecosystem Research Project. Ecological Applications 4:211–25.

Peterson, D.L. 1997. Forest structure and productivity along the Oregon transect. Pages 173–218 in H.L. Gholz, K. Nakane, and H. Shimoda, eds. The use of remote sensing in the modeling of forest productivity. Kluwer Academic Publishers, Boston.

Robinove, C. 1981. The logic of multispectral classification and mapping of land. Remote Sensing of Environment 11:121–30.

Rock, B., T. Hoshizaki, and J.R. Miller. 1988. Comparison of in situ and airborne spectral measurements of the blue shift associated with forest decline. Remote Sensing of Environment 24:109–27.

Rowe, J.S. 1972. Forest regions of Canada. Canadian Forest Service Publication No. 1300. Environment Canada, Ottawa, Ontario, Canada.

Ruimy, A., B. Saugier, and G. Dedieu. 1994. Methodology for the estimation of net primary production from remotely sensed data. Journal of Geophysical Research 99:5263–83.

Running, S., R. Nemani, D.L. Peterson, L.E. Band, D.F. Potts, L.L. Pierce, et al. 1989. Mapping regional forest evapotranspiration and photosynthesis and coupling satellite data with ecosystem simulation. Ecology 70:1090–1101.

Running, S., and E.R. Hunt, Jr. 1993. Generalization of a forest ecosystem process model for other biomes, BIOME-BGC, and an application for global-scale models. Pages 141–57 in J. Ehleringer, and C. Field, eds. Scaling physiological processes: leaf to globe. Academic Press, Toronto.

Running, S.W. 1990. Estimating terrestrial primary productivity by combining remote sensing and ecosystem simulation. Pages 65–86 in R.J. Hobbs, and H.A. Mooney, eds. Remote sensing of biosphere functioning. Springer-Verlag, New York.

Running, S.W. 1994. Testing FOREST-BGC ecosystem process simulations across a climatic gradient in Oregon. Ecological Applications 4:238–47.

Spanner, M., L. Johnson, J. Miller, R. McCreight, J. Fremantle, J. Runyon, et al. 1994. Remote sensing of leaf area index across the Oregon Transect. Ecological Applications 4:258–71.

Spanner, M.L., L. Pierce, D. Peterson, and S.W. Running. 1990. Remote sensing of temperate coniferous forest leaf area index: the influence of canopy closure, understory vegetation and background reflectance. International Journal of Remote Sensing 11:95–111.

Stehman, S.V., and R.L. Czaplewski. 1998. Design and analysis of thematic map accuracy assessment: fundamental principles. Remote Sensing of Environment 64:331–44.

Teillet, P.M. 1986. Image corrections for radiometric effects in remote sensing. International Journal of Remote Sensing 7:1637–51.

Thomas, I.L., V.M. Benning, and N.P. Ching. 1987. Classification of remotely sensed images. Adam Hilger Publ., Bristol, UK.

Trotter, C., J.R. Dymond, and C.J. Goulding. 1997. Estimation of timber volume in a coniferous forest plantation using Landsat TM. International Journal of Remote Sensing 18:2209–23.

Waring, R.H., J.B. Way, E.R. Hunt, L. Morrisey, K.J. Ranson, J. Weishempel, et al. 1995. Imaging radar for ecosystem studies. BioScience 45:715–23.

Waring, R.H., and S.W. Running. 1998. Forest ecosystems: analysis at multiple scales, second edition. Academic Press, San Diego, CA.

Weishempel, J.F., R.G. Knox, K.J. Ranson, D.L. Williams, and J.A. Smith. 1997. Integrating remotely sensed heterogeneity with a three-dimensional forest succession model. Pages 109–33 in H.L. Gholz, K. Nakane, and H. Shimoda, eds. The use of remote sensing in the modeling of forest productivity. Kluwer Academic Publishers, Boston.

White, J.D., S.W. Running, R. Nemani, R.E. Keane, and K.C. Ryan. 1997. Measurement and remote sensing of LAI in Rocky Mountain montane ecosystems. Canadian Journal of Forest Research 27:1714–27.

Wickham, J.D., R.V. O'Neill, K.H. Riitters, T.G. Wade, and K.B. Jones. 1997. Sensitivity of selected landscape pattern metrics to land-cover misclassification and differences in land-cover composition. Photogrammetric Engineering and Remote Sensing 63:397–402.

Woodcock, C.E., and V.J. Harward. 1992. Nested-hierarchical scene models and image segmentation. International Journal of Remote Sensing 13:3167–77.

Woodcock, C.E., J.B. Collins, V.D. Jakabhazy, X. Li, S.A. Macomber, and Y. Wu. 1997. Inversion of the Li-Strahler canopy reflectance model for mapping forest structure. IEEE Transactions of Geoscience and Remote Sensing 35:405–14.

Wulder, M.A. 1998. Optical remote-sensing techniques for the assessment of forest inventory and biophysical parameters. Progress in Physical Geography 22:449–76.

Wulder, M., E.F. LeDrew, M.B. Lavigne, and S.E. Franklin. 1998. Aerial image texture for improved estimation of LAI in mixedwood stands. Remote Sensing of Environment 64:64–76.

Zheng, D., E.R. Hunt, Jr., and S.W. Running. 1996. Comparison of available soil water capacity estimated from topography and soil series information. Landscape Ecology 11:3–14.

14
Spatially Variable Thematic Accuracy: Beyond the Confusion Matrix

KENNETH C. McGWIRE and PETER FISHER

An essential aspect of the increasing sophistication of ecological models is the use of spatially explicit inputs and outputs. Thus, the challenge of documenting the uncertainty of model parameters must expand to include the distribution of error across the surface of maps, satellite images, and other ecological data that are keyed by geographic location. It has become more common to report the overall accuracy of map data sets. Support for such accuracy statements is seen in the descriptive attributes that are defined in file format conventions (e.g., the spatial data transfer standard, SDTS; FGDC 1998). These attributes include documentation of the root mean square error for positional accuracy and error rates associated with the delineation of specific map features. The probability of mapping errors, however, is generally not consistent across the surface of a map data set (Congalton 1988a; Steele et al. 1998), and standard methods have not been adopted for presenting the spatial distribution of error in thematic maps. The confusion matrix is the most commonly accepted method for assessing the accuracy of thematic maps, but it is entirely devoid of spatial context. This chapter addresses shortfalls in various approaches to predicting the distribution of error in thematic maps derived from image data.

Starting with a brief background on the confusion matrix, this chapter will show how attempts to use the error rates reported in the confusion matrix to estimate the spatial distribution of map error are of limited value. Rather than relying on the confusion matrix, some statistical mapping procedures provide estimates of confidence in class assignment at the per-pixel level, and some of the challenges in using such information are identified. Following this, we will examine the potential of using simulation methods that could be combined with Monte Carlo techniques to assess the impact of spatially dependent errors in thematic maps on ecological models.

The most commonly accepted method for documenting the accuracy of thematic maps has been the cross-tabulation of predicted versus actual class membership at a number of independent test points. This cross-tabulation creates a square matrix that is generally referred to as either a confusion, contingency, or error matrix. A number of explanations of the confusion matrix have been provided in remote sensing literature since 1970. Fenstermaker (1994) contains an extensive compilation of this work, and Veregin (1989) provides an excellent synthesis on the topic. Further information on the confusion matrix was provided in Chapter 12. To summarize, a confusion matrix is derived from a number of test samples where the mapped value and actual value of a land cover attribute are determined. The number of samples required is dependent on the expected map accuracy and the required confidence limits. By dividing the number of samples falling on the diagonal of the matrix by the total number of samples, we arrive at the overall percent correctly classified (PCC) for the map. The cross-tabulation of the confusion matrix, however, also provides valuable information on errors of omission (Type I) and commission (Type II) associated with individual map classes. Errors of omission are often indicated by *producer's accuracy* (100%—Type I error rate), which refers to the proportion of cases where a known area of class X is correctly identified in the map product. On the other hand *user's accuracy* (100%—Type II error rate) identifies the frequency with which a location mapped as class X is properly labeled. These two statistics are calculated by tabulating the proportion of correct entries for a particular class either by row or column in the confusion matrix.

In order to ensure that the confusion matrix reflects overall map accuracy, the test samples should be located randomly throughout the study area. Sample points might also be stratified by class to ensure representation of less abundant cover types. Even though these are statistically valid approaches, the confusion matrix itself does not maintain any information on biases in map accuracy that may exist in different parts of a map data set. Congalton documents such spatial autocorrelation in the error of thematic maps derived from remote sensing data (1988a) and demonstrates its potential impact on systematic sample placement schemes (1988b). Labovitz (1984) and Campbell (1981) showed how spatial autocorrelation in spectral reflectance causes areas that are proximate to training areas to be classified more accurately from remotely sensed data. Such autocorrelation may arise from spatial variability in land cover reflectance (e.g., Bartlett et al. 1988) or from other scene characteristics (e.g., haze, illumination, target geometry, and target or background reflectance). The effects of such secondary factors on the accuracy of map products derived from remotely sensed data were described in more detail in Chapter 12.

As a result of unequal error rates between land cover classes and additional spatially autocorrelated error sources, the overall map accuracy reported for a map may not be representative of the error found within a subregion of that product. It is becoming common for land cover maps to be generated for very large areas [e.g., statewide vegetation mapping associated with the U.S. National Biological Service's GAP project (Scott et al. 1993) and the DIS-Cover map of global land cover created in conjunction with the

International Geosphere/Biosphere Program (IGBP) (Belward and Loveland 1995; Loveland et al. 1999)]. The error assessment strategies for such
large-area products are generally designed to provide a statistically valid
estimate of overall map accuracy; however, users often wish to focus only on
a subregion of the entire map product, and it will be dangerous for them to
assume that the accuracy statistics provided for the entire product will be
mirrored in their specific subregion.

   The lack of spatial context in traditional validation methods creates
further difficulties for those who wish to understand better the possible
implications of map errors within their study area. Generally, no independent measure for the true pattern of landscape features exists, beyond that
implied by the imperfect map product itself. Fisher (1996) shows how one
might simulate the error in map products by randomizing class assignments
of pixels in an animation to match the frequency of confusion between
specific class pairs. Even though this approach was developed as a useful
visualization technique, its sole use of random noise may not provide a
realistic representation for the purposes of a computational simulation.
Fisher and Langford (1996) used a similar method to document the effects
of map accuracy on an areal analysis, but spatial dependence of map error
was again not considered. Depending on the situation, error in the map
product may be much more the result of confusions along the boundaries of
map objects or the mislabeling of entire landscape features. There has been
an extensive amount of literature on measures of landscape pattern in the
field of landscape ecology, and many of these metrics are described and
tested by O'Neill et al. (1988) and Riitters et al. (1995). One might use the
spatial configuration of the map units themselves, as indicated by metrics
like contagion or fractal dimension as a model to simulate uncertainty in the
map more realistically. The major problem with such an approach, however,
is its circularity. The error itself may define much of the observed spatial
pattern.

14.1 Example Data Sets

To examine this problem of spatially dependent error in thematic maps we
use three land cover maps covering a portion of the USGS 7.5 minute
quadrangle map for Goleta, California. One of these land cover maps was
derived from a manual photointerpretation of 1:24,000 scale aerial photography (Fig. 14.1a). This photointerpreted map has a documented overall
accuracy of 97% (McGwire 1992), and will be treated as "ground truth" for
the purposes of this chapter. The second map product for Goleta is a digital,
supervised classification that was created from Landsat Thematic Mapper
imagery (Fig. 14.1b). In addition to assigning the land cover for each
pixel, the supervised classification technique also assigns a confidence
value that reflects the likelihood ratio from the Bayesian classifier (a lower

FIGURE 14.1. Goleta maps: (a) photo-interpreted, (b) initial TM classification, (c) manually edited TM classification, (d) differences between (a) and (c) in black.

likelihood ratio value is associated with a more probable assignment). Chapter 12 provides further information on supervised classifications of satellite imagery.

Finally, we also use a version of the supervised classification that was filtered for majority class membership within a 3×3 pixel neighborhood to reduce noise and then corrected for gross errors by manual editing (Fig. 14.1c). Unless otherwise noted, this manually corrected product will be the one referred to when we discuss the TM classification in the following analyses. The pattern of error in the manually edited TM classification (relative to Fig. 14.1a) is displayed in Figure 14.1d by overlaying the classification with the photointerpreted product and identifying all areas that differ. The fact that the error portrayed in Figure 14.1a is not random provides the basis for our discussion.

Confusion matrixes used in this chapter were developed by comparing random locations between the maps developed from the TM imagery and the photointerpreted product. The overall mapping accuracy of the three products is shown below each image in Figure 14.1. These accuracy assessments were performed at the per-pixel level because this is the most common form of verification for remote sensing classifications. In this chapter we assume that no error arises from spatial misregistration between the map derived from aerial photography and the Landsat data. There is opportunity to improve the methods described here by factoring in the positional uncertainty in ground truth observations.

14.2 Spatial Variability in Classification Accuracy

Even if a thematic map is properly documented with a confusion matrix, the confusion matrix does not represent spatial variability in map accuracy. As an area-averaged indicator, the confusion matrix is only truly representative of the full spatial extent from which test samples were derived. Thus, if an analyst were only interested in studying a subregion of some thematic product, then there is a good chance that the information in the confusion matrix would be inaccurate for the purposes of that study. The smaller the subregion, the greater the loss of confidence in the original confusion matrix. This relationship is depicted in Figure 14.2, where the TM classification of the Goleta test site is divided into nonoverlapping subregions of

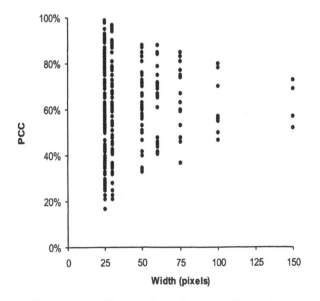

FIGURE 14.2. PCC versus linear dimension of subregion.

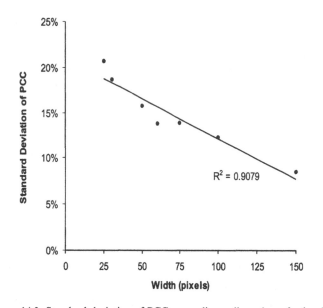

FIGURE 14.3. Standard deviation of PCC versus linear dimension of subregion.

varying size. Map accuracy (PCC) within each subregion is then plotted against the extent of the subregion. Figure 14.3 shows the relationship of the standard deviation of PCC to the extent of the subregion for the Goleta TM classification. Even though a good classification will attempt to represent the full range of variability in land cover types across a scene in an unbiased manner, the general relationships depicted in Figures 14.2 and 14.3 can be expected unless error rates are equal among classes and entirely due to random noise or very localized boundary uncertainty. Even if the confusion matrix indicates equal producer's and user's accuracy for each class, it is typical that this error will be distributed unevenly throughout the map as a result of the mislabeling of specific landscape features or spatially dependent biases arising from unexplained geographical variability.

The problem with the accuracy of subregions being substantially different than that reported for an overall product will likely be of increasing importance to the ecological research community as statewide and national programs develop large-area data sets. For example, the GAP program of the National Biological Survey in the United States is promoting the development of statewide maps of vegetation types derived from Landsat imagery. There will undoubtedly be many who will want to use these data sets to support ecological analyses in selected portions of a state. Despite the fact that the GAP program promotes standardized reporting of map accuracy, the actual map accuracy within a subregion will be uncertain unless one performs a local assessment. Even for very simple map parameters (e.g., the area covered by a particular vegetation type) this situation increases uncertainty in an analysis. Beyond this, of course, map accuracy will vary at finer scales within the selected subregion. Given a known accuracy for the region under study, spatial dependence in map error within that area may have little or no additional effect on some simple aggregate parameters like areal coverage; however, any analyses that are sensitive to the geometry of mapped features, the spatial relationships between features, or the co-occurrence of features from multiple sources may be strongly affected by spatial variability of map accuracy within the study area. As we shall see, the ability to tease out the spatial pattern of errors within a selected area of study unfortunately is limited.

14.3 Area-Weighted User's Accuracy

One might hypothesize that the error occurring within a given subregion might be estimated by an area-weighted average of the user's or producer's accuracy for land cover types that are mapped in that subregion. Figure 14.4a demonstrates the relationship between actual PCC and that predicted using user's accuracy for subregions derived from the Goleta scene. Each of the points in Figure 14.4a represents a 25×25 pixel subregion of the study area. One obvious feature of Figure 14.4a is that there are two different linear patterns, one with no relationship to actual PCC and the other having a strong relationship. The line with no relationship arises from those subregions that combine a number of different land covers so the PCC estimated from user's accuracy converges toward the overall PCC. The line that displays a strong relation is comprised of those subareas in the north that contain an increasing fraction of the shrubland class that has a much high user's accuracy (92%) than the other classes. A further feature of Figure 14.4a is that the estimates of PCC will be fundamentally limited to the minimum and maximum value of user's accuracy, or producer's accuracy if that is used instead. Thus, in the case of relatively uniform producer's and user's accuracy across classes this method would provide no information at all, although there may actually be significant amounts of spatial variability in error rates. A graphical comparison of actual PCC (a) and the user's accuracy estimates (b) is provided in Figure 14.5 for subregions of the Goleta study area. Figure 14.5b shows that the simple assumption that local error arises from differing proportions of classes with different error rates captures the gross pattern arising from dominance of shrubland in the north, but other areas show no apparent differentiation.

This example demonstrates the limited ability to generate information on the actual spatial distribution of error in thematic maps if the test samples from the product validation are not passed to the user along with the confusion matrix. Although we may know the accuracy of each map class and where the map places those classes, this is not enough information to

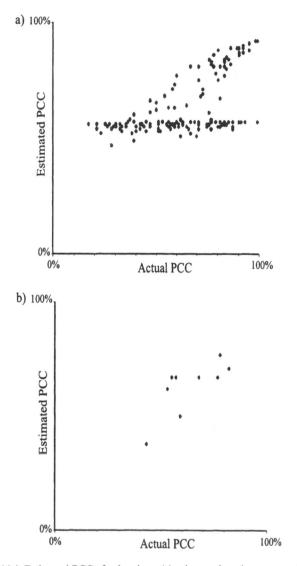

FIGURE 14.4. Estimated PCC of subregions: (a) using producer's accuracy, (b) using test samples.

tell us where much of the error is occurring. Even if we know from the confusion matrix that the errors of omission and commission for our classes of interest are relatively small, there still might be significant concerns regarding where this error occurs. For example, different patterns of error will have varying impacts on studies where the proximity between mapped features is important. A given animal species may require different habitat characteristics within close proximity (i.e., one for shelter, one for forage,

FIGURE 14.5. PCC for subregions (black = 0%, white = 100%): (a) actual, (b) estimated from confusion matrix, (c) estimated from test samples.

and another for reproduction). Randomly distributed error at the pixel level might not call in question the co-occurrence of these features as much as wholesale misclassification of mapped polygons, though both patterns might produce the same error rates.

14.4 Point-Based Estimates

A more direct method of getting at spatial variability in map accuracy might be to re-use the test points that were used to construct the confusion matrix. The study area could be divided into a number of subregions within which local map accuracy could be computed from the test samples. Eventhough this approach goes beyond the traditional confusion matrix by providing some information on the spatial variation in map accuracy, the obvious limitation is that estimates of map accuracy within subregions will be of progressively lower significance as the extent of the subregion decreases. This method will probably provide only qualitative information for fairly broad subregions, unless the goal of characterizing subregions is specifically built into the sampling strategy. Figure 14.4b portrays the relationship between such a point-based estimate and actual PCC for 11% subregions of the Goleta study area. This subdivision size ensured that there were greater than five test samples per subregion. Because sample placement was randomized there was a larger variance in the number of samples per subregion than a spatially stratified design would have provided, resulting in a coarser subdivision than might be expected. If there were a large number of test samples per subregion, a very close linear fit would be expected; however, in this case the estimation error arising from low sample size results in an R^2 of only 0.56 (Fig. 14.4b). As seen in Figure 14.5c, the resulting information on the distribution of map error is quite coarse, and the correspondence between actual and predicted map accuracy is quite low in some subregions when comparing Figure 14.5c to Figure 14.5a. The more detailed the needs of the user with respect to the spatial dependence of map error, the less useful a point-based approach will be.

14.5 Per-Pixel Confidence Maps

Some classification methods (e.g., maximum likelihood or neural networks) are capable of providing confidence estimates for the classification on a per-pixel basis, potentially providing detailed information on the distribution of uncertainty throughout the thematic map. Figure 14.6 plots log-likelihood scores versus PCC for each class using an exhaustive comparison between the original supervised classification (Fig. 14.1b), the per-pixel confidence estimates, and the photointerpreted map of the Goleta study area (Fig. 14.1a).

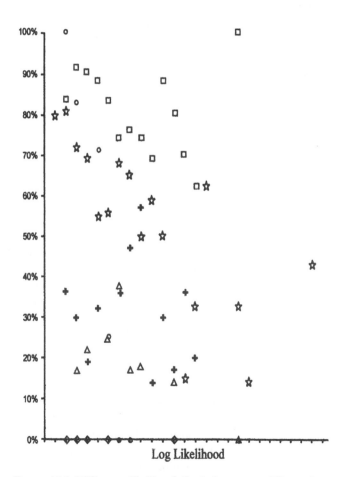

FIGURE 14.6. PCC versus likelihood. Symbols represent different classes.

Note that low log-likelihood values should correspond to a more confident class assignment, and that the likelihood scores were rounded to integer values by the software. The symbols in Figure 14.6 show the class-by-class agreement between the original TM classification and the photointerpreted map for each integer value of the output confidence estimates. This shows that some land cover classes do have a relationship between log-likelihood values and PCC, but the form varies greatly. In addition to this class-specific variation, there are two general limitations to developing relationships between log-likelihood and PCC to predict local map accuracy. First, because the log-likelihood scores are continuous even though the class assignments are discrete, log-likelihood must be binned into ranges within which discrete cases of correct–incorrect can be averaged to create an estimate of PCC. Each of these bins must encompass enough test samples to produce a good estimate of PCC. In other words, a relatively small number of test samples must be binned into an even smaller number of intervals. Thus, the ability to develop good relationships between log-likelihood and map accuracy may be quite limited, and the possibility of developing class-specific relationships is even more unlikely. It is also unlikely that nonlinear methods such as logit modeling would provide a stable solution, given the variability of this relationship between classes and the limited number of samples. The second general limitation is that any postclassification processing performed to increase classification accuracy will distort the relationship between the original confidence value and final class assignment. Such postclassification processing may include common operations such as majority filtering or relabeling based on ancillary map data, as performed on Figure 14.1c.

One of the developments being explored in order to explain the spatial uncertainty in classifications of remote sensing data is "fuzzy" schemes that might allow multiple labels to be assigned to a single pixel (e.g. Foody 1999). The fact that the probability of correctly classifying a pixel is not always directly related to the spectral similarity between that pixel and some model reflectance, however, points out a significant challenge in this area. It will be shown in Chapter 15 that the probabilities of class membership across all classes can be combined to suggest the confidence of class assignment at the per-pixel level; however, the assumption of a monotonic likelihood function that is identical across all classes still underlies this approach. One method of dealing with this difficulty might be taken from McGwire et al. (1996), where pixels from the training fields used in the classifier were reused to translate the per-pixel confidence estimates into a nonparametric function that related directly to map error rates.

The approach of using per-pixel confidence estimates, either directly or in fuzzy approaches, could potentially provide the most detailed information on map accuracy for ecological analyses. The ability to understand the probability of map error at this level would allow a much more precise understanding of the types of confusions that may be taking place in an analysis, particularly when multiple spatial data sources are being combined and co-occurrence is important. For example, as pointed out in Chapter 13, it is important that the type of forest is known when attempting to relate leaf area index to net primary production. It would greatly aid an estimate of uncertainty for such an analysis to know whether hardwood and softwood classes were being confused in areas of higher or lower leaf area index. As demonstrated earlier, however, challenges remain in implementing per-pixel methods of estimating uncertainty. In addition, some common classification methods (e.g., unsupervised clustering) do not lend themselves to per-pixel confidence estimates.

14.6 Spatial Statistics

An alternate approach to portraying the uncertainty derived from specific test samples or individual pixels could be to use spatial statistics to characterize the scales over which errors behave in a nonrandom manner. Even though it does not specifically show where map error occurs, this information can be used to infer the nature of map error within a study area. Congalton (1988a) provides an example of such an analysis using join-count statistics as described by Cliff and Ord (1973). Congalton's analysis was performed using an exhaustive comparison of a classification to a reference map; however, in practice such reference maps do not exist and costs greatly limit the number of ground truth samples.

Join-count statistics can be applied to the test samples generated for a traditional accuracy assessment, but, again, there is some question as to whether there are sufficient samples to obtain a significant answer. Join-count statistics compare the number of agreements and disagreements between binary values across some distance to determine whether a distribution deviates significantly from the null hypothesis of randomness. The expected number of combinations between correct and incorrect pixels and the variance for a random distribution can be compared with observed combinations to calculate the standard normal deviate. The equations for calculating these estimates of the first and second moment may be performed by assuming sampling with, or without, replacement.

Figure 14.7a is a plot of the standard normal deviate for a number of distance intervals, as calculated from the correct–incorrect pairings of 5000 test samples taken from Figures 14.1a and 14.1c. A value less than −2.0 indicates a statistically significant clustering of errors at the 98% confidence level, whereas values greater than 2.0 indicate significant uniformity. The very large number of samples was used to demonstrate the spatial pattern of map error unambiguously. Figure 14.7a shows an extremely high degree of localized clustering in error rates at the finest scales. This clustering of map error is quite apparent in Figure 14.1d. Clustering of map error decreases out to a distance of 55 pixels, beyond which map error becomes much more random. Given that the extent of the Goleta study area is 300 pixels in the x

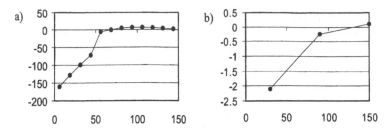

FIGURE 14.7. Spatial pattern of error: (a) 5000 samples, (b) 150 samples.

and y dimensions, distances between samples exceeding 150 pixels would be increasingly concentrated along the boundaries of the study area. It clearly makes sense to ensure that the samples are representative of any randomly located contiguous area, so a maximum intersample distance of 150 pixels was used.

An actual study obviously would not have access to a reference map to generate a large number of samples for the join-count method because we would have used the reference map in the first place. Instead, we would have to rely on the test samples that were used in generating the confusion matrix. One of the difficulties in using this small number of test samples with the join-count technique is in judging the tradeoff between the width of the distance intervals and the available degrees of freedom. For example, a first attempt using test points from the Landsat classification and uniform distance intervals of 30 pixels showed the general pattern of clustering over short distances, but none of the values exceeded 2.0 standard normal deviates. This indicated that there may be insufficient evidence to reject the possibility of random error; however, by extending the distance interval to 60 pixels and thereby increasing the number of samples per interval, there is little detail but greater confidence in the trend (Fig. 14.7b).

Information on the spatial extent of errors in map products may be used to provide a better understanding of the potential magnitude of errors in associated ecological analyses. The join-count statistic, however, provides no guidance on where these errors are likely to be occurring. Attempts have been made to interpolate continuous probability surfaces for map accuracy using spatial statistics. Steele et al. (1998) described a method whereby a surface of mapping uncertainty was interpolated from information on the accuracy with which training pixels were back-classified in Landsat imagery. In this approach, kriging was used in an attempt to provide an optimal interpolation of map error, and the pixel size of the error surface was degraded to an appropriate spatial resolution based on the form of the semivariogram (see Chap. 9 for more information on kriging). This attempt to interpolate map accuracy from the training pixels has several problems, however, including the difficulty of obtaining a properly randomized sampling of training locations and the assumption that the error rate is independent of land cover class. For example, a training field for water in the middle of a pond would have an extremely high back-classification accuracy, so that region of the interpolated surface would be assumed to have a high accuracy. The lake, however, might be surrounded by a highly complicated morass of conifers, hardwoods, meadows, and wetlands where the classifier might perform quite poorly. Although the authors presented a surface of map confidence derived from the training sites, a useful extension would have been to map the estimated error variance for each interpolated pixel. These error variance estimates are provided directly by the kriging technique, based on the number and proximity of samples relative to the modeled semivariance. In other words, it would be possible to have a map of uncertainty for a map of the uncertainty in a map! This still would not deal with the fundamental assumption of the error rate being independent of land cover class, but it would provide more information on how reasonable the interpolation might be in different areas. Despite these limitations, the approach of Steele et al. does quantify the spatial dependence of reflectance characteristics that may contribute, in part, to actual map error.

14.7 Simulation

One of the principal ways of testing the sensitivity of complex models to errors in input data is with a Monte Carlo approach. Instead of working with the abstract distribution of uncertainty across the surface of an imperfect map, we can create multiple representations of the patterns that might underlie that map. A sensitivity analysis could be performed by running an ecological model multiple times, each time randomly altering the distribution of classes in the input map. For the purposes of visualizing map error, Fisher (1996) describes random reassignment of pixels based on frequencies of misclassification documented in a confusion matrix. If the accuracy of the land cover map is any better than random assignment (1.0/#classes), however, then the actual accuracy of these simulations will be less than that of the original product. Discounting the prior probabilities associated with unequal areal extent and mapping accuracy between classes, if one randomly relabeled pixels, then all those selected pixels that were correct would become incorrect. Of those that were incorrect, the number becoming correct would be the inverse of the total number of classes minus one. Thus, the expected accuracy of the changed pixels would be $(1 - \text{PCC})/(\#\text{classes} - 1)$. The resulting overall map accuracy would be expected to be percentage unchanged times the original accuracy plus the percentage changed times the probability of a correct pixel reassignment, or:

$$\text{PCC}^2 + \frac{(1 - \text{PCC})^2}{(n - 1)}$$

For the Landsat-based map used in the preceding sections this method would be expected to reduce the original map accuracy of 64% to an average of

43% for the random simulations. Also, as previously noted, the spatial distribution of map classes and map error is nonrandom (Figs. 14.1d and 14.7), so a randomized reassignment of pixels would not provide a realistic spatial configuration. For example, random pixel reassignment would not allow one to simulate incorrectly labeled patches within the landscape, as are apparent in Figure 14.1d. These two issues highlight an important point with respect to simulating map error in Monte Carlo studies. If an ecological model is sensitive to error in spatially explicit inputs, then the fact that we do not know which pixels are actually incorrect means that sensitivity analyses will typically provide an exaggerated indication of the negative effects of map error!

If available, information on the spatial distribution of error in a map product could be used to provide more realistic simulations, though these would also generally have lower accuracy than would the original product. Eventhough we know that error in a map product almost always has spatial structure, we have seen how difficult it is to extract this type of information from the standard methods for map accuracy sampling. The most likely source of information for alternate reconstructions of land cover would be derived from the spatial structure of the classes portrayed in the map itself. This approach is dangerous, however, because the primary motivation for generating simulations is the knowledge that the map is an imperfect source of information with which to begin. Thus, the maps that pose the greatest need for error simulation will also provide the worst indications of how to meet that need.

Indicator stochastic simulation provides a method for creating multiple realizations of a thematic product based on empirically derived models of spatial dependence (Bierkins and Burrough 1993a). Fisher (1998) demonstrates a similar approach with the use of continuous data using digital elevation models. The indicator stochastic simulation method is based on the geostatistical technique of indicator kriging. Indicator kriging uses a nonparametric conditional probability distribution function (CPDF) that incorporates the spatial connectivity of classes so the class membership of a given location can be predicted from the neighboring class values. Categorical or ordinal data (e.g., a land cover map) are transformed into $n-1$ (number of classes − 1) binary variables $(0, 1)$ that identify whether the observed class is less than or equal to each of the $n-1$ class values, where the class values are an arbitrarily assigned sequence of integers for each land cover class. The CPDF is then derived using semivariograms that are developed from the binary indicator variables.

Bierkins and Burrough (1993b) demonstrate how indicator kriging can be used to generate multiple randomized instances of a categorical soils map from point sampling data. Their approach would have to be modified in the case of simulating thematic map data because the number of ground truth samples are typically an infinitesimally small subset of the total spatial data set, and sole reliance on these samples would discard the wealth of spatial information in the map product itself. Bierkins and Burrough's study (1993b) documents problems in obtaining stable variograms from their data when using as many as 100 field sample points for defining just four soil classes. Most land cover maps have a greater number of classes, and many more points would be required to characterize the spatial connectivity of classes adequately. Thus, random samples from the map product might be required to supplement the original test samples in order to generate stable semivariograms for the kriging technique. Eventhough random samples taken from the map product provide a wealth of data for generating semivariograms, this does raise the issue that map error may be significant in these samples. For example, random samples from land cover classes that have large errors of commission are likely to display broader patterns of spatial connectivity than actually exist.

As with other geostatistical techniques, one of the challenges in using the indicator kriging method is dealing with the underlying assumptions of randomness and statistical stationarity across the map. As seen in Figure 14.1b, patches of a given land cover class are not distributed randomly throughout the study area; instead, they tend to occur in specific subregions. This results in nonstationarity that is seen as a persistent upward trend at longer distance intervals in the semivariograms derived from the transformed indicator variables of the Goleta map (Fig. 14.8). Notice that the

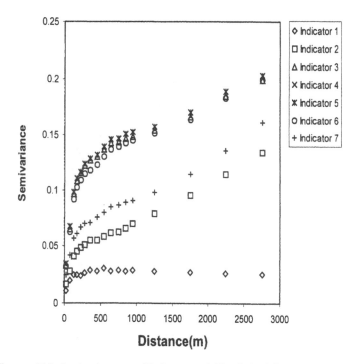

FIGURE 14.8. Semivariograms of indicator variables derived from 10,000 random samples from the Landsat-based map.

nonstationarity in shrubland (Class 2) is superimposed on all subsequent indicator variables because of their interdependence.

Assuming that stable semivariograms could be generated for the indicator variables, our experience with the failure of point-based methods earlier in this chapter also suggests that the indicator kriging method would need additional samples taken directly from the classification in order to provide realistic simulations of the underlying land cover. It is possible to include this information by using additional, unverified map samples as "soft" variables in the indicator kriging method (Gomez-Hernandez and Srivastava 1990a). The question arises, however, as to how many samples of soft variables should be used in the stochastic simulations. If the entire classified map was used, then the output of every simulation would be quite similar to the original map. If no additional samples are taken from the map to supplement the ground truth samples, the resulting simulations would be thoroughly unrealistic and the actual map error would be excessive. Between these extremes, the density of random samples taken from the map product as soft variables will interact with the typical patch size of the land-cover classes to determine the similarity of the simulation to the original map product and actual land cover.

Figure 14.9 demonstrates a random simulation for the Goleta study area that was developed using the ISIM3D program (version 2.5; Gomez-Hernandez and Srivastava 1990b) using a visual estimation of semivariogram parameters that ignored the effects of nonstationarity and anisotropy.

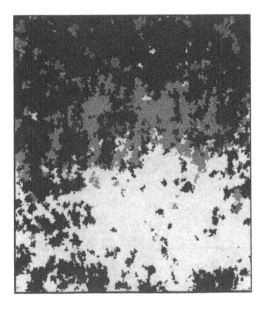

FIGURE 14.9. A random simulation of the Goleta study area using indicator stochastic simulation.

This simulation used an 8% subsample of pixels from the indicator variables of the Landsat classification to serve as soft variables in the kriging method. The simulation in Figure 14.9 has 78% agreement with the Landsat-based map and an actual map accuracy of 61% (only 3% less than the original). The tradeoff between the percentage of points taken from the Landsat classification for use as soft variables and the resulting map accuracy is shown in Figure 14.10. Figure 14.10 also shows how the use of map samples as soft variables affects the degree of similarity to the original map product. An example of a simulation developed without any additional points to serve as soft variables is presented in Figure 14.11, obviously creating an unreasonable result.

It is important to remember that eventhough patterns of uncertainty derived from indicator stochastic simulations may be helpful in a Monte Carlo analysis, they reflect the spatial dependence of classes in the region, not the pattern of map error. Figure 14.12 shows the differences between the Landsat classification and the simulations generated by indicator kriging in Figure 14.9. The pattern seen in Figure 14.12 is clearly different from the actual pattern of error portrayed in Figure 14.1d. Much of this difference appears to arise because class reassignment in the simulation is being performed at the pixel level, whereas actual error also includes the misclassification of larger patches in the landscape. We saw this coarse patchiness in the clustering of map error in the join-count statistic of Figure 14.7, though there were insufficient samples to characterize this pattern confidently (especially if one needed to do so on a class-by-class basis). In order to

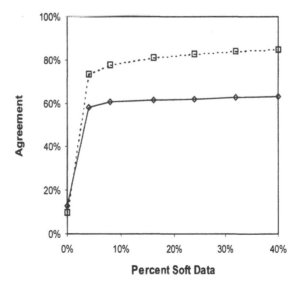

FIGURE 14.10. Trade-off of similarity (dashed) and map accuracy (solid) with percent of pixels used as soft variables.

FIGURE 14.11. A simulated map based only on the 150 ground truth locations.

mimic the patchiness seen in Figure 14.1d with the indicator kriging method, we need to reduce the number of samples used for soft variables. As seen in Figure 14.11, however, this results in a decreasing correspondence with reality.

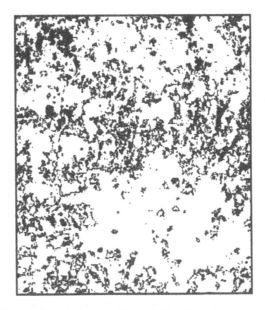

FIGURE 14.12. Differences between Figure 14.8 and the Landsat classification.

The significance of the differences between Figures 14.1d and 14.12 can be seen in ecological analyses where the extent of contiguous patches is important, as in identifying areas of suitable habitat for animal species. For example, if we are trying to find suitable relocation sites for an animal species that requires a certain minimum area of contiguous habitat, then the type of error shown in Figure 14.1d and suggested in Figure 14.7 is actually much more dangerous to the analysis than the fine-scale discrepancies that might be assumed from Figure 14.12. Similar concerns arise if we are concerned about edge effects, as described near the end of Chapter 3.

14.8 Summary

Uncertainty in the accuracy of spatial data that is used in ecological modeling may arise from a number of sources. The breakdown in the relationship between overall map accuracy and the actual accuracy of specific subregions suggests that accuracy assessments for large area maps should be adjusted to match the smallest spatial domain for which the producer expects the product to be used (e.g., county use of a statewide product). This recommendation includes both an increase in the number of test samples, and the distribution of test samples within the expected subregions. We might imagine many examples where a statewide map would be used in a study of ecological impacts in one or two counties. In cases where the subregion being studied is much smaller than the spatial extent that the original validation was designed for, the user of the data will need to perform his or her own tests of map accuracy because the published accuracy statistics are quite possibly irrelevant.

The most information on the spatial distribution of uncertainty within a study area would come from per-pixel confidence estimates, if progress could be made to relate these confidence estimates to actual error rates or at least to consistent monotonic distributions. The join-count statistic and the work by Steele et al. (1998) provide good directions for statistically representing the characteristic scales of spatial variability in map accuracy, eventhough work is needed to apply these to the needs of ecological modelers. For example, we need a better understanding of nonstationarity in the error field and methods of dealing with this given that error rates will vary in a discontinuous manner depending on land cover.

Although methods like indicator stochastic simulation do exist to create multiple representations of a map for Monte Carlo analysis, these methods should be based on the spatial dependence of misclassifications, not on the spatial dependence of the mapped features themselves. Despite the realism imparted to such simulations by incorporating models of spatial variability in land-cover classes, this is not the same as the spatial variability in map error. Confusing these two types of variability will have a direct, negative effect on the accuracy and realism of the resulting simulations. A much more promising direction would be to use the general concept of Steele et al. as the basis for stochastic simulations. Such an approach could build on the work of Bierkins and Burrough (1993a,b) or Fisher (1998), assuming the aforementioned questions about the underlying error field were adequately addressed.

The known spatial dependence of error in map products suggests that the user community needs to rethink the state-of-the-practice methods for reporting map accuracy. This expansion of error documentation methods may include the addition of a spatial error summary in the product metadata, including at the least both the coordinates and the attribute data that were used in the standard product validation. Approaches should be developed to store ancillary per-pixel information from classifiers (e.g., likelihood ratios from Bayesian methods or root mean square error from linear mixture models) in a standardized format with sufficient documentation to ensure proper use. Standardized methods should be developed for generating digital accuracy reports that support more sophisticated use of map products, including the development of realistic simulations for Monte Carlo analyses in models that use spatial data inputs.

Acknowledgments. This work was supported in part by funding from the National Aeronautics and Space Administration (subcontract KK8020 from NAG5-6259) and from NASA through a memorandum of understanding with the National Institutes of Health (grant NIH-5 P01A139808-02).

References

Bartlett, D., M. Hardisky, R. Johnson, M. Gross, V. Klemas, and J. Hartman. 1988. Continental scale variability in vegetation reflectance and its relationship to canopy morphology. International Journal of Remote Sensing 43:595–98.

Belward, A., and T. Loveland. 1995. The IGBP-DIS 1-km land cover project: remote sensing in action. Pages 1099–106 in Proceedings of the twenty first annual conference of the remote sensing society. The Remote Sensing Society, The University of Nottingham, UK.

Bierkins, M., and P. Burrough. 1993a. The indicator approach to categorical soil data. I. Theory. Journal of Soil Science 44:361–68.

Bierkins, M., and P. Burrough. 1993b. The indicator approach to categorical soil data. II. Application to mapping and land use suitability analysis. Journal of Soil Science 44:369–81.

Campbell, J. 1981. Spatial correlation effects upon accuracy of supervised classification of land cover. Photogrammetric Engineering and Remote Sensing 47:355–64.

Cliff, A., and J. Ord. 1973. Spatial autocorrelation. Pion Limited, London, UK.

Congalton, R. 1988a. Using spatial autocorrelation analysis to explore the errors in maps generated from remotely sensed data. Photogrammetric Engineering and Remote Sensing 54:587–92.

Congalton, R. 1988b. A comparison of sampling schemes used in generating error matrices for assessing the accuracy of maps generated from remotely sensed data. Photogrammetric Engineering and Remote Sensing 54:593–600.

Fenstermaker, L. 1994. Remote sensing thematic accuracy assessment: a compendium. American Society for Photogrammetry and Remote Sensing, Bethesda, MD.

FGDC (Federal Geographic Data Committee). 1998. Spatial data transfer standard. FGDC-STD-002-1998. Computer Products Office, National Technical Information Service, Springfield, VA.

Fisher, P. 1996. Visualization of the reliability in classified remotely sensed images. Photogrammetric Engineering and Remote Sensing 60:905–10.

Fisher, P., and M. Langford. 1996. Modeling sensitivity to accuracy in classified imagery: a study of areal interpolation by dasymetric mapping. Professional Geographer 48:299–309.

Fisher, P. 1998. Improving error models for digital elevation models. Pages 55–61 in R. Jeansoulin and M.F. Goodchild, eds. Data quality in geographic information: from error to uncertainty. Hermes, Paris.

Foody, G. 1999. The continuum of classification fuzziness in thematic mapping. Photogrammetric Engineering and Remote Sensing 65:443–51.

Gomez-Hernandez J.J., and R.M. Srivastava. 1990a. ISIM3D: an ANSI C three-dimensional multiple indicator conditional simulation program. Computers and Geosciences 16:395–440.

Gomez-Hernandez J.J., and R.M. Srivastava. 1990b. ISIM3D (version 2.5), an ANSI C program downloaded from ftp://mundo.dihma.upv.es/pub/gcosim3d/.

Labovitz, M. 1984. The influence of autocorrelation in signature extraction—an example from a geobotanical investigation of Cotter Basin, Montana. International Journal of Remote Sensing 50:315–32.

Loveland, T.R., Z. Zhu, D.O. Ohlen, J.F. Brown, B.C. Reed, and L. Yang. 1999. An analysis of the global land cover characterization process. Photogrammetric Engineering and Remote Sensing 65(9):1021–32.

McGwire, K. 1992. Analyst variability in the labeling of unsupervised classifications. Photogrammetric Engineering and Remote Sensing 58:1673–77.

McGwire, K., J. Estes, and J. Star. 1996. A comparison of maximum likelihood-based supervised classification strategies. GeoCarto 11:3–13.

O'Neill, R., J. Krummel, R. Gardner, G. Sugihara, B. Jackson, D. DeAngelis, et al. 1995. Indices of landscape pattern. Landscape Ecology 1:153–62.

Riitters, K., R. O'Neill, C. Hunsaker, J. Wickham, D. Yankee, S. Timmins, et al. 1995. A factor analysis of landscape pattern and structure metrics. Landscape Ecology 10:23–39.

Scott J., F. Davis, B. Csuti, R. Noss, B. Butterfield, C. Groves, 1993. Gap analysis – a geographical approach to protection of biological diversity. Wildlife Monographs 123:1–41.

Steele, B., C. Winne, and R. Remond. 1998. Estimation and mapping of misclassification probabilities for thematic land cover maps. Remote Sensing of Environment 66:192.

Veregin, H. 1989. A taxonomy of error in spatial databases. Technical paper 89-12. National Center for Geographic Information and Analysis, Santa Barbara, CA.

15
Modeling Spatial Variation of Classification Accuracy Under Fuzzy Logic

A-Xing Zhu

A common form of spatial data used by ecologists in their research and management activities is categorical maps (e.g., soil type and vegetation cover–type maps). For example, using these categorical maps landscape ecologists analyze spatial structures of landscape elements and study the ecological implications of these structures (e.g., see Turner and Gardner 1990). Hydroecological modelers derive important landscape parameters (e.g., soil moisture regimes and vegetation types) from these data sets for simulating hydroecological processes over space (e.g., see Running and Hunt 1993; Band 1995; Franklin et al. 1997; Warning and Running 1998). Natural resource managers use these data for resource management and for studying wildlife habitats (e.g., see Brown et al. 1998; Ripple 1994, Chap. 3). Categorical maps are derived through classification, particularly through the classification of remotely sensed imagery (including air photos). For example, vegetation cover–type maps are now commonly derived from the classification of satellite images (Beaubien 1994; Lillesand and Kiefer 1994; Campbell 1996; Jensen 1996; Chap. 13); soil maps were produced through soil polygon delineation based on air photo interpretation (Hudson 1990). With the rapid development of remote sensing technologies, remotely sensed data have become the major source, and image classification techniques have become the major method, for creating these categorical natural resource maps (Chaps. 12, 13). These data sets, however, cannot be created without classification errors because many geographical phenomena do not exist as distinct and discrete classes; instead, they lie along spatial continua and exist as class intergrades (Leung 1987; Mark and Csillag 1989; Lunetta et al. 1991; Zhu 1997a,b, Chap. 16). In addition, remotely sensed data may not capture all the necessary details (Chaps. 5 and 12). Classification errors in these data sets could have serious implications for ecological applications using these data sets (Stoms et al. 1992, Chaps. 2 and 4). It is therefore necessary to provide information about the nature of classification errors in these data sets.

Classification accuracy (the opposite of classification error) is often reported using the global accuracy statistics such as percent correctly classified (PCC) and Kappa (Cohen 1960; Congalton et al. 1983, Chap. 14). Although they give an overall assessment of classification accuracy, these global statistics give no information about spatial variation of classification accuracy in the categorical data sets (Goodchild 1995, Chap. 14). The overall classification accuracy may therefore not be applicable to subregions of a large area map (e.g., a statewide cover-type map). This is because some areas may be classified with accuracy higher than the reported overall value, whereas other areas may be classified with accuracy significantly lower than that value (Chap. 14). On the other hand, information on the spatial variation of data quality can be very important for ecologists in their decision making employing spatial data and GIS-based analysis. First, the provision of information on spatial variation of data quality will allow ecologists to assess areas where data quality does or does not meet the requirements of a given application. Second, this data quality information will also allow ecologists or data producers to allocate data quality improvement efforts more effectively. Third, ecologists can use this data quality information to assess the risk of basing decisions on this type of data set and to quantify the likelihood of alternative outcomes predicted by dynamic models (see Chaps. 1, 2, and 19).

This chapter presents an approach to modeling spatial variation of classification accuracy in categorical maps under fuzzy logic. *Classification accuracy* refers to the closeness of the assigned class label to the true class at a location. *Classification error* is the opposite of classification accuracy, and it means the deviation of the assigned label from the true class at a location. It is difficult, if not impossible, to derive a map of classification accuracy because it would require one to know the true class at every location across the landscape being mapped; however, the ambiguity in assigning a class label to a location may be computed based on the information used in the classification process. It is then possible that this ambiguity (referred to as *classification uncertainty*) may be used to approximate classification accuracy.

This chapter is divided into four major sections. Section 15.1 examines the process of classification and the genesis of classification errors. Section 15.2 provides an overview of a similarity representation scheme based on fuzzy logic for retaining detailed classification information, which is then followed by the discussion of two measures for quantifying classification uncertainty using the similarity scheme. Two case studies are presented in Section 15.4 to illustrate the application of this uncertainty quantification approach for depicting the spatial variation of classification uncertainty. Summaries and implications are given in Section 15.5.

15.1 Classification and Associated Uncertainty

15.1.1 Classification Process

The generation of categorical maps through conventional classification is based on the concept that geographic features/phenomena exist in distinct and discrete classes. The process of classification often consists of two steps: class definition and class assignment (Zhu 1997a,b). During class definition, the characteristics (i.e., class mean, variance/covariance) of a class are defined and are assumed to be distinct from other classes. For example, in remote sensing classification, the reflectance characteristics for each class of objects (e.g., water bodies, coniferous forests, and grasslands) are defined and assumed to be distinctly different from each other. Once the characteristics of each class are defined, the second step is class assignment that is to assign an object at a given location to a particular class by comparing the characteristics of the object to the characteristics of the given class. Class assignment is often performed under crisp logic, which means that each location (e.g., a pixel in remote sensing classification) is assigned to one and only one resource class to which it displays the greatest level of similarity. Because only one class is associated with each location in the categorical map and no indication of the relative strength of class membership is provided, full membership in the allocated class is implied.

15.1.2 Classification Errors

Errors are often introduced by assuming full membership in one class and no membership in other classes in the preceding classification process. These errors can be understood as the exaggeration and ignorance of membership. Suppose that a pixel has membership in more than one class but does not have full membership in any of the classes. When this pixel is assigned to one class, membership in other classes is ignored and information is omitted. At the same time, the pixel assumes full membership in the assigned class for which the pixel does not *fully* "qualify"; thus, membership in that class has been exaggerated. It must be pointed out that the concepts of membership exaggeration and ignorance outlined here pertain to a single classified location.

There are four major scenarios under which classification errors mostly occur, although other scenarios may exist. First, many geographical phenomena do not exist as distinct and discrete classes; instead, they lie along spatial continua and exist as class intergrades (Leung 1987; Mark and Csillag 1989; Lunetta et al. 1991; Zhu 1997a,b, Chap. 16). For example, an area might contain just one homogeneous feature, but this feature is intermediate to the prescribed classes (class intergrade or ecotone) (Wood and Foody 1989; Foody 1992; Fisher 1997). Thus, the feature at the location may bear membership in more than one of the prescribed classes. The intermediate nature will be lost when the location is labeled only as one of the prescribed classes.

The second is that geographic features/phenomena over an area may not be of one type (mixed class or mixed-grade area) (Robinson and Thongs 1985; Foody and Cox 1994; Maselli et al. 1996; Fisher 1997). For example, the area covered by a remote sensing pixel can be occupied by many different cover types (the mixed pixel problem) and it could have membership in all of these cover types. Assigning the area to any one of these cover types will certainly cause classification errors.

Under the third scenario, the feature at a location may be of an undefined class (extragrade) (McBratney and De Gruijter 1992; Jensen 1996). It does not bear much of a membership in any of the prescribed classes. This location may be assigned to an unclassified category (labeled as unclassified) in some classification practice. In many cases, it is assigned to the class with the highest similarity, although this similarity is very weak. In the latter cases, membership of this pixel in the assigned class is seriously exaggerated.

Finally, classification of feature at a given location is based on the observable characteristics of the geographic feature at that location. Due to the limitations of means used to observe these characteristics, it is possible that two areas containing two different types of features may not be distinguished based on the observed information and are then labeled as one class. Much of this scenario will eventually become one of the first three as means for observing geographic features/phenomena improves.

As outlined earlier, many geographical phenomena have membership in multiple classes; thus, classification errors are common in categorical data sets derived from the classification technique. Any classification of the entity at a location would contain some degree of membership exaggeration and ignorance, even if the entity is correctly classified, unless the feature is a typical case of a resource class (Zhu 1997a). Degrees of exaggeration and ignorance also vary from location to location because geographic features/phenomena often exist in spatial continua and as class intergrades.

15.1.3 Exaggeration and Ignorance Uncertainty

Classification errors (membership exaggeration and membership ignorance) due to the mixed grade, intergrade, and extragrade nature of geographic features/phenomena may not be known exactly, but they may be approximated by classification uncertainty. *Classification uncertainty* is defined as the ambiguity in assigning a class label to an area based on the membership values computed from classification processes. There are two aspects of this uncertainty: membership exaggeration uncertainty and membership ignorance uncertainty. *Exaggeration uncertainty* is defined as the exaggeration involved in assuming full membership of an area in the assigned class. It can be perceived as the dissimilarity between the entity at that location and the assigned class. The smaller the membership in the assigned class, the greater is the exaggeration. Exaggeration uncertainty can be used to approximate classification errors due to membership exaggeration. The higher the exaggeration uncertainty, the greater the membership exaggeration error, and the lower the classification accuracy.

Ignorance uncertainty is defined as the membership values of an entity in the prescribed classes other than the assigned one. It can be perceived as the similarity values between the entity and the prescribed classes other than the assigned one. It is related to the fuzziness of an entity compared with the definitions of classes. The fuzzier the entity in relation to the class definitions, the more evenly distributed the membership among the classes, and the greater the ignorance uncertainty. Ignorance uncertainty can be used to approximate classification errors due to membership ignorance. The higher the ignorance uncertainty, the greater the membership ignorance, and the lower the classification accuracy.

15.2 Fuzzy Representation of Spatial Information

Classification uncertainty of mixed-grade, intergrade, and extragrade areas can only be computed if a fuzzy information representation scheme is employed in the classification process. Under a fuzzy representation (Fisher and Pathirana 1990; Wang 1990a; Foody 1992; Lowell 1994; Maselli et al. 1996; Chap. 16) or a similarity representation (Zhu 1997b), the classification of the object at location (i,j) is represented by an n-element vector (referred to as a *similarity vector*, Zhu 1997b), S_{ij} ($S_{ij}^1, S_{ij}^2, \ldots, S_{ij}^k, \ldots, S_{ij}^n$), where S_{ij}^k is the membership value of the local object at (i,j) in class k, and n is the total number of prescribed classes. The membership value can range from 0 to 1 with 0 meaning no membership in class k, whereas 1 means full membership in class k.

The elements in a similarity vector do not have to sum up to unity since they are similarity measures (Zhu 1997b). It is possible that the object at a location may bear high similarity values to many similar classes and the sum of these similarities can exceed unity. The object at another location may be very unique (such as an extragrade object) and it may not bear much similarity to any of the prescribed classes, and the sum of its similarity values can then be less than a unity.

The similarity value, S_{ij}^k, can be approximated operationally by the membership value computed from either a maximum likelihood procedure (Wang 1990a,b; Foody et al. 1992) or a fuzzy classification procedure (Bezdek et al. 1984; Foody 1992) or neural networks (Civco 1993; Foody 1996; Gong et al. 1996; Carpenter et al. 1997; Zhu 2000) or expert system approaches (Zhu et al. 1996). The case studies presented later in this chapter will illustrate two of these approaches: maximum likelihood classification and expert system.

15.3 Measuring Uncertainty Elements and Their Spatial Variation

With the similarity vector describing the entity at a location being populated, ignorance and exaggeration uncertainty associated with classification of that entity can then be quantitatively defined. The classification of the entity at a given location can be achieved by "hardening" the similarity vector describing this entity. The "hardening" (class assignment) is done by assigning the entity to that class in which the entity has the highest membership (Zhu 1997a,b). Zhu (1997a) uses an entropy measure to estimate the ignorance uncertainty associated with this hardening. The entropy measure of a vector is calculated as:

$$H_{ij} = -\frac{1}{\log_e n}\sum_{k=1}^{n}[(S_{ij}^{tk})\log_e(S_{ij}^{tk})] \tag{15.1}$$

where S_{ij}^{tk} is the normalized similarity value (i.e., the sum of the normalized values in a vector is 1.0), n is the total number of candidate classes, and H_{ij} is the entropy associated with the similarity vector for point (i,j) and has a range of 0 to 1. An entropy value of 0 means that the entity at a given location has full membership in one of the prescribed classes and 0 membership in all other classes. Because the membership values in all other classes are 0 ($H_{ij} = 0$), there is no membership ignorance involved in the hardening process and the ignorance uncertainty for this pixel is then 0. When the entropy value of a similarity vector reaches 1, it means the object is similar to all classes at the same degree, and none of the classes would be a good representative for this object. Labeling the object as any one of these classes would induce the highest degree of membership ignorance, and the ignorance uncertainty for this location would then be 1.

Zhu (1997a) defines the exaggeration uncertainty associated with this hardening to be:

$$E_{ij} = (1 - S_{ij}^g) \tag{15.2}$$

where E_{ij} is the exaggeration uncertainty measure and S_{ij}^g is the similarity measure between the entity at (i,j) and class g to which the entity is assigned. S_{ij}^g is a similarity measure expressing the entity's membership saturation in Class g and is not related to other similarity values in the vector. Equation (15.2) should be applied to a vector that is not normalized. Information about exaggeration is otherwise lost through the normalization process (Zhu 1997a).

The two uncertainty measures can be produced for every location being classified. Thus, maps (images) describing the spatial variation of these two uncertainty elements can be produced when a resource categorical map is generated from the similarity vectors over an area. The spatial variation of classification accuracy in the categorical map can then be approximated and depicted by these uncertainty images.

15.4 Case Study Results and Discussion

15.4.1 Soil Mapping Case Study

This study was conducted in the Lubrecht Experiment Forest located about 50 km northeast of Missoula, Montana, for the purpose of mapping 12 identified soil series (classes) (Zhu et al., 1996). The elevation in the area

FIGURE 15.1. Digital elevation model of the study area (light tone means high elevations). Figure from Zhu (1997a), reproduced by permission, the American Society for Photogrammetry and Remote Sensing.

ranges from about 1200 m to about 2000 m, with high elevation in the northeast and southwest and low elevation in the northwest (Fig. 15.1). The area is considered as a semi-humid to semi-arid region with strong moisture contrasts between the low elevation areas and the high elevation areas, and between north-facing and south-facing slopes (Nimlos 1986).

The similarity representation (or fuzzy representation) of soil spatial information over this study area was derived through a knowledge-based inference approach (Zhu et al. 1996). Figure 15.2 illustrates the inference methodology. Data on soil-formative environmental conditions were characterized using GIS techniques and stored in a GIS database. Knowledge on soil–environmental relationships was extracted using knowledge acquisition techniques (Zhu 1999) and stored in a knowledgebase. The fuzzy inference engine then combines the knowledgebase with the GIS database to produce a similarity representation of soil spatial variation.

A soil series map over the study area was created through hardening the preceding similarity representation of soils (Fig. 15.3). The two types of uncertainty (i.e., ignorance and exaggeration) associated with this hardening were estimated using Equations (15.1) and (15.2) for every location over the study area, which produces two uncertainty images for the area (Figs. 15.4 and 15.5).

Ignorance Uncertainty in the Soil Map. One observation is that entropy values are high in the mid-elevation areas (areas around B in Fig. 15.4) where soils are transitional to the soil series prescribed for the low elevations

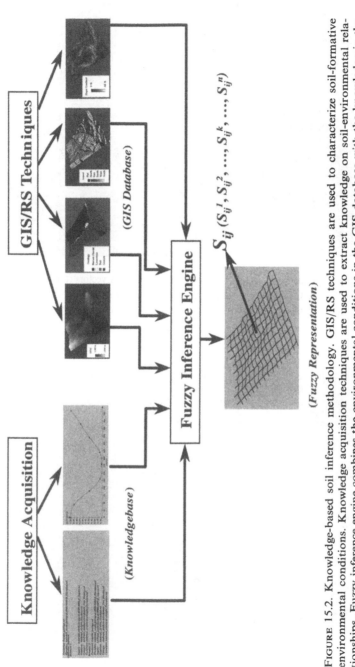

FIGURE 15.2. Knowledge-based soil inference methodology. GIS/RS techniques are used to characterize soil-formative environmental conditions. Knowledge acquisition techniques are used to extract knowledge on soil-environmental relationships. Fuzzy inference engine combines the environmental conditions in the GIS database with the knowledge in the knowledgebase to produce a fuzzy representation of soil spatial information.

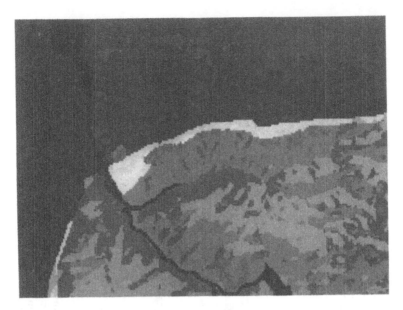

FIGURE 15.3. Soil series map from hardening the similarity vectors (the black areas along the Elk Creek were not included in this study). Figure from Zhu (1997a), reproduced by permission, the American Society for Photogrammetry and Remote Sensing.

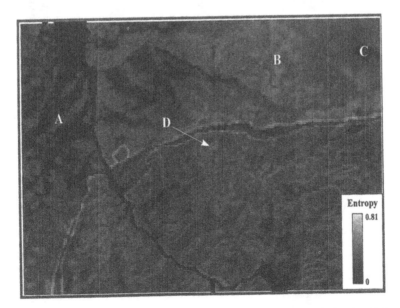

FIGURE 15.4. Entropy map for the Lubrecht area with light tone indicating high entropy values (Note: the black areas along the Elk Creek were not included in this study). Figure from Zhu (1997a), reproduced with permission, the American Society for Photogrammetry and Remote Sensing.

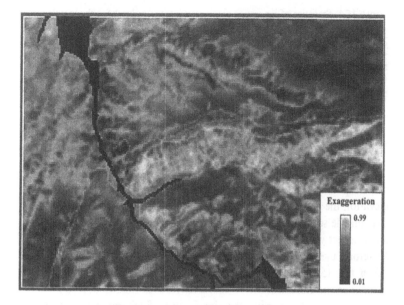

FIGURE 15.5. Spatial distribution of membership exaggeration in class assignment (light tone indicate high degrees of membership exaggeration). Figure from Zhu (1997a), reproduced with permission, the American Society for Photogrammetry and Remote Sensing.

(Area A) and those for the high elevations (Area C). Assigning these transitional soils to any prescribed soil series would imply that these soils can be treated as the same as the prototypes of the prescribed soil series in terms of managerial practices on these soils. This implication could result in misuses of these soil resources because managerial practices on these soils may have to be substantially different from those on the prescribed soil series. The high values in the entropy image over these areas would alert the users that the soils over these areas could be very different from the prescribed classes and that different managerial practices may be needed.

The second observation is that the entropy values at the boundaries of a soil body are often high. This is easy to understand because the soils at the boundary areas often bear high similarities to many different soil types. Classifying these transitional soils to any single soil class would introduce a high degree of uncertainty.

The third observation is that the entropy values on the south-facing slopes seem to be higher than on the north-facing slopes. This pattern is particularly apparent at low elevations and can be explained by the factor that soils on south-facing slopes are more intermittently distributed than are those on the north-facing slopes. Because the study area is a semi-arid to semi-humid region, moisture condition is the dominant factor during the soil-forming process. The moisture condition on these slopes depends on two factors:

TABLE 15.1. Statistics of ignorance measures (entropy values) for the matched and mismatched sites.

Statistics	Matched Sites	Mismatched Sites
Mean	0.54	0.64
Variance	0.047	0.049
Number of Sites	38	26

Source: From Zhu 1997a, reproduced with permission, the American Society for Photogrammetry and Remote Sensing.

slope aspect and elevation. On the north-facing slopes or on slopes of high elevation the moisture condition is often more spatially homogeneous than that on the south-facing slopes at low elevations, where subtle changes in slope aspect would result in a significant difference in moisture conditions; therefore, the soils on the south-facing slopes are less spatially contiguous (D in Fig. 15.4) and soils are more transitional to the prototypes of the prescribed soil classes.

Another way to examine the significance of ignorance uncertainty in revealing the accuracy of soil mapping is to test a hypothesis. If the ignorance uncertainty estimated by the entropy measure is a useful measure of classification accuracy, it can then be hypothesized that on average the locations where soils are misclassified would have a higher entropy value than would those where soils are correctly classified. To evaluate this hypothesis, soil series at 64 field sites were determined. These 64 sites were divided into two groups. The first group (i.e., the matched group) contains all of the sites where the inferred soil series match the observed ones, whereas the second group (i.e., the mismatched group) contains the rest. The entropy values for all of the 64 sites were obtained from the entropy map. The mean and standard deviation of entropy for each of the two groups were calculated and are reported in Table 15.1.

A Student-t test with the assumption that the population variances of the two groups of sites are unequal was used to evaluate the hypothesis. The calculated t value is 1.727 with a degree of freedom of 53. The critical t value greater than which the preceding hypothesis can be accepted with 95% of confidence at the degree of freedom of 40 is 1.684. Because the calculated t value is greater than the critical t, statistically speaking, on average, the entropy value for misclassified locations would be higher than that for correctly classified locations.

Exaggeration Uncertainty in the Soil Map. The image of exaggeration uncertainty (Fig. 15.5) has a spatial pattern very different from that of ignorance uncertainty. Exaggeration is very high for areas at low elevations, particularly for those areas with south-facing slopes. This situation can be created under two scenarios. First, the local soil is really different from any of the prescribed soil series; therefore, it bears low similarity values to all of the prescribed soil series. The second scenario is that the soil similarity vector is not an accurate representation of the local soil, and the low similarity values are the result of low confidence of soil experts in mapping the soils in these areas. In this study, both scenarios exist and both contribute to the high exaggeration values over the low elevation areas. Due to the semi-arid to semi-humid nature of this region, soils at low elevation, particularly those on the south-facing slopes, are more susceptible to moisture stress than are soils at high elevations. These soils are often not well developed and are of class intergrades. They are spatially highly variable (intermittently distributed), which makes it difficult for soil experts to relate these soils to specific environmental conditions. When using environmental indexes to derive soil similarity vectors (Zhu et al. 1996), soil experts would be very conservative in giving high similarity values to these soils. It can be concluded that exaggeration uncertainty reveals different kinds of classification accuracy.

15.4.2 Cover-Type Mapping Case Study

The second study area is the upper McDonald drainage basin, which is located in the west central area of Glacier National Park, Montana. It is about 20,000 hectare in size. The watershed is dominated by steep terrain with a strong relief. Elevation ranges from about 950 m to about 2900 m (Fig. 15.6). Most mountain slopes at low-to-middle elevations are covered with conifer-dominated forests. Forests give way to alpine tundra, and as elevation increases forests are replaced by bedrock outcrops and glaciers (Dutton and Zhu 1995). Direct human disturbance on the natural ecosystems in the upper McDonald drainage watershed has been minimal. The purpose for this case study is to evaluate the spatial variation of classification accuracy of remotely sensed imagery by examining the ignorance and exaggeration uncertainty images.

The similarity vectors, S_{ij}, were populated by membership values computed from a procedure extracted from a maximum likelihood classification method. S_{ij}^k is defined as:

$$S_{ij}^k = \exp\left[-\frac{1}{2}(X_{ij} - M_k)^T C_k^{-1}(X_{ij} - M_k)\right] \qquad (15.3)$$

where C_k and M_k define the characteristics of class k with C_k being the variance–covariance matrix and M_k being the mean vector of spectral reflectance for class k. X_{ij} is the reflectance values (pixel vector) for pixel (i,j). The term $(X_{ij} - M_k)^T C_k^{-1}(X_{ij} - M_k)$ is the Mahalanobis distance between pixel (i,j) and the centroid of class k in the spectral space, which is defined by the spectral bands. This distance can be thought of as a *squared distance function* that measures the distance between the observation and the class mean as scaled and corrected for variance and covariance of the class

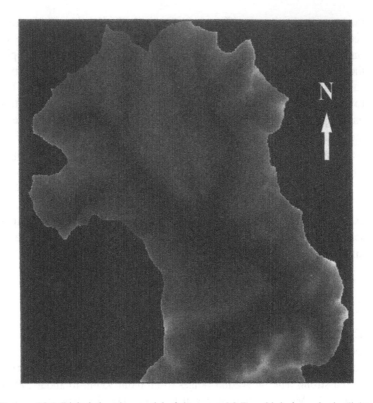

F<small>IGURE</small> 15.6. Digital elevation model of the upper McDonald drainage basin (light tone means high elevations).

(Strahler 1980). It is a measure of how typical the pixel is of class k (Foody et al. 1992).

The similarity is negatively related to this distance. Equation (15.3) has a range of 1–0, and its value depends on the Mahalanobis distance. With a distance of zero (the pixel is at the centroid of class k in the spectral space), the expression returns a value of 1, which means that the pixel is a typical case (member) of class k. The similarity is then unity. When the distance increases (the pixel is moving away from the class centroid), the similarity drops exponentially and eventually reaches 0. This property of the expression makes it a desirable function for computing the similarity value of a pixel to a given class and for populating the similarity vector.

A supervised training process (Lillesand and Kiefer 1994; Campbell 1996; Jensen 1996) was employed to compute C_k and M_k. Although there are six major cover types (i.e., streams, glaciers, alpine tundra, forest, meadows, and bare soils), the signatures of the latter three cover types were used in the training process and the membership values (similarity values) to each of these three major classes were then computed using Equation

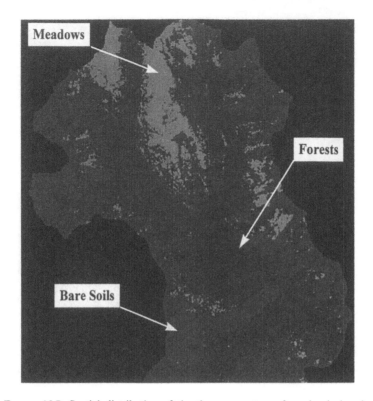

FIGURE 15.7. Spatial distribution of the three cover types from hardening the membership vectors.

(15.3). The other three cover types were considered to be undefined so that the uncertainty values over areas covered by these undefined classes can be examined. The classification of the three cover types through hardening the similarity vectors is shown in Figure 15.7. Figures 15.8 and 15.9 show the ignorance and exaggeration uncertainties in the cover type map.

Ignorance Uncertainty in the Cover-Type Map. The ignorance uncertainty in the centers of major cover types (A in Fig. 15.8) is relatively low. This is expected because each of these areas is covered by just one typical cover type, and the entities over this central region bear little to no resemblance to any other cover types. The membership in the similarity vector concentrates in one category; thus, the entropy is low, indicating the low uncertainty in classification.

The areas labeled as B or Bt in Figure 15.8 are the transitional regions and often have high ignorance uncertainty values. It is easy to understand that the entities over these transitional areas are of mixed grades such as alpine

344   A-Xing Zhu

FIGURE 15.8. Ignorance uncertainty shown as entropy values. Entropy values are low in centers of the cover types (A) and high in the transitional zones (B and Bt). The bright white areas (C) are glaciers.

tundra (Bt in Fig. 15.8), which is a mixture of dwarf trees, grasses, shrubs, and bare soils. This nature of mixed grades makes the signatures of pixels over these areas bear various degrees of similarity to many cover types, and the membership is more evenly distributed across several categories. This even distribution of membership results in high entropy values, which signify the low accuracy of classification over these areas.

The bright white areas (C in Fig. 15.8) are areas covered by glaciers whose spectral signature is very different from those of forests, bare soils, and meadows. Because glaciers were not defined as a class, the similarity vectors over these areas contain zero membership values in each of the three prescribed categories. As a result, the entropy values over these areas are of unity (100.0), which indicate the highest ignorance uncertainty in the classification of entities over these areas. If one compares Figures 15.7 and 15.8, one should notice that these areas are classified as bare soils. This is due to the way the algorithm labels locations with equal membership values in the prescribed classes. When equal membership values in all classes

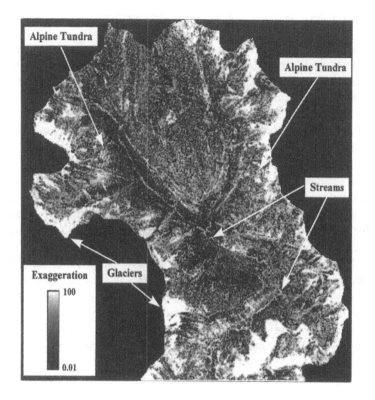

FIGURE 15.9. Exaggeration uncertainty. Exaggeration uncertainty is high around mountain peaks that are often covered by glaciers. It is also high along the streams in the middle of the image but moderate over the transitional areas.

are encountered, the location is labeled as the class that comes up first on the list, which is bare soils in this case.

Exaggeration Uncertainty in the Cover-Type Map. The exaggeration uncertainty portrayed in Figure 15.9 conforms to the basic accuracy pattern shown in Figure 15.8, but further identifies areas of high classification errors due to membership exaggeration. The core areas of the major cover types all have low exaggeration uncertainty values, and the fringes of these major cover types (or the transitional areas) have moderate exaggeration uncertainty. The areas covered with undefined classes (e.g., the glaciers and the streams, see Fig. 15.9) have very high exaggeration values. The cores of these areas have exaggeration of 100, which indicates the total misclassification of the entities at these locations. Although alpine tundra is also an undefined class, areas covered by this class still have moderate exaggeration values due to the fact that the class itself is a mixture of dwarf trees, shrubs, grasses, and bare soils, which makes the exaggeration values close to those in the other transitional areas.

15.4.3 The Quality of the Resource Maps

The spatial variation of classification accuracy of the preceding two categorical maps is clearly portrayed by the two sets of uncertainty maps. The soils at the high elevations in the Lubrecht study area were generally mapped with a relatively high accuracy, whereas the soils at low elevations, particularly those on the south-facing slopes, were mapped at low confidences. Error reduction efforts may be needed to improve the accuracy of soil mapping over these south-facing and low elevation areas. Soils in the middle elevations are transitional to soil series prescribed for the low elevations and those for the high elevations. This transitional nature of mid-elevation soils may require different managerial practice from those applied to either high or low elevation soil series. In addition, additional soil series may need to be designated for these transitional areas so that soils in these areas can be more accurately mapped (i.e., represented).

The uncertainty maps associated with the cover-type map for the upper McDonald drainage basin suggest that the cores of the three major cover types were mapped with a relatively high accuracy. These uncertainty maps also reveal the locations of the transitional areas of these major cover types. The uncertainty maps, more importantly, identify areas of two other major cover types (glaciers and streams) that are not yet included in the class list. These areas were classified at a very low accuracy.

15.5 Summary

This chapter presented an approach to modeling spatial variation of classification accuracy in categorical maps. Classification errors due to membership exaggeration and membership ignorance can be modeled by using a similarity representation of geographic information. The classification errors due to membership exaggeration can be approximated by exaggeration uncertainty, which can in turn be defined as the deficit from full membership. The classification errors due to membership ignorance can be estimated by the ignorance uncertainty, which can be defined as entropy of the similarity vector. Images of these two uncertainty measures can be created along with categorical maps. Spatial variation of classification accuracy can then be depicted by these uncertainty images.

The two case studies illustrated the effectiveness of this approach to modeling spatial variation of classification accuracy. Images of classification uncertainty can be used to identify areas of possible ecotones (e.g., high ignorance uncertainty with moderate exaggeration uncertainty) and to depict areas with undefined classes (i.e., areas of possible serious misclassification) (e.g., high ignorance and exaggeration uncertainty). Areas of both low ignorance and exaggeration uncertainty can be identified as the core areas of the prescribed classes and can be considered to have high classification accuracy.

The implication for ecologists is that spatial variation of classification accuracy is an inherent part of any categorical map. If categorical maps need to be produced, classification uncertainty maps should also be generated so that the spatial variation of classification accuracy in the categorical maps can be assessed. A fuzzy representation scheme can facilitate the computation of the two uncertainty measures and the generation of images of classification uncertainty. By examining the spatial variation of these uncertainty images, areas of different levels of classification accuracy can be identified. With the provision of this data quality information, the fitness of data for use in a given project can first be assessed. Error reduction efforts can then be effective deployed. For example, additional field data collection may be conducted in areas of high classification uncertainty and/or new classes may be designated for areas with undefined classes. The provision of classification uncertainty can also facilitate decision making with uncertainty (see Chap. 18 for details).

It must be pointed out that although the two measures have been shown to be useful in depicting the spatial variation of classification accuracy, the effectiveness of these measures depends on the merit of the similarity vectors, which are used to represent the status of a geographic entity at a given location in relation to the prescribed sets of classes. The means used to populate the similarity vectors, therefore, dictate the significance of these uncertainty measures.

Acknowledgments. This research was supported by a startup grant from the Graduate School at University of Wisconsin-Madison. Mr. Xun Shi's assistance on modifying the computer programs for computing the class memberships is greatly appreciated. The image data for the classification performed in this study were provided by Glacier National Park, Montana.

References

Band, L.E. 1995. Scale: landscape attributes and geographical information systems. Hydrological Processes 9:401–22.

Beaubien, J. 1994. Landsat TM images of forests: from enhancements to classification. Canadian Journal of Remote Sensing 20:17–26.

Bezdek, J.C., R. Ehrlich, and W. Full. 1984. FCM: the fuzzy c-means clustering algorithm. Computers and Geosciences 10:191–203.

Brown, N.J., R.D. Swetnam, J.R. Treweek, J.O. Mountford, R.W.G. Caldow, S.J. Manchester, et al. 1998. Issues in GIS development: adapting to research and policy-needs for management of wet grasslands in an environmentally sensitive area. International Journal of Geographical Information Science 12:465–78.

Campbell, J.B. 1996. Introduction to remote sensing, second ed. The Guilford Press, New York.

Carpenter, G.A., M.N. Gjaja, S. Gopal, and C.E. Woodcock. 1997. ART neural networks for remote sensing: vegetation classification from Landsat TM and terrain data. IEEE Transactions on Geoscience and Remote Sensing 35: 308–25.

Civco, D.L. 1993. Artificial neural networks for land-cover classification and mapping. International Journal of Geographical Information Systems 7:173–86.

Cohen, J. 1960. A coefficient of agreement for nominal scales. Educational and Psychological Measurement 20:37–46.

Congalton, R.G., R.A. Mead, and R.G. Oderwald. 1983. Assessing landsat classification accuracy using discrete multivariate analysis statistical techniques. Photogrammetric Engineering and Remote Sensing 49:1671–78.

Dutton, B., and A.X. Zhu. 1995. Soils of the McDonald Drainage Glacier National Park Montana. A research report submitted to Glacier National Park, National Park Service, West Glacier, MT.

Fisher, P., and S. Pathirana. 1990. The evaluation of fuzzy membership of land cover classes in the suburban zone. Remote Sensing of Environment 34:121–32.

Fisher, P. 1997. The pixel: a snare and a delusion. International Journal of Remote Sensing 18:679–85.

Foody, G.M. 1992. A fuzzy sets approach to the representation of vegetation continua for remotely sensed data: an example from lowland heath. Photogrammetric Engineering and Remote Sensing 58:221–25.

Foody, G.M., N.A. Campbell, N.M. Trodd, and T.F. Wood. 1992. Derivation and applications of probabilistic measures of class membership from the maximum likelihood classification. Photogrammetric Engineering and Remote Sensing 58: 1335–41.

Foody, G.M, and D.P. Cox. 1994. Sub-pixel land cover composition estimation using a linear mixture model and fuzzy membership functions. International Journal of Remote Sensing 15:619–31.

Foody, G.M. 1996. Relating the land-cover composition of mixed pixels to artificial neural network classification output. Photogrammetric Engineering and Remote Sensing 62:491–99.

Franklin, S.E., M.B. Lavigne, M.J. Deuling, M.A. Wulder, and E.R. Hunt, Jr. 1997. Landsat TM-derived forest covertype for use in ecosystem models of net primary production. Canadian Journal of Remote Sensing 23:91–99.

Gong, P., R. Pu, and J. Chen. 1996. Mapping ecological land systems and classification uncertainties from digital elevation and forest-cover data using neural networks. Photogrammetric Engineering and Remote Sensing 62:1249–60.

Goodchild, M.F. 1995. Attribute accuracy. Pages 59–79 in S.C. Guptill and J.L. Morrison, eds. Elements of spatial data quality. Pergamon, Oxford, UK.

Hudson, B.D. 1990. Concepts of soil mapping and interpretation. Soil Survey Horizons 31:63–73.

Jensen, J.R. 1996. Introductory digital image processing, second edition. Prentice Hall, NJ.

Leung, Y. 1987. On the imprecision of boundaries. Geographic Analysis 19:125–51.

Lillesand, T.M., and R.W. Kiefer. 1994. Remote sensing and image interpretation. John Wiley & Sons, New York.

Lowell, K. 1994. An uncertainty-based spatial representation for natural resources phenomena. Pages 933–44 in T.C. Waugh and R.G. Healey, eds. Advances in GIS research: proceedings of the sixth international symposium on spatial data handling, Taylor & Francis, London.

Lunetta, R.S., R.C. Congalton, L.K. Fenstermaker, J.R. Jensen, K.C. McGwire, and L.R. Tinney. 1991. Remote sensing and geographic information system data integration: error sources and research issues. Photogrammetric Engineering and Remote Sensing 57:677–87.

Mark, D.M., and F. Csillag. 1989. The nature of boundaries on "area-class" maps. Cartographica 26:65–78.

Maselli, F., A. Rodolfi, and C. Conese. 1996. Fuzzy classification of spatially degraded Thematic Mapper data for the estimation of sub-pixel components. International Journal of Remote Sensing 17:537–51.

McBratney, A.B., and J.J. De Gruijter. 1992. A continuum approach to soil classification by modified fuzzy k-means with extragrades. Journal of Soil Science 43:159–75.

Nimlos, J.T. 1986. Soils of Lubrecht experiment forest. Miscellaneous publication no. 44, Montana forest and conservation experiment station, Missoula, MT.

Ripple, W.J. 1994. The GIS applications book—examples in natural resources: a compendium. American Society for Photogrammetry and Remote Sensing, Bethesda, MD.

Robinson, V.B., and D. Thongs. 1985. Fuzzy set theory applied to the mixed pixel problem of multispectral land cover databases. Pages 871–86 in B. Opitz, ed. Geographic information systems in government, Deepak Publication: Hampton, VA.

Running, S.W., and E.R. Hunt, Jr. 1993. Generalization of a forest ecosystem process model for other biomes, BIOME-BGC, and an application for global-scale models. Pages 141–58 in J.R. Ehleringer and C.B. Field, eds. Scaling physiological processes: leaf to global. Academic Press, New York.

Stoms, D.M., F.W. Davis, and C.B. Cogan. 1992. Sensitivity of wildlife habitat models to uncertainties in GIS data. Photogrammetric Engineering and Remote Sensing 58:843–50.

Strahler, A.H. 1980. The use of prior probabilities in maximum likelihood classification of remotely sensed data. Remote Sensing of Environment 10:135–63.

Turner, M.G., and R.H. Gardner. 1990. Quantitative methods in landscape ecology. Springer-Verlag, New York.

Wang, F. 1990a. Improving remote sensing image analysis through fuzzy information representation. Photogrammetric Engineering and Remote Sensing 56:1163–69.

Wang, F. 1990b. Fuzzy supervised classification of remote sensing images. IEEE Transactions on Geoscience and Remote Sensing 28:194–201.

Warning, R.H., and S.W. Running. 1998. Forest ecosystems, second edition. Academic Press, New York.

Wood, T.F., and G.M. Foody. 1989. Analysis and representation of vegetation continua from Landsat Thematic Mapper data for lowland heaths. International Journal of Remote Sensing 10:181–91.

Zhu, A.X., L.E. Band, B. Dutton, and T.J. Nimlos. 1996. Automated soil inference under fuzzy logic. Ecological Modelling 90:123–45.

Zhu, A.X. 1997a. Measuring uncertainty in class assignment for natural resource maps under fuzzy logic. Photogrammetric Engineering and Remote Sensing 63:1195–202.

Zhu, A.X. 1997b. A similarity model for representing soil spatial information. Geoderma 77:217–42.

Zhu, A.X. 1999. A personal construct-based knowledge acquisition process for natural resource mapping. International Journal of Geographical Information Science 13:119–41.

Zhu, A.X. 2000. Mapping soil landscape as spatial continua: the neural network approach. Water Resources Research 36(3):663–77.

16
Alternative Set Theories for Uncertainty in Spatial Information

PETER FISHER

Spurred by the importance attached to the subject by the National Center for Geographical Information and Analysis through its Initiative 1 (Goodchild and Gopal 1989) much research on uncertainty in spatial information in general, and in natural resource information in particular has been completed. Edited volumes by Burrough and Frank (1996) and Goodchild and Jeansoulin (1998) bear witness to the continuing interest, as do a host of journal papers published in such journals as *Photogrammetric Engineering and Remote Sensing* and the *International Journal of Geographical Information Science* (formerly *Systems*). Much of that research has been motivated by applications grounded in empirical analysis, so it has been more concerned with the results than with the set theoretical basis of the work. At the same time, there have been significant advances and controversy in the understanding of the basis of the set theory itself from applied mathematicians and logicians (Pawlak 1994).

I will present a set theoretical view of uncertainty in geographic information and particularly in natural resources in this chapter. The discussion is based on applications of GIS to natural resources with illustrations from a variety of natural resource contexts. No single application is examined in detail, and much of the writing is speculative. Even experimental implementations of the set theories discussed are not necessarily available yet, so no comparative analysis is attempted but will be the subject of further work.

This exploration of existing set theories is not necessarily grounded in a belief that any one method is better than another; indeed, they are broadly complementary. There is, however, an obligation on the user of any set theory (and we all are) to evaluate it against others as part of establishing whether it is appropriate in a particular application. The chapter concludes by identifying the concept of higher-order uncertainty, which may be associated with all set theories.

16.1 Measurements and Sets

The primary observation in much spatial information is the measurement of a parameter (i.e., the color, elevation, pH, species frequency, etc.). It is inevitable within any database that some of these observations contain error (although it may be small) in the value recorded; thus, for all such measurements there is a probability that they are correctly measured. The error can be estimated in a number of ways (e.g., by estimation or by repeated measurements) (Taylor 1982). This enables the assignment of a confidence interval; a range of values within which we can be certain (at some probability level) that the real value lies. Measurement error is clearly defined. It involves the desire to measure a specific property, and its success can be estimated. Such error estimates can be mapped out over space, and the resulting errors can be propagated through multiple transformations (Heuvelink 1998). Much geostatistical analysis is based in the estimation and interpolation of single parameters over space. The errors associated with those estimates are recorded in the variance estimates, or in the variety of estimates inherent in a stochastic simulation of the property (Journel 1996; Chainey and Stuart 1998; Chap. 10 this volume).

Analysis of single properties is only one type of analysis, and it is not the approach most generally adopted in the analysis of spatial data within GIS. It is, however, without doubt the most precise form of analysis. People are more generally interested in mapping ideas across space, and ideas are best expressed as sets. In the context of spatial information, sets correspond to map categories. At the simplest the set can be defined by a range of values of a single parameter, yielding perhaps those areas with high, medium, and low values of that parameter (e.g., soil pH, canopy density, elevation, etc.). Sets may alternatively be mapped from a whole variety of properties as a result of a multivariate classification or taxonomy. It is possible to define a mapping unit more specifically as that contiguous area on the ground over which some set of properties are considered similar.

In many cases properties may work as indicators of conditions (plant and animal indicator species), or by threshold values (in Soil Taxonomy). Map categories can also be based on the opinion of experts, and many are a combination of some number of these. Soil maps, for example, show the extent of categories based on threshold values in multiple parameters to define taxonomic class. The categories are then projected over space to identify the mapping units that are an expression of the expert opinion of the soil surveyor. However the sets are defined, they can be addressed by a number of different set theories that conform to our different understandings and perceptions of the nature of the sets.

16.2 Boolean Sets

In much use of geographic information systems (GIS) the fundamental spatial unit of analysis has historically been based on information extracted from paper maps. These areas are conceptually and fundamentally Boolean. For any theme (e.g., vegetation) a number of classes are identified (map categories), the spatial extent of which are then indicated on the map (map units). It follows that for a particular class or unit (both sets) any location in the map space is either a member of the set or it is not. No other assignment is acceptable. This model of space is intriguing because it is based in the production of the map, not in the sciences contributing to the production of the map and using the information. It is the result of the need to simplify the data and clarify the pattern in the map. This is the principal paradigm that has been applied in mapping and has been referred to as the paradigm of cartographic production (Fisher 1996, 1998).

In essence any location has a membership value for the Boolean set that is either 1 or 0: $\mu_x(x) \in \{0, 1\}$, where 1 indicates that the object belongs to the set, and 0 that it does not. This coding is frequently implicit within GIS, where a binary theme such as a mask map coded as 0 or 1 is the only explicit example of a Boolean set within GIS.

It is intriguing that this paradigm has pervaded the production of spatial information even when that information is derived directly from measurements such as in the mapping of land cover and land use information classes from raw satellite imagery (Campbell 1987). Doubt is of course possible in the assignment of any object or location to a class. That doubt can be expressed in a number of ways, but in the mainstream interpretation of satellite imagery it is exploited in the maximum likelihood classifier, which assigns any location to the class it is most likely to belong to, minimizing the possibility of it being misclassified.

As noted earlier, concepts of probability are relatively well understood, and have certainly dominated the discussion of uncertainty in ecological analysis. With classes being assigned to Boolean sets, the Boolean logic can be applied with the traditional operators based around union, intersection, and negation.

Implicit in the use of the Boolean model of space is the assumption that all boundaries are crisp and abrupt. The transition between one class and another in some measurement (uni- or multivariate or subjective) specifically is abrupt; furthermore, the boundaries in space (and time, if appropriate) are also abrupt. Furthermore, in the literal implementation of the model it is implicit that any one location can belong to one and only one set.

In the context of spatial information, the use of Boolean sets can be conceptually problematic, and other set theories may be more appropriate.

16.3 Rough Sets

Given that spatial information can be represented by sharp boundaries (as is implicit in the Boolean model of space), the process of generating geographic information (digitizing) has always been problematic, especially the generation of raster (grid-cell) information from vector (point and line). The vector–raster transformation has been reviewed by a number of authors, but the essential problem is the decision as to which class a particular cell should be assigned. The problem is related to the granularity of the information, and this is fundamental to rough set theory, articulated by Pawlak (1982, 1986). The granularity of a measurement essentially allows three different classes of information to be identified: cases within the set, cases outside, and those on the boundary. These sets are directly identifiable in the vector-raster transformation (Fig. 16.1). Granularity can occur in attribute space

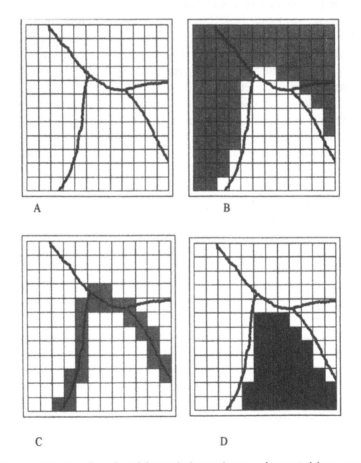

A   B

C   D

FIGURE 16.1. The effect of spatial granularity can be seen when rasterizing a vector coverage (A). For any polygon, three different areas can be distinguished: the area outside a polygon (B), the boundary area (C), and the area within the polygon (D).

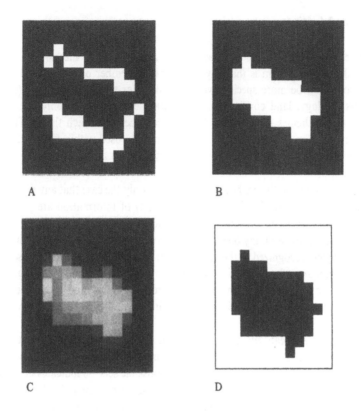

FIGURE 16.2. The combined effect of spatial and attribute granularity shows in the extraction of polygon areas from a gray-scale image (e.g., single of multiband band satellite image) (A). Again, three distinct areas can be distinguished: that outside a polygon (B), the boundary area (C), and the area within the polygon (D).

as well as in locational space, however, and it is thus possible to identify the three sets in the extraction of land cover information from satellite imagery (Fig. 16.2). This is indeed done, if not very explicitly in the identification of mixed pixels as a class in parallelepiped and maximum likelihood classification of satellite imagery (Campbell 1987). Some have found rough sets hard to distinguish from fuzzy sets, but the distinction is clear and grounded in granularity and discernibility (Pawlak 1985; Dubois and Prade 1990).

As an alternative to the $\{0, 1\}$ Boolean model of space, rough set theory provides a three-valued model. The sets defined are arguably no more than three separate Boolean sets, but the power of rough sets comes in the discernibility of classes, through the exploration of alternative granularities. Whether there are advantages or not in the use of rough sets in the analysis of uncertainty in spatial information is to be explored in the future (Worboys 1998).

16.4 Multisets

A second assumption of Boolean sets as implemented in the cartographic production paradigm is that any location is a member of one set and one set only. To be more specific, within a particular theme of mapping (e.g., soils, geology, land cover, etc.) one location can belong to only one set, but across themes naturally the location has one set in each theme. Thus, the multiset model is actually the one commonly used within GIS, but the separation into themes (e.g., coverages, layers, etc.) allows users to ignore this.

Within any one theme, however, it is commonly the case that any detailed examination shows that many classes (or sets) of information are present and those classes are irregularly discontinuous across space. Even the number of sets present at any one location is not predictable (Fig. 16.3). This is most widely recognized in the habitats of tropical rain forests, but the same is true of many other vegetation environments, including woodlands or parklands. The forest or woodland is sufficient recognition of land cover type for some work, but a finer distinction of environmental types is necessary for others.

Whereas Boolean sets are distinguished for any location by the integer values 0 or 1, multisets are coded by any number of integers between zero and an upper limit which is the number of possible sets—0 indicates none of these is present; $\mu_x(x) \in \{0, 1, 2, \ldots\}$. New models of spatial information such as ArcInfo's Spatial Database Engine and the associated shapes database mean that multisets are becoming convenient to implement within GIS, although that is not the reason for implementing that data structure. Multiple coverages and exhaustive overlay of the multisets are both approaches that can be used to store spatial multisets, but it is not possible to extract them directly or to manipulate them.

16.5 Fuzzy Sets

The Boolean model assumes that it is possible to assign an object or location to a specific Boolean set. This is not necessarily the case. There are many instances where the conceptualization of the set is not clear cut. Thus, many landscape features (e.g., hilltops, mountains, valleys, etc.) are not well defined (Fisher and Wood 1998). Vegetation associations similarly are not clear cut, and habitat types are equally poorly defined (Moraczewski 1993a,b), and Zhu (Chap. 15), among others, has articulated the case for soil information being poorly defined. In a situation where it is desirable to assign an object or phenomenon to a class, but the class is not well defined, then it is better described by fuzzy sets. These define a membership (in the real number range from 0 to 1) that describes the degree to which a phenomenon may match the characteristics of a prototype concept $\mu_x(x) \in [0, 1]$.

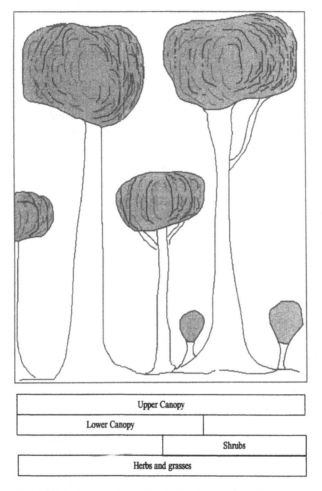

Upper Canopy

Lower Canopy

Shrubs

Herbs and grasses

FIGURE 16.3. Within the woodland class subunits are discontinuous over space, giving rise to multisets of woodland habitats, including the upper canopy, the lower canopy, the shrub layer, and the herbs and grasses.

It is interesting that just as rough sets were initially confused with fuzzy sets (Pawlak 1985), so were fuzzy sets not clearly distinguished from probability a few years earlier (Zadeh 1980). There is, however, no *conceptual* question of fuzzy set membership being a probability. One estimates the probability of a measurement or a Boolean outcome being correct, whereas the other measures the degree to which a phenomenon matches the proto-typical version of that phenomenon. Reference to the classic Sorites Paradox from Ancient Greek philosophy clarifies the question of the conceptualiza-tion of a fuzzy set in terms of vague concepts (Williamson 1994; Fisher 1999, 2000). There are many examples of fuzzy set theory applied to geographic information (Fisher 1999, 2000).

16.6 Mixed Set Models

A number of possible shortcomings of the Boolean model of space are identified and some alternatives to that set theory have been outlined (i.e., rough sets, multisets, and fuzzy sets). There is no implication that this is an exhaustive list of set theories that may be applicable to spatial information; however, it is possible to mix them in any combination. Thus, there are fuzzy multisets (Pawlak 1994) where the real number interval is used to encode the degree to which a location or object belongs to each of a set of candidate sets. The membership in each is stored as a list of real numbers: $\mu_x(x) \in [0, +\infty]$. Dubois and Prade (1990) have similarly discussed the distinction between fuzzy rough sets and rough fuzzy sets. This distinction is related to sources of granularity and vagueness in different contexts.

16.7 Higher-Order Levels of Uncertainty

Just as the identification of a Boolean set is susceptible to error (expressed as the probability of the identification being correct) so too with a fuzzy set. The membership is estimated at a location but that estimate may be susceptible to error; therefore, a probability of the fuzzy set membership being correct can also be estimated. This is a second level of uncertainty, and still higher levels of uncertainty can be recognized. Indeed, in vagueness theory, it is fundamental that if there is uncertainty over a concept, then there is also a vagueness about the vagueness—a fuzzy membership of the fuzzy concept for which a fuzzy membership is estimated (Williamson 1994). Higher levels of uncertainty have largely been ignored in processing of spatial information, except in the case of the probability of the correct identification of a Boolean set. It is assumed that enough uncertainty is modeled in the first order, again showing its confusion with probability. Work by Lees (1996) and Hughes et al. (1999) is looking at higher levels of uncertainty in satellite imagery as it relates to fuzzy sets.

16.8 An Example

Vegetation mapping can be used as an example of the application of some of these alternative set theories. In a region it may be possible to identify two broad vegetation classes (e.g., grassland and woodland). For one purpose (e.g., regional mapping) these can be mapped as Boolean map categories where every location in the region can be assigned to one or the other of these vegetation types. There is a finite probability at any individual location that the location is incorrectly identified as belonging to that class.

When the data are converted to a raster format, there is a rough boundary between the two classes such that we cannot be certain of the correct identification of the area in the vicinity of the boundary. The extent of this boundary set is dependent on the pixel size of the raster data set. As the pixel size increases the boundary set occupies an increasingly large part of the region. If we are using the frequency of trees per hectare to identify the difference between woodland and grassland, there may also be a boundary set in that attribute, and that boundary set may occupy a large area if there is an extensive ecotone between grassland and woodland. If it is an agricultural landscape, then effectively there may be no such ecotone. However the boundary set is estimated, it is the same in the woodland to grassland transition as it is in the reverse.

Upon closer examination (perhaps for another application) it is observed that the grassland and the woodland are actually mixed, and particularly in the region of the ecotone, where one vegetation types blends into the other, and that the three-class model of the rough set (i.e., within the class, in the boundary, and outside the class) is insufficient to model the landscape. This context is better modeled as a fuzzy set. Membership of 1 is recorded for the category grassland where the plants present match the central concept of grassland, and 0 is recorded for woodland, and vice versa when the species match the concept of woodland. The plant association elsewhere belongs to some extent to the woodland and to some extent to the grassland; therefore, a number is recorded in the range $[0, 1]$ for each vegetation type. The constraint is frequently placed that the sum of the memberships in a single location should sum to unity, but this is not a necessary constraint. At any location there is a probability of the fuzzy set membership estimated being correct. For example, the correctness could be estimated in the normal way by repeat measurements.

If this fuzzy space is modeled in a raster GIS environment, then there is again a spatial granularity to the model. Thus, we could also distinguish a rough fuzzy set associated with the boundary and the core area.

Further examination shows that there are interesting habitats in the region (both the grassland and the woodland). Within the woodland we can distinguish a high canopy where certain birds nest. Other birds prefer to nest in lower trees and saplings. Some species of mammal roam on both the ground and the trees within the woodland, whereas still others roam across both the grassland and the woodland, and a third group of mammals and some birds favor the open grassland. We now have a number of different categories of habitat (five in all), some of which co-exist at the same geographic location and others of which are distinct. This is conveniently modeled as multisets. These can be approximated by using nonexhaustive Boolean coverages of a region such that some parts of the region are not covered by anything. If the spatial resolution of the data set is changed (rasterization), then, again, rough sets occur at the boundaries of the sets. Notice that the boundaries of the different categories of rough sets are not co-existing in this case, unlike in the earlier Boolean case.

The degree to which particular species stay within the habitats described is not clear cut; however, there is a degree to which animals favor one of the habitats, but it is not exclusive. This is therefore better modeled as a group of fuzzy multisets, each having core central concept areas, as well as gradational boundaries.

In looking at any landscape, this range of possible set models is very real. The most suitable model is dependent on many factors, including data, application, scale, etc.

16.9 Summary

The analysis of uncertainty is frequently a pragmatic compromise. None of the alternatives to Boolean sets identified in this chapter are widely researched, nor are they implemented within standard software packages or geographic information systems. Some are used implicitly without any awareness either that the approach is in use or that there is mathematics of the set theory available. A failure to understand the set theoretic basis of the uncertainty being examined is perhaps the most fundamental of uncertainties. The confusion of probability and fuzzy sets has been identified widely in the literature, but here the replacement of multisets with minimal intersections of Boolean sets, as well as the implicit use of rough sets in several areas, is identified. The application science has frequently been ahead of the formal mathematics, but the onus remains on the application scientist to understand the theoretical context of their analysis better.

Much research is being expended on fuzzy set theory and its application to land cover and vegetation mapping from satellite imagery (Chap. 15), but it should be realized that this is only one of the available alternatives.

Finally, the understanding and modeling of uncertainty ultimately depends on only one thing: the conceptualization of the information by the investigator, and the defensibility of that conceptualization. If a Boolean concept of the information can be defended, then the appropriate way to assess uncertainty is by probabilities. Alternative set theories may be necessary when vagueness or granularity is inherent in the process, and important in the context of use. It is likely that in any context more than one conceptualization of the problem is possible and that different models of uncertainty may consequently be defensible. In this situation the consequences of different models may be explored. Recognition of one model of uncertainty is very likely to lead to higher levels of uncertainty, but it is not yet evident that this will necessarily yield useful outcomes.

Although a variety of set theories are available, positive advantages of each have not been established. It is clear, however, that fuzzy sets that have been suggested as the principal alternative to Boolean sets are not the only contender and that further research is required to establish the suitability of these different set models for spatial information.

Acknowledgments. Many people have contributed to my continuing education in the ideas discussed here, but I would particularly like to thank my collaborators on the EU-funded FLIERS (ENV4-CT96-0305) and REVIGIS projects, including Lucy Bastin, Martin Brown, Marianne Edwards, Andrew Frank, Mike Hughes, Ioannis Kanellupoulos, Robet Jeansoulin, Martien Molenaar, Nicos Silleos, John Stell, Graeme Wilkinson, and Michael Worboys. I must also thank Mike Goodchild and Carolyn Hunsaker for inviting me to the NCEAS workshop, and the editors for comments that have undoubtedly improved the manuscript.

References

Burrough, P.A., and A. Frank, eds. 1996. Geographic objects with indeterminate boundaries. Taylor & Francis, London.

Campbell, J.B. 1987. Introduction to remote sensing. Guilford, New York.

Chainey, S., and N. Stuart. 1998. Stochastic simulation: an alternative interpolation technique for digital geographic information. Pages 3–24 in S. Carver, ed. Innovations in GIS 5. Taylor & Francis, London.

Dubois, D., and H. Parde. 1990. Rough fuzzy sets and fuzzy rough sets. International Journal of General Systems 17:191–209.

Fisher, P.F. 1996. Concepts and paradigms of spatial data. Pages 297–307 in M. Craglia and H. Couclelis, eds. Geographic information research: bridging the Atlantic. Taylor & Francis, London.

Fisher, P.F. 1998. Is GIS hidebound by the legacy of cartography? Cartographic Journal 35:5–9.

Fisher, P.F., and J. Wood. 1998. What is a mountain? or The Englishman who went up a Boolean geographical concept but realised it was fuzzy. Geography 83:247–56.

Fisher, P.F. 1999. Models of uncertainty in spatial data, volume 1. Pages 191–205 in P. Longley, M. Goodchild, D. Maguire, and D. Rhind, eds. Geographical information systems: principles, techniques, management and applications. John Wiley & Sons, New York.

Fisher, P.F. 2000. Sorites paradox and vague geographies. International Journal of Fuzzy Sets and Systems 113:7–18.

Goodchild, M., and S. Gopal, eds. 1989. Accuracy of spatial databases. Taylor & Francis, London.

Goodchild, M., and R. Jeansoulin, eds. 1998. Data quality in geographic information: from error to uncertainty. Hermes, Paris.

Heuvelink, G. 1998. Error propagation in environmental modelling with GIS. Taylor & Francis, London.

Hughes, M., J. Bygrave, L. Bastin, and P.F. Fisher. 1999. High order uncertainty in spatial information: estimating the proportion of cover types within a pixel. Pages 319–23 in K. Lowell and A. Jaton eds. Spatial accuracy assessment: land information uncertainty in natural resources. Ann Arbor Press, Chelsea, MI.

Journel, A. 1996. Modelling uncertainty and spatial dependence: stochastic simulation. International Journal of Geographical Information Systems 10:517–22.

Lees, B. 1996. Improving the spatial extension of point data by changing the data model. In Proceedings of the third international conference/workshop on integrating GIS and environmental modeling. NCGIA Santa Barbara, CD-ROM.

Moraczewski, I.R. 1993a. Fuzzy logic for phytosociology. 1. Syntaxa as vague concepts. Vegetatio 106:1–11.

Moraczewski, I.R. 1993b. Fuzzy logic for phytosociology. 2. Generalization and prediction. Vegetatio 106:13–20.

Pawlak, Z. 1982. Rough sets. International Journal of Computer and Information Sciences 11:341–56.

Pawlak, Z. 1985. Rough sets and fuzzy sets. Fuzzy Sets and Systems 17:99–102.

Pawlak, Z. 1986. Rough sets: theoretical aspects of reasoning about data. Kluwer, Dordrecht, The Netherlands.

Pawlak, Z. 1994. Hard and soft sets. Pages 130–35 in W.P. Ziarko, ed. Rough sets, fuzzy sets and knowledge discovery. Springer-Verlag, Berlin.

Taylor, J.R. 1982. An introduction to error analysis. Oxford University Press, New York.

Williamson, T. 1994. Vagueness. Routledge, London.

Worboys, M. 1998. Computation with imprecise geospatial data. Computers, Environment and Urban Systems 22:85–106.

Zadeh, L.A. 1965. Fuzzy sets. Information and Control 8:338–53.

Zadeh, L.A. 1980. Fuzzy sets versus probability. Proceedings of the IEEE. 68:421.

17
Roles of Meta-Information in Uncertainty Management

KATE BEARD

Meta-information is information that describes information. Bretherton (1994) defined it as information that makes data useful. The importance of meta-information becomes most apparent in situations where the data user is not the data collector and is not likely to be familiar with the characteristics and idiosyncrasies of the data. As more geospatial information becomes available on-line for use by ecologists, meta-information becomes essential to understand and minimize the uncertainty in using the data. Many ecologists collect their own data, but their analysis may require integration with data collected by other scientists. In a context in which it is becoming ever-more common to share and integrate heterogeneous digital data sets, there are several questions for ecologists:

- How effectively can ecologists use other scientists' data?
- How effectively can ecologists integrate their data with other scientists' data?
- How can ecologists better understand the uncertainty in their own data analysis?
- How can they communicate the uncertainty in their data and analysis to others?
- How can they understand the uncertainty in another scientist's analysis?

Meta-information can address each of these questions and is critical to the overall management of uncertainty. Uncertainty is not a characteristic of data or information, but rather of a state of knowledge regarding the data. By providing information about data or information to the ecologist, meta-information has the potential to reduce some amount of uncertainty directly. Key sources of uncertainty include imprecision, incompleteness, vagueness, and ambiguity in language, concepts, observations, interpretations, and representations. This chapter reviews sources of uncertainty and identifies the roles meta-information can play in mitigating uncertainty. This chapter reviews current concepts of meta-information and discusses how these can be extended to improve uncertainty management. *Uncertainty management* refers to improving comprehension and understanding of imperfections in data or information.

17.1 Sources of Uncertainty

Uncertainty is inherent in ecological analysis and synthesis. Sources of uncertainty include nondeterministic environmental processes; conceptual vagueness and ambiguity in taxonomies and ecological terms (e.g., old growth forest); sampling and data collection problems that lead to incomplete, incorrect, or imprecise observations; use of surrogate variables; constraints on digital representations (e.g., resolution of a raster data set); and data processing steps (e.g., interpolation, image classification). This section highlights sources of uncertainty by considering an example.

Suppose an ecologist wishes to investigate the spatial patterns of avian species richness. Due to a number of factors, qualification or quantification of avian species richness is subject to uncertainty. First, the concept is subject to ambiguity and error. Many studies have used breeding bird surveys for such measurements (Jones 1996). The quality of breeding bird surveys can vary substantially. The accuracy of the species identification varies with the experience of the observer, the survey routes vary from year to year, and observations occur only along roads, which biases the counts against rare birds who prefer remote habitats, to name just a few problems (Robbins et al. 1986). The raw bird counts are subject to these types of imperfections, which may then be exacerbated by the process of converting the count data to an operational spatial variable of species richness and a mapped representation (e.g., a map generated by some form of interpolation process such as kriging).

Suppose the map of avian species richness is made available for use by others. The question is what should be documented so that the generated map of species richness can be used effectively by other scientists or policy makers. Most users would wish to know, for example, how many breeding bird survey routes were used, covering what time period, the spatial distribution of the routes, if routes were run for all years, how avian species richness was calculated, what interpolation method was used, and what the resulting resolution was. Meta-information cannot reduce the uncertainty in the individual surveys, but if each survey is well documented and derivative products are well documented, substantial uncertainty can be reduced or avoided in subsequent analyses. In many cases, sources of uncertainty can be reduced by comprehensive meta-information and easy access to it.

17.2 Description of Meta-Information

The term *meta-information* originated in the database design community and generally referred to data dictionaries or definitions and explanations of the database schema (Mark and Roussopoulos 1986; Lilywhite 1991; Newman 1991). The definition of the term has broadened and meta-information has received more attention as a result of several standards efforts. In the United States the geospatial standards effort resulted in the Content Standards for Digital Geospatial Metadata (FGDC 1994; FGDC 1997). The objectives of this standard are to provide a common set of terms and definitions for the documentation of digital spatial data sets. The metadata content elements included in the standard were determined on the basis that metadata serve the following four roles:

- Availability: Information to determine if information exists for a geographic location.
- Fitness for use: Information to determine if data meet a specific need.
- Access: Information needed to acquire a set of data.
- Transfer: Information needed to process and use a set of data.

The geospatial content standard has almost 200 elements that cover identification and citation information, spatial referencing, formats, data quality, status, and distribution information. In the United States, all federal agencies are expected to document their data sets with metadata that is compliant with the standard. The Global Change Master Directory's DIF (DIF 1999) format and the variety of standards used for documenting data within the NASA Distributed Active Archive Centers (DAACs) are examples of other metadata efforts. Several international geospatial standards (ISO/TC211 1998, CEN 1995) have also been developed. Other more general metadata standards efforts related to information access and distribution include USMARC (US machine-readable catalog) and the Dublin Core. USMARC (1994) provides very detailed bibliographic descriptions for all types of information from maps to photographs to music. The Dublin Core was designed with web searches in mind and identifies a minimum subset of metadata that serves primarily the role of determining the availability of information resources (Weibel et al. 1995).

A number of metadata standards efforts related to the ecological sciences have appeared. These include the National Biological Service's (now BRD) "Content Standard for National Biological Information Infrastructure Metadata" (NBII) (NBS 1995), a standard proposed by Michener et al. (1997); metadata for GAP analysis (Cogan et al. 1994); and a variety of Long-Term Ecological Research (LTER) program metadata formats in use throughout the LTER network.

The National Biological Service was the first to develop a profile as an extension to the FGDC geospatial content standard to cover elements of particular interest to biologists and ecologists. One extension to the standard requires documentation of methodology. This requires a scientist to describe the physical methods used to gather the data, the experimental design, sample frequency, treatments or strata, statistical and spatial sampling design, and sample completeness, representativeness, and biases. For example, in a bird survey, relevant elements would include the methods used to detect species occurrences (casual sightings, transects, focal point surveys, vocalizations, mist nets); whether or not evidence of breeding activity was required; descriptions of the habitat strata in a stratified design; and known biases (e.g., nonterritorial birds being undersampled, and some juveniles not identified to species). The profile also calls for documentation of taxonomic sources, procedures, and treatments; and information describing types of specimens collected (e.g., herbarium specimens, blood samples, photographs, individuals, or batches).

In terms of uncertainty management, the data quality section of the digital geospatial metadata standard is generally the most pertinent. The data quality section covers positional accuracy, attribute accuracy, completeness, consistency and lineage information. For example the Horizontal Positional Accuracy Value element of the standard asks for an estimate of the accuracy of the horizontal coordinate measurements in the data set as expressed in (ground) meters. The NBII metadata as a profile extension to the standard includes all the geospatial data quality sections plus additional elements.

Much of the current geospatial metadata is being generated for pre-existing geospatial data sets and typically for interpreted and second- or later-generation representations (e.g., a soil map) rather than original observations (e.g., field soil samples). For most pre-existing maps the complete processing record or lineage is not known and cannot be documented. For example, topographic maps usually have no meta-information describing stereo-compilation instructions given to the stereo operators, whether they were correctly and consistently followed, whether the capture of features was complete, or whether features were generalized, and, if so, to what degree. Most of this sort of detailed processing history has not been compiled for distribution to users. For many spatial data sets, the metadata is generated long after data collection; thus, it is necessarily of limited and of minimal value for uncertainty management. Comprehensive meta-information that is generated as data are collected and updated as data are processed (Beard 1996) will be better able to support uncertainty management. The next sections describe the roles meta-information can play in supporting uncertainty management.

17.3 Use of Meta-Information for Uncertainty Management

Meta-information is most typically provided as an accompanying text document only loosely associated with the data it describes. To be more useful, meta-information should be accessible in close association with the information it describes. Achieving the full benefits of meta-information requires an integrated database approach in which the meta-information is accessible in conjunction with the data. In an integrated database environment there are two possible approaches to utilizing meta-information for uncertainty management. In the first approach, meta-information does not directly document uncertainty; rather, it serves as supporting information for uncertainty analysis. In the second approach, measures of imperfections or reliability are computed and stored directly as meta-information. In the second case the meta-information can be a specific uncertainty measure [a root mean square error (RMSE) measure or a misclassification matrix] or, alternatively, an uncertainty measuring process (Goodchild et al. 1998). In both cases, the database allows for query, analysis, and display of meta-information in conjunction with the data. The first approach includes three distinguishable subcategories:

- Contextual information as meta-information.
- Lineage information as meta-information.
- "Ground truth" as meta-information.

17.3.1 Approach 1: Meta-Information as Supporting Information for Uncertainty Analysis

Contextual Information as Meta-Information. Meta-information in the form of contextual information does not indicate the reliability of the information directly; rather, it provides a basis to interpret reliability. For example, consider the meaning of water quality samples taken as part of a shellfish monitoring program. The samples are tested and measured for fecal coliform. As part of the sampling procedure, the current weather conditions (cloudy, overcast, rainy, sunny) are observed, and water temperature and salinity are concurrently measured. These concurrently measured variables can be considered as either data or metadata. In the context of shellfish monitoring they may be considered meta-information if they provide a context in which to evaluate the measured fecal coliform values. The weather, temperature, and salinity values do not indicate problems with a fecal coliform value independently, but in combination they might. There is less certainty in a measured fecal coliform value absent the contextual information on weather, temperature, and salinity. Certain conditions must apply for information to be considered contextual information. Contextual information includes related variables measured at the same or approximately the same time and location. Contextual information for the water quality sample could include the amount of precipitation in the watershed for the week prior to sampling.

To make contextual information effective as meta-information for uncertainty analysis it must be easy for users to analyze it in relation to the data. Fast search and retrieval should allow users to find contextual information quickly, and a set of exploratory tools should allow users to visualize and explore patterns in the data and meta-data. A combination of data, contextual information, and exploratory tools, for example, could allow users to investigate whether an unusual data value is in error or possibly unusual for good reason. Using the shellfish example, heavy rains in a watershed prior to sampling could be an explanation for a high fecal coliform value. For example, graphs of contextual meta-information (e.g., scatter plots linked to maps, correlograms, or other statistics of correlation and spatial dependence) (Cressie 1991; Anselin 1998) could help ferret out potential errors in the data. Contextual meta-information is currently not part of a meta-data standard, but it could be very useful for ecological as well as other scientific analysis.

Lineage as Meta-Information. Investigation of the reliability of spatial data and analysis requires more information than is generally available with most current digital geospatial data sets. To assess imperfections and their potential impact on analysis it is necessary to know how data were collected, the assumptions, concepts, and objectives behind the observations, and the processing history behind any digital representation. Suppose a representation of digital bathymetry has been acquired and a user wishes to perform a rigorous assessment of its fitness for use. The user would ideally need to know the path the ship made in collecting the soundings, the instrument used to take the soundings, the navigational equipment used to determine location (e.g., LORAN C, GPS), how soundings were processed to generate bathymetry, and if the bathymetry was subsequently generalized. A complete processing record constituting an information set's lineage forms an essential foundation for reliability assessment.

Current lineage information that is compliant with the geospatial meta-data content standard may describe a generic processing sequence, but not one that applies to a specific data set. For example, the meta-information for all USGS digital orthophoto quadrangles has identical data quality sections that describe a general data collection and processing sequence. Table 17.1 is an excerpt from the data quality section that describes all digital

TABLE 17.1. Excerpt from data quality report for digital orthophoto quads.

The DOQ horizontal positional accuracy and the assurance of that accuracy depend, in part, on the accuracy of the data inputs to the rectification process. These inputs consist of the digital elevation model (DEM), aero-triangulation control and methods, the photo source camera calibration, scanner calibration, and aerial photographs that meet National Aerial Photography Program (NAPP) standards. The vertical accuracy of the verified USGS format DEM is equivalent to or better than a USGS Level 1 or 2 DEM, with a root mean square error (RMSE) of no greater than 7.0 m. Field control is acquired by third-order Class 1 or better survey methods sufficiently spaced to meet National Map Accuracy Standards (NMAS) for 1:12,000-scale products. Aerial cameras have current certification from the USGS, National Mapping Division, Optical Science Laboratory. Test calibration scans are performed on all source photography scanners.

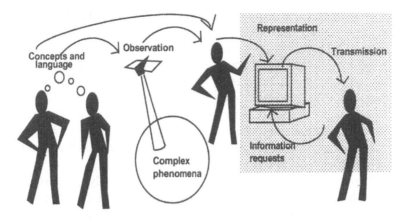

FIGURE 17.1. Schematic of an information system.

orthophoto quads. The described process may not necessarily apply to each individual quadrangle. This approach works for very standardized data collection operations, but there is often sufficient variation in collection and processing histories that a single generic processing description is not adequate. Lineage records can be generalized for standard products, but they should indicate important deviations in individual data sets where they occur.

Geospatial meta-information is currently predominantly associated with the lineage of digital representations (Lanter 1991). A comprehensive meta-information environment should include concepts and measurements in a lineage trail in addition to describing the processes applied to digital representations. Figure 17.1 shows a schematic of an information system that indicates concepts and observations as fundamental parts of information development that should be included in the lineage trail. The gray box in the figure represents a computational system. Currently lineage information is typically only documented for representations stored within the computational system. A more complete lineage record should begin with documentation of concepts such as species richness or biodiversity or a specific taxonomy, and these should then be incorporated within the computational system. Within a computational system, the concepts could be directly associated with observations and representations that arise from them. Concepts may have their own meta-information such as their definition, who defined them, and their objective. A household is an example of a concept that might be defined differently for census taking versus market analysis. A wetland is another concept that is usually defined differently by soil scientists, ecologists, and others. Such concepts are important in sharing digital spatial representations, but the particular meaning of a concept behind a representation is typically lost or disassociated from a representation once it has left the original data collector.

Following the sequence of Figure 17.1, prior to a data collection effort, a concept is usually operationalized and a measurement process specified. For example behind a field soil survey lies the concept of soil taxonomies and soil classes. The operational definition of a soil class is determined by several field characteristics including color, texture, and position in the landscape. Soil scientists make transects in the field, observe the pertinent soil characteristics, and assign and delineate soil classes accordingly. In soil mapping, the field observations are typically discarded and the lineage record for a digital soil map only begins with a description of the conversion of a hard copy soil map into digital form.

For ecological analysis as well as most other scientific analyses that require reliability assessment, lineage meta-information should include documentation of concepts and observations or measurements and their connection to digital geospatial representations. The FGDC geospatial metadata standard does not include detailed specification for concepts, but the NBII has moved in this direction and the FGDC is working on related standards and has recently developed a vegetation classification standard (FGDC 1997). The stated purpose of the National Vegetation Classification Standard is to build accuracy, consistency, and clarity in the structure, labeling, definition and application of a systemic vegetation taxonomy for the United States. Such formal taxonomies are useful but still do not address the need to relate a specific concept to a data set. Suppose a map of old growth forest is created. Since there are several possible definitions of old growth forest, the specific working definition of old growth forest used in the generation of that particular map should be documented, clearly linked to the map, and made accessible for review by any subsequent users.

A comprehensive lineage structure should allow for integration of concepts, observations, and geospatial representations all within an integrated database environment. Currently the lineage structure within the metadata content standard includes the elements shown within the first column of Table 17.2. The second and third columns of the table include excerpts from a lineage report from the NBS Global Change database.

Even though processing steps are well conveyed, detailed source information is not. To illustrate how digital representations (maps) can be connected to associated concepts and measurements, two examples of representations with meta-information are shown in Table 17.3. One is the documentation for a digital map of old growth forest and the other is documentation for a map of soil texture interpolated from a dense set of field sample points.

These examples include elements not explicitly included in the FGDC metadata standard. The concept and observation set elements make connections between concepts, observations, and representations. These connections can be modeled by a relation of the general form $R = \{[\text{information\_object\_id}, [\text{property}, \text{value}]]\}$. Such a relation can track the concept definition specified for a representation. For example, the relation Concept Definition $= \{[\text{representation\_id}, [\text{concept def}, \text{ogfdef.txt}]]\}$ makes an

TABLE 17.2. Lineage elements from the content standard for digital geospatial metadata.

Lineage elements	Example elements from NBS Colorado	
Source information		
Source citation		
Source scale denominator		
Type of source media		
Source time period of content		
Source citation abbreviation		
Source contribution		
Process step	Process step: 1	Process step: 2
Process description	Process description: vegetation map combined with an aspect map derived from USGS DEM (30 meter cell size)	Process description: created 200 meter strip along vegetation boundary types
Source used citation abbreviation		
Process date	Process date:	Process date:
Process time	Process time	Process time
	Source identity: unknown	Source identity: unknown
Process contact	Process contact	Process contact
	Contact information:	Contact information:
	Contact person primary	Contact person primary
	Contact organization: national biological survey	Contact organization: national biological survey
	Contact person: Nancy Thorwardson	Contact person: Nancy Thorwardson
	Contact organization primary	Contact organization primary
	Contact position	Contact position
	Contact mail address: unknown	Contact mail address: unknown
	Contact voice telephone: unknown	Contact voice telephone: unknown
	Contact facsimilie telephone: unknown	Contact facsimilie telephone: unknown
	Contact electronic mail address: unknown	Contact electronic mail address: unknown
	Contact instructions: none	Contact instructions: none

TABLE 17.3. Meta-information for connecting concepts and observations with representations.

Old growth forest map	Kriged soil texture map
Concept: old growth forest definition #	Concept: soil texture definition
Variable(s): old growth forest	Variable(s): soil texture
Variable type: derived—image interpretation	Variable type: observed
Observation set: analog photo set { }	Observation set: soil sample set #
Creation process: classification	Creation process: kriging

explicit connection between a representation (i.e., old growth forest map) and the definition used to construct the map. The file reference for the concept definition (ogfdef.txt) is a pointer to a text file containing the definition of old growth forest specified as:

- Area: 3 ha minimum.
- Age: 50% of maximum age, meaning at least 175–225-years old, depending on the species.
- Undisturbed: Minimal signs of human disturbances (e.g., burning, logging, or plowing).
- Definition Source: Dunwiddie and Leverett 1996.

A similar relation tracks the observations used to create a representation. For example, the relation Has Observations = {[representation_id, [observation_set_id, id#]]} can be constructed. For the old growth forest map, the observations consist of a set of aerial photographs, and the meta-information for the old growth forest map simply points to this set of photos. If a user wishes to know more about the photos, their meta-information can be searched and displayed (see Table 17.4). Table 17.4 shows an example of a subset of meta-information for a set of photos.

Other elements in Table 17.3 indicate that the presence of old growth forest is the mapped variable. The variable type listed as derived means that the old growth forest was not the variable directly measured. By checking the observation set meta-information shown in Table 17.4 we can see that the observed variable was visible light. The presence of old growth forest was interpreted from this variable and the creation process is the interpretation process applied to the photos to generate the map. This can be contrasted with the soil texture map in which soil texture is the map variable, but it is also the variable that was directly observed in the field.

Digital spatial representations are typically constructed from a set of observations and one or more processing steps. An example would be kriging a set of point observations or classifying a remotely sensed image. In an active geographic information system where users are performing analysis there will be a steady stream of new representations being created. A new representation is considered to be one generated by some operation (e.g.,

TABLE 17.4. Sample meta-information for observations.

Analog photo
Ground coverage: 10 × 10 m
Location of coverage: −68.21, 44.35
Time of coverage 5/9/96 15:21
Sampling type: exhaustive
Spatial resolution: {na}
Variable(s): visible light
Measurement device: Camera

rectification, classification, difference, slope, aspect) or the same operation with different parameters (supervised classification with 10 class vs. 7 classes). Properties describing characteristics of a creator operation can be modeled by the same general relation described earlier with a specific case CREATOR = {[representation_id, [creation_operation, IDRISI_slope]]} and a general relation of the form R = {[property_1, value_1], [property_2, value_2]} with specific case R = {[creation_operation, IDRISI_slope], [algorithm, x]}. These two sample relations indicate a new slope coverage was created using IDRISI software and an algorithm listed as X. As each new representation is created its lineage record can be automatically updated. Representations may share several common meta-information elements but they differ by the operation that generated them and its parameters. An operation (e.g., resampling) may change one or more representation characteristics (e.g., spatial resolution).

In addition, relationships between representations can be managed by a predecessor relation. This relation would allow lineage tracking for any representation by providing information to chain back to any previous representation or back to the original representation if necessary. It can be modeled by a relation of the general form R = {[information_object_id, [property, value]]}. For the specific predecessor relation this becomes PREDECESSOR = {[representation_id, [predecessor_id, id#]]}.

"Ground Truth" as Meta-Information. Information collected for validation purposes can also be considered meta-information. In keeping with the first category of meta-information, "ground truth" is not a direct measure of uncertainty; rather, it is used to evaluate uncertainty. Examples of validation data sets can be higher accuracy GPS control points or image interpretation performed by an expert to evaluate an image classification. Information about how validation sets are collected can be referenced as part of a data set's meta-information. For example, in using photointerpretation to validate an image classification it is important to know what sampling scheme was used, whether points or areas were sampled, and the scale of the photography. In fact, validation data sets should be treated as any other geospatial data set. For example, GPS control points can be a data set unto themselves, in another context they may be a validation set for aerotriangulation data. Any geospatial data set covering the same location and measuring the same variable potentially could be a validation data set. Validation data sets could be ones generated by data producers to check their own data. A validation data set could also be generated by a data user to validate a producer's data set independently. To make a data set a validation data set simply requires the construction of a new relation and assurance that the validation set meets appropriate criteria (i.e., independently collected). A validation data set relation can be modeled by the same basic relation given above where R = {[information_ object_id, [property, value]]}. The relation to describe the connection between a data set and a

validation data set is: Has Validation Set = {[representation_id, [valida-tion_set_id, id#]]}.

By linking contextual, lineage, and validation data-sets more closely with the data, it will be easier for users to assess the data of interest more effec-tively. None of the preceding meta-information examples are direct measures of uncertainty but they can be queried and analyzed to construct measures of uncertainty. This level of meta-information and the analysis required to assess uncertainty may not be of interest to the casual data user but would be available to a user interested in constructing an in-depth reliability analysis.

17.3.2 Approach 2: Meta-Information as Measures of Uncertainty

Meta-information can also consist of explicit measures of uncertainty or an encapsulated uncertainty measurement process (e.g., simulation, Goodchild et al. 1998). Information management systems have typically treated all information in the system as certain (Motro and Smets 1997). For example, a vegetation class receives one presumed certain value (e.g., hardwood forest). One possibility to represent uncertainty is to store multiple values. As an example, a set of possible values {hardwood forest, mixed wood forest, softwood forest} could indicate that a polygon may have any one of the three values. The counterpart for continuous variables would be a value plus or minus some amount. The 7-m RMSE for USGS Digital Elevation Models (DEMS) (USGS 1993) is an example of this type of measure. The alternative to storing multiple values as an indication of imprecision is to store a measure of imprecision with each value. In this case, a land-cover classification would receive an assignment of probability for each class (e.g. {hardwood forest 0.6, mixed-wood forest 0.3, softwood forest 0.1}) where the probability values are meta-information that directly describes the uncertainty of the classes.

Such measures of uncertainty are typically generated from the first category of meta-information and may in fact be used to replace the first category. For example, "ground truth" information could be collected to evaluate a classified remotely sensed image. A misclassification matrix could be constructed and linked to the image or a simple statistic such as percent correctly classified (PCC) could be stored. With this information stored in the database in association with the data set, the ground truth data set and associated meta-information could then be discarded. Such an approach could reduce the information storage overhead, but would limit the flexibil-ity of users to investigate a data set more thoroughly, and comprehend its uncertainties better.

In storing measurements of uncertainty directly it is essential to be clear about what information units the uncertainty measures attach to because there are several different levels of granularity among geospatial objects.

Several measures of uncertainty are global measures (e.g., the producer's and user's accuracy and PCC from a misclassification matrix and the RMSE for DEMs). Such global measures are not locally useful because, for example, the 7-m RMSE does not describe the errors present for a specific geographic location where terrain may be very rugged but is a presumed average for a region based on a set of well defined points. In addition, PCC measures accuracy of a class, not the accuracy of a class assigned to a specific polygon.

Procedures that generate localized measures of uncertainty will generally be more useful. Stochastic conditional co-simulation (Kyriakidis et al. 1999) is a procedure that generates a local accuracy map that can be stored as meta-information. Conditional co-simulation creates alternate equiprobable realizations of a spatial variable. The realizations reproduce "ground truth" data at measurement locations. Local probability distributions for true values at observed locations are constructed from simulated values at these nodes. Local mismatch likelihood between a reported and "ground truth" value is modeled by determining the probability for the reported value at any location to over- or underestimate the true values at reported locations. A local accuracy map is constructed as the difference between reported values and the mean of the simulated values. This local accuracy map can be stored as meta-information for a data set using the relations described earlier. For example, the relation: Has Accuracy Map = {[representation_ id], [accuracy_map_id, id#]} may be created.

Uncertainty measures for geospatial data sets are now commonly generated by simulation (Goodchild et al. 1992; Englund 1993; Journel 1996; Mower 1998). These simulations can be used to generate multiple realizations of a spatial variable where the collection of realizations is an implicit measure of the uncertainty in a single representation. Producers of geospatial data sets may run such simulations and store either multiple realizations or mean and standard deviations as meta-information. Good-child et al. (1998) have proposed encapsulating a simulation process with a data set as meta-information. Under this approach, the simulation process could be run interactively by an ecologist or any other analyst to gain an appreciation of the overall uncertainty within a data set.

17.4 Summary

The role of meta-information is becoming more appreciated in a digital environment, but the potential benefits of meta-information have not yet been fully realized. Meta-information for uncertainty management can take many forms. There are trade-offs in storing explicit measures of uncertainty versus providing more extensive meta-information that can be analyzed to investigate uncertainty in greater detail. Data producers are the logical ones to generate measures of uncertainty to store as meta-information for a data set because they are most familiar with the data. Producers are now beginning to provide a fixed set of standard measures such as RMSE, but these may not be adequate for ecologists and other scientists who desire spatially specific estimates of reliability. The tradeoff can be a readily available measure of uncertainty that is not particularly useful against a larger volume of meta-information that must be synthesized, but which is eventually more pertinent to an individual scientist. A workable solution may be some combination of the two approaches. In either case, substantial advances in uncertainty assessment will not be possible without more comprehensive meta-information. Data producers must generate more comprehensive meta-information and use it to compute more refined uncertainty measures or pass the first type of meta-information on in a useable form for interested scientists to carry out their own uncertainty evaluation. Future research needs to be directed toward more comprehensive meta-information collection and management strategies within an integrated database environment, and uncertainty analysis tools tailored for an information/meta-information combination.

Acknowledgment. The support of the National Imagery and Mapping Agency for this work under grant NMA202-97-1-1021 is gratefully acknowledged.

References

Anselin, L. 1998. Interactive techniques and exploratory spatial data analysis. Pages 253–66 in P. Longley, M.F. Goodchild, D.J. Maguire, and D.W. Rhind, eds. Principles of geographic information systems. John Wiley & Sons, New York.

Beard, M.K. 1996. A structure for organizing metadata collection. In Proceedings of the third International Conference on Integrating GIS and Environmental Modeling Sante Fe, NM. January 12–26. Santa Barbara, CA:NCGIA. http://www.ncgia.ucsb.edu/conf/sante-fe_cd_rom/main.html.

Bretherton, F. 1994. A reference model for metadata: a strawman. http://www.llnl.gov/liv_comp/metadata/papers/whitepaper/bretherton.ps.

CEN/TC 287, N 404, 1995. Working draft on geographic information: definitions. draft for discussion URL:http://forum.afnor.fr/afnor/WORK/AFNOR/GPN2/Z13C/PUBLIC/DOC/.

Cogan, C., and T. Edwards, Jr., 1994. Metadata standards for GAP analysis. Idaho Cooperative Fish and Wildlife Research Unit, University of Idaho, Moscow, ID.

Cressie, N. 1991. Statistics for spatial data. John Wiley & Sons, New York.

Directory Interchange Format (DIF) Writer's Guide, Version 7. 1999. Global change master directory. National Aeronautics and Space Administration [http://gcmd.nasa.gov/difguide].

Englund, E. 1993. Spatial simulations: environmental applications. Pages 432–37 in M. Goodchild, B. Parks, and L. Steyart, eds. Environmental modeling with GIS. Oxford University Press, New York.

Federal Geographic Data Committee. 1994. Content standards for digital geospatial metadata. Federal Geographic Data Committee, Washington, DC. June 1994.

Federal Geographic Data Committee. 1997. Content standards for digital geospatial metadata. Federal Geographic Data Committee, Washington, DC. (revised April 1997).

Goodchild, M.F., G. Sun, and S. Yang. 1992. Development and test of an error model for categorical data. International Journal of Geographical Information Systems 6(2):87–104.

Goodchild, M.F., A. Shortridge, and P. Fohl. 1999. Encapsulating simulation models with geospatial data sets. Pages 123–30 in K. Lowell and A. Jaton, eds. Spatial Accuracy Assessment: Land Information Uncertainty in natural Resources. Ann Arbor Press, Chelsea, MI.

ISO/TC211 1998. URL:http://www.statkart.no/isotc211/.

Jones, M.T. 1996. Avian spatio-temporal dynamics and the core-satellite species hypothesis. Ph.D. Thesis. University of Maine, Orono, ME.

Journel, A.G. 1996. Modeling uncertainty and spatial dependence: stochastic imaging, International Journal of GIS 10(5)517–22.

Kyriakidis, P.C., A.M. Shortridge, and M.F. Goodchild. 1999. Geostatistics for conflation and accuracy assessment of digital elevation models. International Journal of GIS 13(7):677–707.

Lanter, D.P. 1991. Design of a lineage-based meta-database for GIS. Cartography and Geographic Information Systems 18(4)255–61.

Lilywhite, J. 1991. Identifying available spatial metadata: the problem. Pages 3–11 in D. Medyckyj-Scott ed. Metadata in the geosciences. Group Publications Ltd., Londonborough.

Mark, L., and N. Roussopoulos. 1986. Metadata management. IEEE Computer. December 1986, 26–36.

Michener, W.K. et al. 1997. Non-geospatial metadata for the ecological sciences. Ecological Applications 7(1):330–42.

Motro, A., and P. Smets. 1997. Uncertainty management in information systems: from needs to solutions. Kluwer Academic Publishers, Boston, MA.

Mower T.H. 1999. Accuracy (Re) Assurance: Selling Uncertainty Assessment to the Uncertain. Pages 3–10 in K. Lowell and A. Jaton, eds. Spatial Accuracy Assessment: Land Information Uncertainty in natural Resources. Ann Arbor Press, Chelsea, MI.

National Biological Service. 1995. Content standard for national biological information infrastructure metadata. Ftp://ftp.nbs.gov/pub/nonspmet.gra/nonspstd.wp.

Network Development and MARC Standards Office. 1988. USMARC format for bibliographic data. Library of Congress, Cataloging Distribution Service, Washington, DC.

Newman, I. 1991. Data dictionaries, information resource dictionary systems and metadatabases. Pages 69–83 in D. Medyckyj-Scott, ed. Metadata in the geosciences. Group Publications Ltd., Londonborough.

Robbins, C.S., D. Bystrak, and P.H. Geissler. 1986. The breeding bird survey: its first fifteen years 1965–1979. U.S. Fish and Wildlife Service, Washington, DC. Resource Publication 157. 196 pp.

USGS. 1993. Standards for digital elevation models, technical report national mapping program. Technical Instructions. U.S. Department of Interior, Washington, DC.

USMARC Format for Bibliographic Data. 1994. Network Development and MARC Standards Office; Cataloging Distribution Service, Library of Congress, Washington, DC.

Weibel, S., J. Godby, and E. Miller. 1995. OCLC/NCSA Metadata workshop report. Technical report, OCLC/NCSA. 1995. http:www/oclc.org:5046/conference/metadata/dublin_core_report.html.

18
Uncertainty Management in GIS: Decision Support Tools for Effective Use of Spatial Data Resources

RONALD EASTMAN

One of the largest challenges facing environmental scientists and managers is the effective use of existing information and the strategic targeting of new data acquisitions. For example, in the context of environmental management in developing nations, existing data are commonly characterized as poor in quality, out of date, irregular in coverage, and unusable for strategic planning. The cost and time frame required to fully rectify this situation, however, is generally prohibitive, and decisions must be made regardless of the information resources available. Thus, the challenge is to make effective use of information that is available and to limit expenditures to only those data collection activities that have the greatest potential for the reduction of uncertainty.

To date, geographic information systems (GIS) have largely been used as a technology for *complexity management*, which is a means for coping with the enormous volume of data associated with geographic information and the extensive calculations needed to rectify and analyze these data in a decision-making context. For example, a typical raster[1] GIS suitability analysis might involve more than 30 million data elements and close to 200 million calculations! The fact that this is a trivial problem for a GIS attests to the success with which it can cope with such complexity. Furthermore, the speed and consistency with which it is able to do so, and its facility in providing presentation-quality output, make this a very attractive technology. Despite this extraordinary capability, however, solving the problem of complexity does not imply that we have therefore achieved effective decision making with spatial data.

Uncertainty pervades the decision-making process. Have we considered all the relevant variables? Have we assigned them proper weight? What measurement error exists? Have we aggregated this evidence in an appropriate

---

[1]*Raster* GIS software systems represent geographic data by recording the condition or state of the landscape for each cell in a fine matrix that covers the region of interest. This contrasts with *vector* systems that record the boundary or course of distinct features as a set of connected points.

manner? What if we are wrong? How likely is it that we are wrong, and thus what are the limits to the conclusions that can be reached? These are the kinds of questions that complexity management does not address, but which form the focus of a new horizon that might be called *uncertainty management* in GIS.

18.1 Uncertainty Management

Uncertainty is not simply a flaw to be avoided or ignored; rather, it is an inherent attribute of data and data manipulation processes. To appreciate this more fully, it is of value to consider the nature of the decision-making process. By one definition, a decision can be viewed as a choice between alternatives (Eastman et al. 1995) (i.e., alternative actions, alternative sites for a facility, alternative combinations of features, and so on). A decision, however, can equally be understood as the process of allocating features to one or more decision sets. For example, we may wish to allocate lands to one of two alternatives—protected versus developed lands. We may similarly wish to determine to which of several land-cover types pixels belong, based on multispectral reflectance of electromagnetic energy. These decisions each involve an allocation of lands into one or more decision sets. Thus, the decision process itself can be seen to contain three basic components (Fig. 18.1), each of which is a significant site for uncertainty—evidence, the decision set, and a relation (i.e., decision rule) that associates the two.

18.1.1 Evidence

The measured attributes of features constitute evidence by which set membership is determined. Thus, to decide which areas would be inundated by a global rise in sea level, one might collect elevation data—evidence from

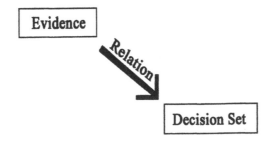

FIGURE 18.1. The three basic components of a decision: evidence, the set of selected choices, and the relation that is established between the two (i.e., the decision rule). Each is a significant source of uncertainty in the decision process.

which the decision can be made. This is clearly a major potential source of uncertainty because all measurements are subject to error. There is a significant concern here with measurement error, its expression (see Goodchild and Gopal 1989), assessment (e.g., Congalton 1991), and how it propagates through a decision rule (e.g., Heuvelink 1993; Burrough 1986). This kind of uncertainty is usually represented by a root mean square error (RMSE) in the case of quantitative data, or proportional error (percent incorrect) in the case of qualitative data, and relies upon classical probability theory and statistical inference for its assessment and propagation; however, the spatial character of measurement error and its impact on propagation is of equal concern. For example, spatially autocorrelated errors in ground height measurements will have considerably less impact on slope calculations than errors that are independent over the characteristic distance over which slopes are determined.

18.1.2 The Decision Set

The second basic element of a decision is the decision set itself. Even though cases exist where the decision set is unambiguous in its character (e.g., whether an area will be inundated or not by a sea level rise), decision sets quite frequently lack such definitive boundaries in their definition. For example, we may have a set defined as "lands suitable for development" that we characterize as areas on shallow slopes; however, what constitutes shallow? If we establish a hard criterion (e.g., slopes of 8% or less), does this imply that slopes of 8.1% are definitively unsuitable, or that those of 7.9% are unquestionably good? In reality, there is no hard boundary. Rather, we have a gradation from slopes that are clearly suitable to those that are clearly unsuitable.

Criteria such as this may be characterized as *fuzzy* rather than *crisp*. Fuzzy set theory (Zadeh 1965) has provided a very robust means of handling such cases, with direct implications for the development of decision rules that can relate evidence to set membership. It has consequently been the subject of much interest in GIS (e.g., Burrough 1989). Fuzziness in the character of the set, however, is not the only source of ambiguity in the decision process, and it is all too often that any source of "fuzziness" is handled as a fuzzy set. For example, in the case of error in the measurement of elevations, there is ambiguity as to the true height of a location, and thus whether it will or will not be inundated by a sea level rise. This, however, is not a fuzzy set. The set itself (the set of inundated locations) is crisp—locations will either be under water or they will not! Our uncertainty is only in the measured height of features, and thus the likelihood that they belong to the decision set—a problem for which the logic of classical set probability theory is more appropriate for the development of a decision rule. Regardless, both issues lead to statements of membership in the decision set that are intermediate between true and false—statements that more generally fall under the concept of *fuzzy measures*. Fuzzy measures refer to any set function that is monotonic with respect to set membership (Dubois and Prade 1982), including classical *probabilities*, the *beliefs* and *plausibilities* of Dempster-Shafer theory (Gordon and Shortliffe 1985), and the *possibilities* of fuzzy sets. We therefore need to agree on aggregation procedures that can accommodate these very different sources of uncertainty and allow for the combination of evidence in establishing the relation between the evidence and the decision set. Chapter 15 provided another example of the use of fuzzy logic.

18.1.3 The Relation (Decision Rule)

The final basic element of a decision is the specification of the relationship between the evidence and the decision set. This is known as a *decision rule*, and involves the aggregation of evidence and a procedure for allocating candidates to the proper decision set. Uncertainty here arises from at least three sources:

1. The first, and perhaps most widely recognized, is *model specification error*. We tend to think of this largely in terms of whether complete and appropriate evidence has been gathered to establish membership in the decision set. Of particular concern is the explanatory value of attributes and the balance between the reduction of specification error versus the compounding of measurement error as additional variables are considered (e.g., Alonso 1968); however, there are also substantial issues related to the procedures used for the aggregation of evidence (Eastman and Jiang 1996). For example, one might define areas suitable for development as being those on shallow slopes *and* near to roads. If membership in the set of shallow slopes is 0.6 and proximity to roads is 0.7, what is the membership in the decision set? Is it the 0.42 of probabilities, the 0.6 of fuzzy sets, the 0.78 of Bayes, the 0.88 of Dempster-Shafer, or the 0.65 of linear combination? The introduction of the ordered weighted average procedure in GIS (Eastman and Jiang 1996; Yager 1988) illustrates the diversity of perspectives that can be introduced into the aggregation procedure and their implications for risk aversion and the degree of tradeoff between factors.

2. The second case where uncertainty arises is in cases where the evidence does not directly and perfectly imply the decision set under consideration. In the examples of inundated lands there is a direct relationship between the evidence and the set under consideration; however, there are also cases where only indirect and imperfect evidence can be cited. For example, we may have knowledge that water bodies absorb infrared radiation. Thus, we might use the evidence of low infrared reflectance in a remotely sensed image as a statement of the *belief* that the area is

occupied by open water. This is only a belief, however, because other materials also absorb infrared energy. The use of such indicator variables and surrogate measures is common, and thus a significant source of uncertainty.

Statements of belief in the degree to which evidence implies set membership are very similar in character to fuzzy set membership functions. They are not definitions of the set itself, however, but simply statements of the degree to which the evidence suggests the presence of the set (however defined). Thus, the logic of fuzzy sets is not appropriate here, but, rather, that of *Bayes* (Bonham-Carter et al. 1988) and *Dempster-Shafer* theory (Gordon and Shortliffe 1985).

3. The third area of uncertainty concerns *ignorance*. Ignorance can arise from a variety of sources, including incompleteness of evidence or from the presence of unknown classes.

Dempster-Shafer theory (see Gordon and Shortliffe 1985) provides an elegant procedure for the accommodation of incomplete evidence by making a distinction between statements of *belief* in a hypothesis and the *plausibility* of that hypothesis. Belief is based on hard evidence for the hypothesis, whereas plausibility expresses the degree to which there is a lack of evidence *against* the hypothesis. For example, we may have only limited belief that a particular disease vector exists in an environment because hard evidence is scarce, but simultaneously feel that the plausibility of that vector species is high because there is little evidence to suggest that it is not present. In such a case, our ignorance is high because of this discrepancy between our belief in its presence and the plausibility that it might be encountered.

With regard to the presence of unknown classes (e.g., the mapping of vegetation into predefined classes that do not include one or more species that actually exist in the area being mapped), little work has been undertaken. This has profound effects on the logic of procedures such as Bayesian probability analysis, which assumes that the classes being considered are exhaustive. For example, Bayesian probability procedures commonly employed in remote sensing may conclude with confidence that a location belongs to a particular class, even though there is virtually no true evidence to support it, as long as it does not look like any other class being considered. It is for this reason, for example, that Eastman (1997) has developed a modified form of Dempster-Shafer analysis for the classification of remotely sensed imagery that explicitly handles the possibility of an unknown land cover category.

It is clear from this brief consideration that many sources of uncertainty exist in the decision process; however, uncertainty *can* be managed. With the proper tools, it can be characterized and used quite effectively to place bounds on a decision, establish the location and nature of new data that should be collected, and gain important new insights into the nature of the features being considered. Several examples follow to illustrate this.

18.2 Measurement Error and Decision Risk

The first example is concerned with estimating the impacts of accelerated sea level rise on rice agriculture in Vietnam (Eastman and Gold 1997), and serves to illustrate the impact of measurement error on the decision making process. Figure 18.2 (see color insert) illustrates the case study region—an area in the vicinity of the Cua-Lo estuary near Vinh in north-central Vietnam. This is an area dominated by paddy rice agriculture—a net export crop of considerable economic value. The problem, then, is to delineate areas that would be affected by an estimated sea level rise of 0.48 m by the year 2100. A key element of the analysis included the development of a digital elevation model based on French colonial maps from the early part of this century, supplemented with height inferences derived from Landsat TM satellite imagery, and subsequent characterization of inherent measurement error. In this case, it was estimated that the RMS error of derived heights was in the order of 0.30 m. Thus, at any one location, we need to accept the fact that the stated height is no more than a best guess drawn from a range of possible values.

The typical approach to solving this decision problem in GIS would be to threshold the elevation model at 0.48 m, thereby separating areas that would be flooded from those that would not. Given that the heights in the elevation model are uncertain, however, what do we conclude about an area with a height of 0.56 m? It would nominally appear that this area is above the projected level of the sea. Given measurement error, however, there is a possibility that the stated height is incorrect, and that the true height is actually below this new level of the sea. Thus, the best that we can say is the probability that we think the area would be flooded. To facilitate this, we have implemented in the IDRISI system a soft thresholding tool, named PCLASS, that can estimate these probabilities by evaluating the area under the normal curve subtended by the original threshold. Figure 18.3 (see color insert) illustrates the result.

We have clearly moved away here from the traditional hard decisions of GIS, yet, at some point, a hard decision needs to be made. There is no right decision here. Any hard decision we make will be associated with some likelihood that the decision will be wrong. This image fortunately tells us this directly because the probability image is essentially a mapping of *decision risk* (Eastman 1996); more specifically, it expresses the likelihood that an area would be flooded if we were to say that it would not. This decision risk surface can then be thresholded at a specified risk level to achieve the final decision. For example, Figure 18.4 (see color insert) shows the areas of potential inundation at a 5% decision risk level (i.e., there is no more than 1 chance in 20, which is equivalent to a 95% confidence level, that an area designated as above sea level would in fact be inundated).

The important feature of this analysis is that it makes use of existing data to provide an assessment that is directly related to data quality. In this case, the elevation data came from French colonial maps from 1906. Even though the quality of these maps is the major contributor to decision risk, the concern of the decision group was to establish the upper limit of vulnerability. Thus, the data functioned quite effectively. Furthermore, the uncertainty management tool used established a clear relationship between risk and the quality of information. In this particular example, it is very unlikely that all of these areas would be affected. Given the quality of the data, however, we must accept the possibility that any of them might. To the environmental manager, the implication becomes immediately clear—we require better information to narrow down the possibilities to a more likely set. This procedure can fortunately also be used to estimate the degree of improvement that an investment in new data would produce. For example, in this case, doubling the precision of the elevation model to an RMS of 0.15 m (a major investment requiring new aerial photography and photogrammetric interpretation) would result in a change of only 14% in the area expected to be flooded (as estimated by simply running the procedure a second time with this lower RMS).

18.3 Decision Set Uncertainty

The Vietnam case study illustrates quite well that membership in the final decision set can be other than simply true or false. Thus, in cases where a feature can be allocated to one of several decision sets (e.g., land use classes) it is quite possible that the feature may possess some degree of probability of belonging to more than one class. This discrepancy is ordinarily ignored in favor of the class with the highest probability; however, there is useful information here that can say a great deal about the feature involved. To illustrate this, consider the following case study from Southern Africa.

In the country of Malawi, there have been recurrent discussions on the need for regular inventory coverage of high-resolution remotely sensed data, but the costs are prohibitive; however, the country does receive daily 1-km resolution AVHRR coverage. Even though these data are coarse, they are only coarse spatially. Their temporal resolution is extraordinary. Thus, we have explored the possibility of using uncertainty management tools for subpixel classification of the AVHRR imagery to compensate for the spatial degradation.

The specific procedure is a soft classifier named BAYCLASS in the IDRISI software system. BAYCLASS calculates separate images of the probability that each pixel belongs to each of a set of land cover classes. In the absence of significant error in the definition of training sites, it is reasonable to interpret this information as evidence of the relative proportions of these land-cover types within each pixel. For example, Figure 18.5 (see color insert) shows a typical land cover classification where class membership is assigned based on the maximum likelihood, and one of the set of images produced by BAYCLASS expressing the proportion of each pixel covered by bush savannah. In effect, then, we are enhancing the spatial resolution of these data by producing a subpixel classification directly derived from the uncertainty data. This represents a significant enhancement of available data for a country such as Malawi.

18.4 Uncertainty in Relating Evidence to the Decision Set

As a further illustration of the potential of using uncertainty management, consider the analysis of land degradation in southern Mauritania, as an illustration of the problem of relating evidence to the decision set. During the 1980s in particular, this region experienced significant declines in precipitation and increasing pressure from grazing and human populations. In the monitoring of impacts, however, direct measures of degradation are few. To overcome such a situation, Eastman (1997) developed an approach to maximize the utility of local knowledge and expert opinion in the mapping and aggregation of indicator variables based on a weight-of-evidence approach.

In this particular example (Thiam 1997) a series of indicator variables were developed using biophysical data (e.g., satellite-based measurements of vegetation biomass) and socioeconomic and land utilization data gathered by means of a village-level questionnaire (e.g., grazing distances from village wells, and browse and building material species). These were each used to develop a set of maps that expressed *belief* in the presence of degradation using a software module in IDRISI that can map fuzzy measure membership functions. Local knowledge, assessed by means of the questionnaire, was used as the primary input for the shape of the membership function, whereas expert opinion (i.e., that of the principal investigator of the study) was used for the scaling of these membership statements into expressions of belief. For example, Figure 18.6 (see color insert) shows the belief images derived for degradation as a result of gathering fuelwood, and that arising from grazing around bore holes (wells). A common feature of many of these indicator variables is the fall-off in belief (in the presence and severity of land degradation) as distance increases from the settlement or source of water.

A particularly problematic aspect of this kind of data is that absence of belief does not necessarily constitute disbelief in the hypothesis under consideration. It often arises simply because we do not have adequate information. For example, a village may be somewhat unfamiliar with an area for which they are asked for information. As a result, the indicator variable may be constructed to indicate that they have little reason to believe in the presence of degradation at that location. This statement of belief, however, arises mostly out of ignorance, and thus cannot be used to imply that degradation is absent. Traditional approaches to both classical and fuzzy sets have no basis for dealing with this problem—and yet such problems dominate cases where local knowledge and expert opinion are used. As a consequence, the procedure that has been used to aggregate these data makes use of a special approach known as *Dempster-Shafer theory*, which is a variant of Bayesian probability theory that provides the theoretical foundation for a remarkably powerful weight-of-evidence approach to the aggregation of expert and local knowledge (Gordon and Shortliffe 1985).

Dempster-Shafer theory makes a strong distinction between *belief* (based on hard evidence in support of hypothesis) and *plausibility* (the degree to which there is a lack of evidence against the hypothesis), and provides a mechanism for aggregating statements of belief and disbelief that can yield composite measures of these concepts. For example, Figure 18.7 (see color insert) shows a composite measure of belief in degradation derived from 10 indicator variables for the study in southern Mauritania, based on the procedures developed in the IDRISI GIS software system (Eastman 1997). Figure 18.8 (see color insert) shows its corresponding measure of plausibility, whereas Figure 18.9 (see color insert) shows the difference between the two—a statement of uncertainty known as a *belief interval* that arises from ignorance in the model.

The richness of these statements is particularly striking; however, the utility they provide in determining how the model might be improved is especially interesting. For example, examining Figure 18.9, this spatial portrayal of ignorance allows us to understand that uncertainty is highest at the margins of the grazing zones around wells and boreholes. It became immediately clear that this arose because the questionnaire did not address the issue of whether settled and nomadic tribes intermingle their grazing areas in these zones. Thus, the procedure has indicated where uncertainty is highest (and thus where extra data gathering would be most productive), as well as the nature of the information that needs to be gathered. This insight would have been lost in traditional analyses, and without the ability to provide a spatial portrait of the uncertainty present.

In the context of developing nations, the Dempster-Shafer approach offers a major opportunity to make effective use of information resources that have all too frequently gone unrealized. Because of its depth in portraying the essential uncertainties in expert and local knowledge, it provides strong guidance for the targeting of further data collection, as well as in the development of sound conclusions that are substantiated by the evidence at hand.[2] This procedure, however, also offers significant opportunities for the effective utilization of expert opinion in conjunction with indicator variables in ecological research.

18.5 The Role of Spatial Representation

In each of these examples, it is clear that significant value exists in the characterization of uncertainty arising from the decision process. It is important to note, however, that in many cases full value is realized only when this uncertainty is spatially characterized. For example, in the case study concerned with sea level rise, improvement in the areal precision of impact assessments associated with a given level of decision risk is dependent on the degree of change in the resolution of the measurements, as well as on the configuration of the land. Areas of shallow relief will yield large differences in areal estimates for a given change in resolution, whereas those of steep relief will experience little change. Thus, it is clear that a greater return will be achieved in these lower-relief areas.

As a second example, the implication of uncertainties in the classification of remotely sensed imagery is very much dependent upon the spatial configuration of subpixel mixtures. For example, if one finds support for a mixture of forest and wetland in distinct polygonal features, one would act upon this information quite differently from a case where such mixtures are only found at the common boundaries of uniform single-class stands (where mixtures would be expected). The latter suggests a problem only with the resolution of the imagery, whereas the former suggest that a new mixed class (*forested wetland*) would need to be added.

As a final example, the analysis of ignorance associated with the Mauritanian case study suggested a highly cost-effective solution to the reduction of uncertainty in the study; however, this solution (i.e., asking new questions about the degree of mutual use of areas peripheral to boreholes by both settled and nomadic tribes) was entirely dependent upon a visual appreciation of the spatial configuration of the ignorance determined.

---

[2]This example, although dramatic, does not illustrate the full potential of the Dempster-Shafer weight-of-evidence approach. An additional, and very attractive feature is its organization of a basic set of hypotheses into a hierarchy of combinations, for which evidence may also be productively specified. For example, with four hypotheses, evidence can be supplied for or against any individual hypothesis: [A], [B], [C], or [D] or any upper level combination: [AB], [AC], [AD], [BC], [BD], [CD], [ABC], [ABD], [ACD], or [BCD]. Evidence supportive of one of these upper level combinations lends weight to the possibility that any of its constituent members might be true, without being able to distinguish between them. Such partial evidence is ordinarily rejected on the basis that it is inconclusive; however, this type of evidence is remarkably common, and Dempster-Shafer theory allows such partial evidence to be productively employed.

In each of these examples, most (if not all) of the true gain associated with a characterization and analysis of uncertainty has resulted from its spatial characterization. In contrast, it has perhaps been more common to develop nonspatial summary indicators of uncertainty associated with individual map layers (e.g., RMSE or proportional error) and then to gauge their propagated effects through decision rules. These examples illustrate, however, that *an analysis of uncertainty in the decision-making process should not be handled simply as a metadata problem*, but rather as a parallel spatial analysis with important implications that arise entirely as a result of the spatial configuration of the uncertainty involved.

18.6 Summary

Uncertainty clearly pervades the decision-making process. It is equally clear, however, that with the development of appropriate software tools, this uncertainty can be spatially characterized, allowing effective use of existing information. Through these examples, it can be seen that by incorporating concepts of uncertainty, it is possible to recast decisions into the context of decision risk. Doing so allows one to establish spatial portraits of the value of information, thus providing important directions in the gathering of new data. It can also be seen that uncertainty can lead to fresh understandings of the character of phenomena, thereby supplementing current knowledge. Furthermore, it is evident that new uncertainty management techniques can greatly facilitate the merging of information from indicator variables with expert opinion, allowing the production of quantifiable information in situations where direct evidence is scarce. It is vital to stress, however, that such tools need to be fully spatial in their analysis of uncertainty arising from the decision process. To fail to do so greatly diminishes the potential value of uncertainty management in GIS.

References

Alonso, W. 1968. Predicting best with imperfect data. Journal of the American Institute of Planners 34:248–55.

Bonham-Carter, G.F., F.P. Agterberg, and D.F. Wright. 1988. Integration of geological datasets for gold exploration in Nova Scotia. Photogrammetric Engineering and Remote Sensing 54(11):1585–92.

Burrough, P.A. 1986. Principles of geographical information systems for land resources assessment. Clarendon Press, Oxford.

Burrough, P.A. 1989. Fuzzy mathematical methods for soil survey and land evaluation. Journal of Soil Science 40:477–92.

Congalton, R.G. 1991. A review of assessing the accuracy of classifications of remotely sensed data. Remote Sensing and the Environment 37:35–46.

Dubois, D., and H. Prade. 1982. A class of fuzzy measures based on triangular norms. International Journal of General Systems 8:43–61.

Eastman, J.R., W. Jin, P.A.K. Kyem, and J. Toledano. 1995. Raster procedures for multi-criteria/multi-objective decisions. Photogrammetric Engineering and Remote Sensing 61(5):539–47.

Eastman, J.R. 1996. Uncertainty and decision risk in multi-criteria evaluation: implications for GIS software design. Proceedings, International Institute for Software Technology Expert Group Workshop on Software Technology for Agenda'21: Decision Support Systems. Section 8.

Eastman, J.R., and H. Jiang. 1996. Fuzzy measures in multi-criteria evaluation. Pages 527–34 in Proceedings, second international symposium on spatial accuracy assessment in natural resources and environmental studies. May 21–23, Fort Collins, CO.

Eastman, J.R. 1997. Decision support. Chapter 9 in IDRISI for Windows User's Guide, version 2.0. Clark University, Worcester, MA.

Eastman, J.R., and S. Gold. 1997. Assessing the impacts of sea-level rise in Vietnam. GeoInfo Systems 7(1):38–43.

Goodchild, M.F., and S. Gopal. 1989. Accuracy of spatial databases. Taylor and Francis, London.

Gordon, J., and E.H. Shortliffe. 1985. A method for managing evidential reasoning in a hierarchical hypothesis space. Artificial Intelligence 26:323–57.

Heuvelink, G.B.M. 1993. Error propagation in quantitative spatial modelling: applications in geographical information systems. CIP-Gegevens Koninklijke Bibliotheek, Den Haag.

Thiam, A. 1997. Geographic information systems and remote sensing methods for assessing and monitoring land degradation in the Sahel: the case of southern Mauritania. Ph.D. Dissertation, Graduate School of Geography. Clark University, Worcester, MA.

Yager, R. 1988. On ordered weighted averaging aggregation operators in multicriteria decision making. IEEE Transactions on Systems, Man, and Cybernetics 8(1):183–90.

Zadeh, L.A. 1965. Fuzzy sets. Information and Control 8:338–53.

Part IV
Epilog

19
Epilog

Carolyn T. Hunsaker

> A picture is worth a thousand words, but a map or figure *is not*
> necessarily reality.

19.1 Why Spend Time on Spatial Uncertainty?

For this book, we defined *uncertainty* as "the state of knowledge about a relationship between the world, and a statement about the world." There were several reasons I began working on the topic of spatial uncertainty, and they help reinforce the summary points I want each reader to remember. These include bringing disparate spatial data sets together for analysis with GIS, work with landscape pattern metrics and regional ecological risk assessment, and an exposure to the diverse aspects of geography. I worked for many years on regional or landscape analyses that require spatial data for large geographic areas. Scientists are often happy just to be able to find digital data for vegetation, soils, and topography, let alone worry about their accuracy. Available data are usually at different spatial resolutions and were developed for different purposes, so one is almost immediately faced with some uncertainty issues, although it may not be obvious.

At Oak Ridge National Laboratory, I had the privilege to work with many outstanding landscape ecologists in the application of pattern metrics to monitoring and assessing ecological condition. Even though much effort has been expanded in the development and understanding of pattern metrics, little attention has been given to the sensitivity of such metrics to errors and uncertainties in spatial data. Although we realized spatial uncertainty was an issue (Hunsaker et al. 1990), our papers did not quantify uncertainty (Hunsaker et al. 1994; Riitters et al. 1995; O'Neill et al. 1996). Landscape pattern metrics have been proposed as being useful to monitor change over time. They have also been used as predictor variables in multivariate statistical models and as indicators of habitat condition. How, then, can we effectively use these metrics if we do not have some idea of their sensitivity to error in the spatial data they are based on or their variability over time? At Oak Ridge I also worked with some of the leaders in the development of ecological risk assessment techniques. Most of the early work focused on toxicological effects, but several of us later wrote a series of papers on regional ecological risk assessment (Hunsaker et al. 1990; Suter 1990; and Graham et al. 1991). It was readily obvious that the degree of effort put into estimating variance and uncertainty in toxicity tests was almost nonexistent for spatial data that would be used in a regional risk assessment. A probabilistic risk assessment of habitat condition for a terrestrial animal could require spatial data on vegetation, predator locations, and climate. Of these, only the climate data are likely to have variance estimates.

My early dependence on digital data from both cartographic products and remote sensing technologies, along with the use of GIS, exposed me to the breadth of thinking in the discipline of geography. Involvement in the early conferences on environmental modeling with GIS (Goodchild et al. 1993; Goodchild et al. 1996) and with the University Consortium for Geographic Information Science (UCGIS) (http://www.ucgis.org/) also provided impetus for this research topic. I seldom saw ecologists acknowledge uncertainty in their use of spatial data. Geographers recognized this as an important topic (National Center for Geographic Information Analysis, NCGIA, and UCGIS), but they offered little in the practical realm of how to address the issue. In 1995 I had the privilege of spending time at the NCGIA, University of California, Santa Barbara, to investigate what techniques were available for ecologists to address uncertainty in spatial data. During that year my discussions with Frank Davis, Mike Goodchild, Karen Kemp, Jack Estes, and many others led me to realize that there was much work to be done on this topic. A proposal to the newly formed National Center for Ecological Analysis and Synthesis (NCEAS) resulted in the development of a working group and a workshop in 1997. This book is the first product from that NCEAS working group and associated workshop.

The goal of this book was to explore the many sides of uncertainty with respect to spatial data and to raise awareness of its importance, especially among ecologists. To adequately address this topic, however, we had to involve geographers, including cartographers, physical and cognitive scientists, and GIS and remote sensing experts; scientists that model natural systems; statisticians; and a diverse array of ecologists. Although we have targeted ecological applications, this book will be useful to all types of scientists who work with spatial data in natural resources. We wish the techniques were well defined and straightforward, but the quantification of uncertainty in spatial data is not yet a world of simple recipes and prescriptions. As we said earlier in the book, coping with uncertainty in spatial ecology requires a willingness to adapt to the circumstances of a problem, rather than to follow a prescription.

Ludwig et al. (1993) highlight the important role of uncertainty in resource exploitation and conservation. They list five principles of effective resource management, and one of them is "confront uncertainty." They state that "effective policies are possible under conditions of uncertainty, but they must take uncertainty into account." Our hope is that this book exposes the reader to the diverse approaches from which they can chose: fuzzy set theory and probability theory, quantification of uncertainty and its propagation through an analysis or reduction/elimination of uncertainty, or simple acknowledgment of uncertainty. We provide many examples using different types of data that ecologists often use. The obligation to deal with uncertainty rests equally with data producers, software and hardware vendors, system integrators, and end-users alike (see Hunter, http://www. ncgia.ucsb.edu/giscc/cc_outline.html). Hunter suggests four key reasons why the uncertainty debate has grown in importance: mandatory data quality reporting in some jurisdictions, the need to protect reputations, as a means of safeguarding against litigation, and as part of the basic scientific quest for knowledge. He also presents a strategy for handling uncertainty:

- Consideration of the type of application and the nature of the decision to be made.
- Determining the error in the information product and comparing it with the error specifications for the task at hand.
- Either reducing uncertainty or absorbing it (the former usually involves technical approaches, while the latter usually involves institutional methods).

Ecological risk assessment is an approach that acknowledges error and uncertainty and strives to account for it in all components of an assessment (i.e., input data, exposure model, and effects model). Ecological risk assessment provides a probability distribution of an event happening and an associated confidence level. No matter what the strategy is for handling uncertainty it still should be quantified. For those that seek to reduce uncertainty in spatial data it is unlikely uncertainty will be totally removed. For vegetation data, given unlimited money and time, uncertainty could be substantially reduced. The reporting of error in different ways (Chap. 14) for remotely sensed data would facilitate decisions and efforts on error reduction.

The discussion of spatial uncertainty in this book has been structured around four stages: data, display, analysis and modeling, and decision making. Although a majority of the text addresses data and analysis/modeling stages, we have introduced issues about display and decision making. To date, GIS have largely been used as a technology for complexity management; however, Eastman (Chap. 18) suggested the need for a new effort— uncertainty management in GIS. He gives some excellent examples of decision support tools for effective use of spatial data resources. I hope you conclude that error and uncertainty are not just bad. Acknowledging them promotes active quality control. It is useful to know how error and uncertainty occur, how they can be managed and reduced, how knowledge of errors and their propagation can be used to improve understanding of pattern or process, and how they can help optimize sampling and identify weak parts of spatial analysis.

References

Goodchild, M.F., B.O. Parks, and L.T. Steyaert, eds. 1993. Environmental modeling with GIS. Oxford University Press, New York.

Goodchild, M.F., L.T. Steyaert, B.O. Parks, C. Johnston, D. Maidment, M. Crane, et al., eds. 1996. GIS and environmental modeling: progress and research issues. GIS World Books, Fort Collins, CO.

Graham, R.L., C.T. Hunsaker, R.V. O'Neill, and B.L. Jackson. 1991. Regional ecological risk assessment. Ecological Applications 1(2):196–206.

Hunsaker, C.T., R.L. Graham, L.W. Barnthouse, R.H. Gardner, R.V. O'Neill, and G.W. Suter. 1990. Assessing ecological risk on a regional scale. Environmental Management 14(3):325–32.

Hunsaker, C.T., R.V. O'Neill, S.P. Timmins, B.L. Jackson, D.A. Levine, and D.J. Norton. 1994. Sampling to characterize landscape pattern. Landscape Ecology 9(3):207–26

Ludwig, D., R. Hilborn, and C. Walters. 1993. Uncertainty, Resource Exploitation, and Conservation: Lessons from History. Science 260:17, 36.

O'Neill, R.V., C.T. Hunsaker, S.P. Timmins, B.L. Jackson, K.B. Jones, K.H. Riitters, et al. 1996. Scale problems in reporting landscape pattern at the regional scale. Landscape Ecology 11(3):169–80.

Riitters, K.H., R.V. O'Neill, C.T. Hunsaker, J.D. Wickham, D.H. Yankee, S.P. Timmins, et al. 1995. A factor analysis of landscape pattern and structure metrics. Landscape Ecology 10(1):23–39.

Suter, G.W., II. Endpoints for regional ecological risk assessments. Environmental Management 14:9–23.

Index